T0188364

Identification and Classical Control of Linear Multivariable Systems

Many systems are interactive multivariable systems. To design controllers, a transfer function matrix model is required. In this book, the methods for identifying such models are given for linear multivariable systems by both open loop and closed loop methods. The closed loop reaction curve method and optimization method fall under closed loop identification methods. In open loop identification methods, methods of identifying transfer function model for stable first order plus time delay model and critically under-damped second order system are presented.

An extension of the method for identifying transfer function matrix for an unstable system is also discussed in this book. The main contribution is in the selection of the initial guess values for the model parameters for the optimization method.

Design of multivariable proportional integral (PI) controllers based on steady state gain matrix (SSGM) is described for stable and unstable systems. The method of identifying SSGM for multivariable systems is given for stable and unstable systems. The method is also extended for stable and unstable non-square systems. The method of designing multivariable PI controllers for non-square systems by genetic algorithm is also explained.

This book has been structured for an elective course for undergraduate and a core course for post-graduate students of chemical engineering, instrumentation and control engineering, and electrical engineering.

Dhanya Ram V. is Assistant Professor, Department of Chemical Engineering, at the National Institute of Technology, Calicut. She has 11 years of teaching and research experience. Her research interests are in process control, besides modeling and simulation.

M. Chidambaram is Professor Emeritus, Department of Chemical Engineering, at the National Institute of Technology, Warangal. He has also held various posts at IIT Madras, where he headed the chemical engineering department; was a member of the faculty at IIT Bombay; and was a director of the National Institute of Technology, Tiruchirappalli. He is the co-author of *Relay Autotuning for Identification and Control* (2014) and author of *Mathematical Modelling and Simulation in Chemical Engineering* (2017) published by Cambridge University Press.

Identification and Classical Control of Linear Multivariable Systems

Dhanya Ram V.
M. Chidambaram

CAMBRIDGE
UNIVERSITY PRESS

CAMBRIDGE
UNIVERSITY PRESS

University Printing House, Cambridge CB2 8BS, United Kingdom

One Liberty Plaza, 20th Floor, New York, NY 10006, USA

477 Williamstown Road, Port Melbourne, vic 3207, Australia

314 to 321, 3rd Floor, Plot No. 3, Splendor Forum, Jasola District Centre, New Delhi 110025, India

103 Penang Road, #05-06/07, Visioncrest Commercial, Singapore 238467

Cambridge University Press is part of the University of Cambridge.

It furthers the University's mission by disseminating knowledge in the pursuit of education, learning and research at the highest international levels of excellence.

www.cambridge.org
Information on this title: www.cambridge.org/ 9781316517215

© Dhanya Ram V. and M. Chidambaram 2022

This publication is in copyright. Subject to statutory exception and to the provisions of relevant collective licensing agreements, no reproduction of any part may take place without the written permission of Cambridge University Press.

First published 2022

Printed in India by Avantika Printers Pvt. Ltd.

A catalogue record for this publication is available from the British Library

Library of Congress Cataloging-in-Publication Data

Names: V., Dhanya Ram, author. | Chidambaram, M., author.
Title: Identification and classical control of linear multivariable systems
    / Dhanya Ram V., M. Chidambaram.
Description: Cambridge, United Kingdom ; New York, NY, USA : Cambridge
    University Press, 2022. | Includes
bibliographical references and index.
Identifiers: LCCN 2022016943 (print) | LCCN 2022016944 (ebook) |
    ISBN 9781316517215 (hardback) | ISBN 9781009043472 (ebook)
Subjects: LCSH: Linear control systems. |
Multiple criteria decision making. | BISAC: TECHNOLOGY &
ENGINEERING / Chemical & Biochemical
Classification: LCC TJ220 .V33 2022 (print) |
LCC TJ220 (ebook) | DDC 629.8/32–dc23/eng/20220524 LC record available at
https://lccn.loc.gov/2022016943
LC ebook record available at
https://lccn.loc.gov/2022016944

ISBN 978-1-316-51721-5 Hardback

Cambridge University Press has no responsibility for the persistence or accuracy of URLs for external or third-party internet websites referred to in the is publication,and does not guarantee that any content on such websites is, or will remain,accurate or appropriate.

Contents

Preface

This book, *Identification and Classical Control of Linear Multivariable Systems*, has been structured for an elective course for undergraduate students and for a core course for post graduate students of Chemical Engineering, Instrumentation and Control Engineering, and Electrical Engineering.

Many systems are described by Multi Input and Multi Output (MIMO) systems. To design controllers for such systems, we require the identification of a transfer function matrix of the system. If the system is mildly interactive, then we can design decentralized Proportional and Integral (PI) controllers based on diagonal transfer functions with appropriate detuning of the PI controllers. If the interaction is significant, then a centralized PI control system is to be designed. The design of the controller becomes complicated if the system is unstable in nature.

Classical Control Theory studies the physical systems and the control design in the frequency domain, while Modern Control Theory studies it in the time domain. Systems in the frequency domain are expressed in transfer functions (via Laplace Transforms), while the time-domain systems are described in state-space representations (a set of differential equations). In Chapter 1, a basic review is given of the Classical Control Theory and the Modern Control Theory.

In Chapter 2, basics of open loop and closed loop identification of transfer function models of Single Input and Single Output (SISO) systems are reviewed. An open loop method for identifying First Order Plus Time Delay (FOPTD) model and Critically Damped Second Order Plus Time Delay (CSOPTD) transfer function model is proposed. The closed loop reaction curve method and optimization method are discussed. A review of the techniques available for design of PI controllers for transfer function models for SISO systems is brought out.

In Chapter 3, the concept of relative gain array for the measure of interactions in a multivariable system is given. The need to detune the diagonal controllers' settings is brought out and also the method of tuning decentralized PI controllers by relay auto tune method. Simple methods of designing centralized PI controllers and the analyses of robust stability and robust performances are discussed.

In Chapter 4, a Closed Loop Reaction Curve (CRC) method for the identification of a stable MIMO system is discussed. The system under consideration is controlled by decentralized PI/PID controllers. The closed loop responses and interactions are modelled by the closed loop reaction method. Using the relation between the open loop and the closed loop transfer function matrices, the open loop transfer function matrix is obtained in Laplace domain. In Chapter 5, the method for identifying a critically damped SISO model is proposed. In Chapter 6 the CRC method is extended for identifying the model for critically damped MIMO systems.

For any nonlinear least square optimization method, the selection of initial guess values plays an important role in computational time and convergence. In Chapter 7, a simple and generalized method for obtaining reasonable initial guess values for the model parameters of FOPTD transfer function model when the system is controlled by decentralized PI controllers is discussed. The method to obtain the upper and lower bounds for the model parameters for the optimization routine is also presented. This method gives a quick and guaranteed convergence. The standard optimization routine is used for solving the optimization problem.

The method to get the guess values for the FOPTD model parameters when the system is controlled by centralized PI controllers is discussed in Chapter 8. In Chapter 9, the method to get the guess values for the SOPTD model parameters when the system is controlled by decentralized PI controllers is proposed. Due to increased number of parameters, the convergence is not obtained. The need for using step-up followed by step-down in the set point change is brought out to identify the model parameters. The method of identifying CSOPTD model by optimization method is proposed, with two examples, using conventional step change in the set point. Since the number of parameters in CSOPTD model are less than those in SOPTD model, this method does not have any convergence problem.

In Chapter 10, a generalized technique to obtain the initial guess values for individual transfer function processes of an unstable multivariable system required for the nonlinear least square method is discussed. To determine the lower and the upper bounds of the model parameters to be used in the optimization technique, a simple method is given. The method to identify the model parameters of a MIMO unstable system under a decentralized control system is also proposed. The method is extended to an unstable MIMO system under centralized controllers. In Chapter 11, a method is given to design multivariable PI/PID controllers for stable and unstable multivariable systems. This method needs only the SSGM. The method is based on the static decoupler design followed by SISO PI/PID controllers design and combining the resulting decoupler and the diagonal PID controllers as the centralized controllers. The performance of the controllers is compared with that of the reported centralized controllers based on the multivariable transfer function matrix.

In Chapter 12, a method to identify the SSGM for an unstable TITO system from the input and output variables at steady state condition under closed control is proposed. The method proposed by Davison is modified to design MIMO centralized PI controllers for unstable systems. Two simulation studies are given.

In Chapter 13, simple centralized controller-tuning methods, namely the Davison method and the Tanttu and Lieslehto method, are extended to non-square systems with Right Half-Plane (RHP) zeros. The proposed methods are applied by simulation on two examples: a coupled pilot plant distillation columns and a crude distillation unit. The performances of the square and non-square controllers are compared. In this chapter, Genetic Algorithm (GA) optimization technique is also applied for tuning of centralized and decentralized controllers for linear non-square multivariable systems with RHP zeros.

In Chapter 14, the Davison method is modified for unstable non-square multivariable systems, to design single-stage multivariable PI controllers using only the SSGM of the system. Since the overshoots in the closed loop responses are large, a two-stage P-PI control system is proposed. A method is presented to identify the SSGM of a non-square multivariable unstable system under closed loop control. Effects of disturbances and measurement noise on the identification of SSGM are also studied. In Chapter 15, recent methods of designing multivariable PI controllers are reviewed.

There are several good books available on the system identification and control of multivariable systems (for example, Zhu, 2001; Ikonen and Najim, 2001; and Skogestad and Postlethwaite, 2005, 2014). These books consider identification and control of only stable systems, whereas this book considers identification of the unstable systems also. Non-square stable and unstable systems are considered in this book. The design of multivariable controllers based on SSGM is proposed. A simple method designing controllers based on transfer function matrix model is reviewed.

Many simulation examples are given in each chapter to illustrate the methods and several problems are given at the end of each chapter for students to apply these methods. Chapters 1, 2, 3, 4, and 7 can be used for an elective course for undergraduate students. Chapters 5, 6, 10, 11, and 12 can be used for a graduate course.

It is a pleasure to acknowledge the contributions of the people who helped the authors in writing this book. This book owes much to the numerous authors of original research papers and textbooks. The authors gratefully acknowledge the assistance provided by their graduate students P. Ganesh, K. L. N. Sarma, Dinesh Sankar Reddy, S. Pramod, I. Anand, and C. Rajapandiyan.

The authors are thankful for the support received from the departments of Chemical Engineering at IIT Madras, NIT Calicut, and NIT Warangal. The authors thank Cambridge University Press and the editors involved in the process for their professionalism and for successfully publishing this book.

The first author dedicates this work to her parents, guru, Arun, Abhinand Krishna, Arjun, and Amirtha, and the second author dedicates the work to the Almighty.

Acknowledgements

Papers

[Chapter 2]
Narasimha Reddy, S. and Chidambaram, M. (2020). Model identification of critically damped second order plus time delay systems. *Indian Chemical Engineer*, **62**(1), 67–77.

[Chapter 3]
Nikita, S. and Chidambaram, M. (2018). Case studies of improved relay auto tuning of PID controllers for SISO systems. *Indian Chemical Engineer*, **60**(4) 438–456.

[Chapter 4]
Dhanya Ram, V. and Chidambaram, M. (2014). Closed loop reaction curve method for identification of TITO systems. *IFAC Proceedings Volumes (IFAC-Papers Online)*, **47**(1), 989–996.

[Chapter 5]
Dhanya Ram, V. and Chidambaram, M. (2014). On-line controller tuning for critically damped SOPTD system. *Chemical Engineering Communications*, **202**(1), 48–58.

[Chapter 6]
Dhanya Ram, V. and Chidambaram, M. (2017). Closed loop identification of two input two output critically damped second order systems with delay. *Indian Chemical Engineer*, **57**(2), 79–100.

[Chapter 7]
Rajapandiyan, C. and Chidambaram, M. (2012a). Closed-loop identification of multivariable systems by optimization method. *Industrial and Engineering Chemistry Research*, **51**(3), 1324–1336.

[Chapter 8]
Dhanya Ram, V. and Chidambaram, M. (2016). Identification of centralized controlled multivariable systems. *Indian Chemical Engineer*, **58**(3), 240–254.

[Chapter 9]
Rajapandiyan, C. and Chidambaram,M. (2012b). Closed-loop identification of second order plus time delay (SOPTD) model of multivariable systems by optimization method. *Industrial and Engineering Chemistry Research*, **51**(8), 9620–9633.

[Chapter 11]
Dhanya Ram, V. and Chidambaram, M. (2015). Simple method of designing centralized PI for multivariable systems based on SSGM. *ISA Transactions*, **56**, 252–260.

[Chapter 12]
Dhanya Ram, V., Rajapandiyan, C. and Chidambaram, M. (2014). Steady state gains identification and control of multivariable unstable systems, *Chemical Engineering Communications*, **202**(2), 151–162.

[Chapter 13]
Sarma, K. L. N. and Chidambaram, M. (2005). Centralized PI/PID controllers for non-square systems with RHP zeros. *Journal of the Indian Institute of Science*, **85**(4), 201–214.

Ganesh, P. and M. Chidambaram, M. (2010). Multivariable controller tuning for non-square systems with RHP zeros by genetic algorithm. *Chemical and Biochemical Engineering Quarterly*, **24**(1), 17–22.

[Chapter 14]
Dhanyaram, V. and Chidambaram, M. (2018). SSGM based multivariable control of unstable non-square systems. *Chemical Product and Process Modelling*, **13**(1), 1–14.

[Appendix A]
Dhanya Ram, V., Karlmarx, A. and Chidambaram, M. (2014). Identification of unstable second order transfer function model with a zero by optimization method. *Indian Chemical Engineer*, **58**(1) 29–39.

Presentations in Conferences

[Chapter 10]

Tripathi, Saurabh and Chidambaram, M. (2017). Identification of unstable centralized control systems, Int Conf on Frontiers in Engineering, Applied Sciences & Technology, (FEAST'17) held at NIT-Trichirappalli on 31st March & 1st April, 2017. 21–16.

Tripathi, Saurabh and Chidambaram, M. (2017). Identification of unstable TITO systems by optimization technique, TIMA, Chennai. January 2017.

[Chapter 13]

Kalpana, D. and Chidambaram, M. (2019). Auto-tuning of decentralized PI controllers for a non-square system, Paper presented in VIT, Vellore, 2019.

Abbreviations

BLT	Biggest Log-Modulus Tuning
CRC	Closed Loop Reaction Curve
CSOPTD	Critically Damped Second Order Plus Time Delay System
ETF	Equivalent Transfer Function
FOPTD	First Order Plus Time Delay System
GPM	Gain and Phase Margin
IAE	Integral of Absolute Error
IMC	Internal Model Controller
ISE	Integral of Squared Error
ISP	Industrial Scale Polymerization
ITAE	Integral of Time-Weighted Absolute Error
MIMO	Multi Input Multi Output
MRNC	Model Reference Nonlinear Controller
NC	Narasimha Reddy and Chidambaram
ODE	Ordinary Differential Equation
PI	Proportional–Integral
PID	Proportional–Integral Derivative
RGA	Relative Gain Array
RHP	Right Half Plane
SISO	Single Input Single Output
SOPTD	Second Order Plus Time Delay System

SSA   Sum of Absolute Error

SSGM   Steady State Gain Matrix

TITO   Two Input Two Output

MRC   Model Reference Control

MRRC   Model Reference Robust Control

NI   Niederlinski Index

N.I.   Number of Iterations

RK   Rangaiah and Krishnaswamy

SK   Sundaresan and Krishnaswamy

SSQ   Sum of Squares (between the closed loop model responses and the closed loop process responses for a step change in the corresponding set point values)

Notations

a	Input step magnitude
a_1	Coefficient of denominator dynamics
a_2	Coefficient of denominator dynamics
B	Defined by Eq. (1.11b)
C	Defined as $(3t^+ - t^*)/2$
A_m	Gain margin
φ_m	Phase margin
D	Transfer function matrix of the decoupler
f	$=(1 - \text{fractional response}) = (1 + \theta)\exp(-\theta)$
$g + p, g_c, g_m$	Process, controller, and model transfer function
$G(s), G_p$	Transfer function matrix of the process
G, G_c	Transfer function of the process and the controller
G_{pc}	Closed loop transfer function of the process
G_{cd}	Controller transfer function matrix (diagonal) for the decoupled system
G_m	Transfer function of the model
G_v	Disturbance transfer functions matrix
k_c	Proportional gain of the controller
$k_{c,p}, k_{I,p}$	Elements of $K_{C,d}$ and $K_{I,d}$ respectively
$k_{c,ij}, k_{I,ij}$	Elements of the centralized controller K_C and K_I
k_p	Steady state gain of the process
k_{pm}, k_p	Model and process steady state gains

k_{pc}	Closed loop process gain
$k_{pij,I}$	Process gain relating i^{th} output and j^{th} input for the inner loop
$k_{pij,o}$	Process gain relating i^{th} output and j^{th} input for the outer loop
K	$= k_c k_p$
K_C	Proportional gain of the centralized controller
K_I	Integral gain matrix of the centralized controller whose elements are $k_{c/\tau I}$
$K_{C,d}$	Proportional gain matrix (diagonal) for the decoupled system
$K_{I,d}$	Integral gain matrix (diagonal) for the decoupled system
$K_{C,O}$	K_C for the outer loop
$K_{I,O}$	K_I for the outer loop
K_P	Process gain matrix
L	Time delay defined as $(3t^+ - t^*)/2$ in Chapter 2
M	Closed loop gain matrix of the inner loop controlled system
M_s	Maximum magnitude of the sensitivity function
n	Number of inputs or outputs
p	System zero
P_{11}, P_{22}	Elements of the diagonal P matrix
P	Proportional controller
PI	Proportional integral controller
PID	Proportional integral derivative controller
q	Defined by Eq. (4.19e)
s	Laplace variable
t	Time (seconds)
t_1	Time taken to reach fractional step response of (0.14 for SOPTD model; 0.35 for FOPTD model)
t_2	Time taken to reach fractional step response of (0.55 for SOPTD model;
t_3	Time taken to reach fractional step response of 0.91 for SOPTD model
t^+	Time taken to reach fractional step response of 0.35 for FOPTD or CSOPTD model
t^*	Time taken to reach fractional step response of 0.73 for CSOPTD model
t_m	Time to reach the first undershoot

t_p	Time taken to reach the peak value
t_{p1}, t_{p2}	Time to reach the first and second peak respectively
t_s	Settling time
ΔT	Period of oscillation (Fig. A.1 in Appendix A), complementary sensitivity function
U	Manipulated variable matrix $= \begin{bmatrix} U_1 & U_2 \end{bmatrix}$
U_1	$[u_{1s}, u_{2s}]^T$, for change in the set point of y_1
U_2	$[u_{1s}, u_{2s}]^T$, for change in the set point of y_2
U_1	$[u_{1s}, u_{2s}u_{3s}]^T$, for change in the set point of y_1
U_2	$[u_{1s}, u_{2s}\ u_{3s}]^T$, for change in the set point of y_2
u	Manipulated variable
u_s	Steady state value of u
u_1, u_2, u_3	Manipulated variables
V	Disturbance variable vector
v_1, v_2	Load variables (entering along the input u_1 and u_2 respectively)
v_1	Defined by Eq. (2.30)
v_2	Defined by Eq. (2.31)
v_1, v_2, v_3	Load variables (entering along the input u_1, u_2 and u_3 respectively)
Y	Closed loop response matrix, output variable vector
y_1, y_2	Output variables
y_1	Fractional step response of 0.14 for SOPTD model (Chapter 2)
y_2	Fractional step response of 0.55 for SOPTD model (Chapter 2)
y_3	Fractional step response of 0.91 for SOPTD model (Chapter 2)
y_∞	Value of y at steady state
y_{cii}, y_{cij}	Closed loop response, closed loop interaction respectively
y_{cm}, y_m	Closed loop model response
y_r, y_{mr}	Process and model set point
y_{p1}	First peak in the closed loop response (Fig. A.1 in Appendix A)
y_{p2}	Second peak in the closed loop response (Fig. A.1 in Appendix A)
y_{m1}	First valley in the closed loop response (Fig. A.1 in Appendix A)
y_r	Set point value of y
y_p	Peak value of y
y_{11}, y_{21}	Response and interaction in y_1 and y_2

y_{12}, y_{22}	Response and interaction in y_1 and y_2
y_{c11}, y_{c21}	Closed loop response and interaction in y_1 and y_2 for step change in set point of y_1
y_{c12}, y_{c22}	Closed loop response and interaction in y_1 and y_2 for step change in set point of y_2
Δt	Half period of oscillation shown in Fig. 2.6

Greek Letters

$\propto$	Filter (lead) time constant, real part of dominant pole of the closed loop system
α	$[(t_3 - t_2)/(t_2 - t_1)]$ (for Chapter 2)
β	Imaginary part of dominant pole of the closed loop system, filter (lag) time constant
δ	Tuning constant for K_C calculation in Davison method
δ_1	Tuning parameter for the proportional gain of the centralized controller
δ_2	Tuning parameter for the integral gain of the centralized controller
δ_3	Tuning parameter for the derivative gain of the centralized controller
ε	Tuning constant for K_I in Davison method
$\zeta, \zeta_\in$	Damping coefficient of the closed loop system
ζ_1, ζ_2	Defined by Eq. (2.32) to Eq. (2.33)
η	$= \tau_I \tau_D$
θ	$(t - L)/\tau$
θ^+	$(t^+ - L)/\tau$
θ_m, θ	Model and process time delay
θ_c	Time delay of the closed loop system
σ	Singular value, standard deviation of noise
τ	Process time constant
$\tau_1 \tau_2$	Time constants of overdamped second order systems
τ_e	Effective time constant of the closed loop system
τ_{eo}	Effective time constant of the open loop system
$\tau_{i,j}$	Time of the system relating the output i to the input j
τ_m	Model time constant
τ_I	Integral time constant

τ_D	Derivative time constant
$\tau_{1,d}, \tau_{2,d}$	Time constants of the desired decoupled systems P
$\tau_{d1,d}, \tau_{d2,d}$	Time delays of the desired decoupled systems P
v_1, v_2	Defined as in Eq. (2.30) and Eq. (2.31)
φ_m	phase margin
φ	$= (t - \psi)$
ψ	time delay in the closed loop response
Δ_I	input uncertainty matrix
Δo	output uncertainty matrix
Δt	Sampling time

1

Models, Control Theory, and Examples

The Classical Control Theory studies the physical systems and the control design in the frequency domain, while the Modern Control Theory is in the time domain. Systems in the frequency domain are expressed in transfer functions (via Laplace Transforms), while the time-domain systems are described in state–space representations (a set of first order differential equations). State–space is commonly used to model Multiple Input and Multiple Output (MIMO) systems, like spacecrafts, aircrafts, automobiles, marine vessels, and so on. In this chapter, a basic review of the Classical Control Theory and the Modern Control Theory will be done.

1.1 Classic Control Theory vs Modern Control Theory

The Classical Control Theory and the Modern Control Theory are the two control theories available. The classical control method (introduced before 1950) considers (Seborg et al. 2006) root locus, and Bode, Nyquist, and Routh–Hurwitz methods, which make use of the transfer function model. Multiple Input and Multiple Output (MIMO) systems were considered one loop at a time. Modern control method refers to stat–space methods developed in the late 1950s and early 1960s. Here the system models are considered in the time domain. Analysis and design are carried out in time domain by using computers and advanced numerical methods.

During 1960s, the system representation and analysis were carried out in state variable form (representation by n first-order differential equations). An excellent review of classical control of multivariable systems is given by Skogestad and Postlethwaite (2005). Excellent review on Modern control theory is given by Gopal (2014) and Ogata (2017).

System models can be developed by two distinct methods. Analytical modelling consists of a systematic application of basic physical laws to system components and the inter connection of these components. Experimental modelling, or modelling by synthesis, is the selection of mathematical relationships which seem to fit observed input–output data.

The computational effort in analysing and designing controllers by the Modern Control Theory is increased only marginally for higher order systems. To design the controllers, all the state variables are to be available for measurement. In the Classical Control Theory, the interactions among the feedback loops are to be evaluated and taken into account in designing the controllers. Given the system description in the state–space form, the derivation of the transfer function matrix model is unique, whereas given the transfer function matrix, the derivation of the state space model description is not unique (Gopal, 2014).

1.2   State Space Model Description

A dynamic system is described by the two state variables x_1 and x_2 and the state equations Eq. (1.1) and Eq. (1.2)

$$\frac{dx_1}{dt} = f_1(x_1, x_2, u_1, u_2, d_1, d_2) \tag{1.1}$$

$$\frac{dx_2}{dt} = f_2(x_1, x_2, u_1, u_2, d_1, d_2) \tag{1.2}$$

Where, u_1, u_2 are manipulated inputs and d_1 and d_2 are disturbances affecting the process.

The nominal steady state operating point(s) can be obtained by making the accumulation terms 0 (i.e., $\frac{dx_1}{dt} = 0$ and $\frac{dx_2}{dt} = 0$) and solving the nonlinear equations Eq. (1.3) and Eq. (1.4) for x_1 and x_2 at the given process conditions (u_1, u_2, d_1, d_2).

$$f_1(x_1, x_2, u_1, u_2, d_1, d_2) = 0 \tag{1.3}$$

$$f_2(x_1, x_2, u_1, u_2, d_1, d_2) = 0 \tag{1.4}$$

To solve these nonlinear equations by using any numerical method, initial guesses for x_1 and x_2 are to be given. Depending on the initial guess values the method converges to the nearest possible solution.

In this book, the nonlinear equations are solved by using *fsolve* of Matlab. The numerical solution of the nonlinear equations by using Newton–Raphson method (Chidambaram and Padmasree, 2016) is given in the next section.

1.3 Solution of Simultaneous Nonlinear Algebraic Equations

Example 1

$$-x_1 + Da(1-x_1)\exp\left(\frac{\gamma x_2}{1+x_2}\right) = 0 \qquad (1.5)$$

$$-(1+\beta)x_2 + BDa(1-x_1)\exp\left(\frac{\gamma x_2}{1+x_2}\right) = 0 \qquad (1.6)$$

Where, $B = 0.4$, $\beta = 0.3$, $\gamma = 20$, and Da $= 0.072$.

Write a function program in the Matlab editor debugger window:

function g?**cstr_solve**(x)

beta $= 0.3$;

B $= 0.4$;

gamma $= 20$;

Da $= 0.072$;

$g(1) = -x(1) + \text{Da} * [1 - x(1)\exp[\text{gamma} * x(2)/1 + x(2))]$;

$g(2) = -(1 + \text{beta}) * x(2) + B * \text{Da} * [1 - x(1)] * \exp[\text{gamma} * x(2)/(1 + x(2))$;

save this file as **cstr_solve.m**

In the Matlab command window type the following commands

$>>$ y = fsolve('cstr_solve',[0.1 0.01],foptions)

The answer is:

y = [0.1440   0.0443]

i.e., x(1) = 0.144 and x(2) = 0.0443

Instead of 0.1 and 0.01 as initial guess for $x(1)$ and $x(2)$ if we give 0.4 and 0.4, we get:

$>>$ y = fsolve('cstr_solve',[0.4 0.4],foptions)

The answer is:

y = [0.1440   0.0443]

1.4   State Space Representation of Ordinary Differential Equation (ODE) Model

To check the stability of the steady state operating points and to derive transfer function model for the process at the steady state operating point, the following linearization procedure is used.

Linearization of the nonlinear equations Eq. (1.5) and Eq. (1.6) by Taylor series (Stephanopoulos, 1984) around the nominal steady state operating point $(x_{10}, x_{20}, u_{10}, u_{20}, d_{10}, d_{20})$ gives Eq. (1.7) and Eq. (1.8):

$$\frac{dx_1}{dt} = f_1(x_{10}, x_{20}, u_{10}, u_{20}, d_{10}) + \left(\frac{\partial f_1}{\partial x_1}\right)_0 (x_1 - x_{10})$$

$$+ \left(\frac{\partial f_1}{\partial x_2}\right)_0 (x_2 - x_{20}) + \left(\frac{\partial f_1}{\partial u_1}\right)_0 (u_1 - u_{10})$$

$$+ \left(\frac{\partial f_1}{\partial u_2}\right)_0 (u_2 - u_{20}) + \left(\frac{\partial f_1}{\partial d_1}\right)_0 (d_1 - d_{10}) \tag{1.7}$$

$$\frac{dx_2}{dt} = f_2(x_{10}, x_{20}, u_{10}, u_{20}, d_{20}) + \left(\frac{\partial f_2}{\partial x_1}\right)_0 (x_1 - x_{10})$$

$$+ \left(\frac{\partial f_2}{\partial x_2}\right)_0 (x_2 - x_{20}) + \left(\frac{\partial f_2}{\partial u_1}\right)_0 (u_1 - u_{10})$$

$$+ \left(\frac{\partial f_2}{\partial u_2}\right)_0 (u_2 - u_{20}) + \left(\frac{\partial f_2}{\partial d_2}\right)_0 (d_2 - d_{20}) \tag{1.8}$$

Where, all the derivatives are evaluated at the nominal steady state operating point, i.e., $(x_{10}, x_{20}, u_{10}, u_{20}, d_{10}, d_{20})$. The deviation variables are defined by Δ as in Eq. (1.9a) to Eq. (1.9f).

$$\Delta x_1 = x_1 - x_{10} \tag{1.9a}$$

$$\Delta x_2 = x_2 - x_{20} \tag{1.9b}$$

$$\Delta u_1 = u_1 - u_{10} \tag{1.9c}$$

$$\Delta u_2 = u_2 - u_{20} \tag{1.9d}$$

$$\Delta d_1 = d_1 - d_{10} \tag{1.9e}$$

$$\Delta d_2 = d_2 - d_{20} \tag{1.9f}$$

Substituting these deviation variables of Eq. (1.9a) to Eq. (1.9f) into Eq. (1.7) and Eq. (1.8), we get Eq. (1.10a) and Eq. (1.10b).

$$\frac{d\Delta x_1}{dt} = a_{11}\Delta x_1 + a_{12}\Delta x_2 + b_{11}\Delta u_1 + b_{12}\Delta u_2 + c_1\Delta d_1 \tag{1.10a}$$

$$\frac{d\Delta x_2}{dt} = a_{21}\Delta x_1 + a_{22}\Delta x_2 + b_{21}\Delta u_1 + b_{22}\Delta u_2 + c_2\Delta d_2 \tag{1.10b}$$

Where,

$$a_{11} = \left(\frac{\partial f_1}{\partial x_1}\right)_0, \quad a_{12} = \left(\frac{\partial f_1}{\partial x_2}\right)_0, \quad a_{21} = \left(\frac{\partial f_2}{\partial x_1}\right)_0,$$

$$a_{22} = \left(\frac{\partial f_2}{\partial x_2}\right)_0; \quad b_{11} = \left(\frac{\partial f_1}{\partial u_1}\right)_0,$$

$$b_{12} = \left(\frac{\partial f_1}{\partial u_2}\right)_0, \quad b_{21} = \left(\frac{\partial f_2}{\partial u_1}\right)_0, \quad b_{22} = \left(\frac{\partial f_2}{\partial u_2}\right)_0;$$

$$c_1 = \left(\frac{\partial f_1}{\partial d_1}\right)_0, \quad c_2 = \left(\frac{\partial f_2}{\partial d_2}\right)_0.$$

The equations Eq. (1.10a) and Eq. (1.10b) are linearized approximation of nonlinear equations around the nominal steady state operating point in terms of deviation variables. Hereafter the notation Δ is omited. Eq. (1.10a) and Eq. (1.10b) are written in the form of matrix equation as Eq. (1.11).

$$\dot{x} = Ax + Bu + Cd \tag{1.11a}$$

Where,

$$x = \begin{bmatrix} x_1 \\ x_2 \end{bmatrix}, \quad u = \begin{bmatrix} u_1 \\ u_2 \end{bmatrix}, \quad A = \begin{bmatrix} a_{11} & a_{12} \\ a_{21} & a_{22} \end{bmatrix},$$

$$B = \begin{bmatrix} b_{11} & b_{12} \\ b_{21} & b_{22} \end{bmatrix} \quad C = \begin{bmatrix} c_1 & 0 \\ 0 & c_2 \end{bmatrix} \quad \text{and} \quad d = \begin{bmatrix} d_1 \\ d_2 \end{bmatrix} \tag{1.11b}$$

If all the eigen values of A are negative or have negative real parts, then the steady state operating point is stable. If at least one eigen value is positive or has positive real part, then the steady state operating point is unstable. If the eigen values are complex conjugates with real part 0, then the system is oscillatory around that steady state operating point.

1.5   Transfer Function Derivation

If y is the output of the process and is described either as x_1 or x_2 or both, then it can be expressed as in Eq. (1.12).

$$y = Dx \tag{1.12}$$

Where, $D = \begin{bmatrix} 1 & 0 \\ 0 & 0 \end{bmatrix}$ for x_1 as output, $\begin{bmatrix} 0 & 0 \\ 0 & 1 \end{bmatrix}$ for x_2 as output and $D = \begin{bmatrix} 1 & 0 \\ 0 & 1 \end{bmatrix}$ if both x_1 and x_2 are outputs.

The transfer function model for the system is obtained by using the relation in Eq. (1.13a).

$$\boldsymbol{y}(s) = [\boldsymbol{D}(s\boldsymbol{I} - \boldsymbol{A})^{-1}\boldsymbol{B}]\boldsymbol{x}(s) \tag{1.13a}$$

The procedure can be extended to any number of state variables. Given the matrices A, B and D as in Eq. (1.13b),

$$A = \begin{bmatrix} -1 & -1 \\ 6.5 & 0 \end{bmatrix}; \quad B = \begin{bmatrix} 1 & 1 \\ 1 & 0 \end{bmatrix}; \quad D = \begin{bmatrix} 1 & 0 \\ 0 & 1 \end{bmatrix} \tag{1.13b}$$

we get the transfer function matrix as Eq. (1.14a),

$$G(s) = \begin{bmatrix} \dfrac{(s-1)}{(s^2 + s + 6.5)} & \dfrac{s}{(s^2 + s + 6.5)} \\ \dfrac{(s+7.5)}{(s^2 + s + 6.5)} & \dfrac{6.5}{(s^2+s+6.5)} \end{bmatrix} \tag{1.14a}$$

Where,

$$[y_1 \quad y_2]^{\mathrm{T}} = G(s)[u_1 \quad u_2]^{\mathrm{T}} \tag{1.14b}$$

In Matlab, a command *ss2tf* is used to get transfer function model from the state space representation.

Example 2

>>ss2tf(A,B,D,E,1)

Here E is a null matrix with the same dimension as B. This results in transfer function model of x_1(s) to u_1(s) and x_2(s) to u_1(s) in matrix form:

>>ss2tf(A,B,D,E,2)

This results in transfer function model of x_1(s) to u_2(s) and x_2(s) to u_2(s) in a matrix form.

1.6 Numerical Solution of Simultaneous Nonlinear Algebraic Equations

The method is required for the numerical solution of a nonlinear algebraic equation Eq. (1.15).

$$f(x) = 0 \qquad (1.15)$$

Taylor's series of $f(x_{i+1})$ around a value of x_i gives Eq. (1.16).

$$f(x_{i+1}) = f(x_i) + f'(x_i)(x_{i+1} - x_i) + \text{higher order derivatives} \qquad (1.16)$$

Here f' is the first derivative of f with respect to $x(f' = \partial f/\partial x)$. Neglecting the second and higher order derivatives and letting $f(x_{i+1}) = 0$, we get the iteration equation as Eq. (1.17).

$$x_{i+1} = x_i - [f(x_i)/f'(x_i)] \qquad (1.17)$$

The iterative process is to be continued till convergence is achieved. The steps involved are:

> Initial guess of x_i:
> Calculate x_{i+1} by Eq. (1.17).
> We need to check the following stopping conditions:
>
> (1) abs$[f(x_{i+1})] \leq$ given value, (say, 10^{-4})
>
> (2) check abs$(x_{i+1} - x_i) \leq$ abs$(x_i \times 10^{-3})$
>
> (3) or check the iteration not to exceed the maximum iteration ($N_{\max}$, say 100)
>
> If stopping conditions (1) or (2) are not met, iterate again. If the stopping condition (1) is to be given importance, then make the second condition more stringent (small value, use $x_i \times 10^{-5}$ rather than $x_i \times 10^{-3}$).

Similarly, for the simultaneous nonlinear algebraic equations $F(X) = 0$, we can write the iteration equation in vector form as Eq. (1.18).

$$X_{i+1} = X_i - J^{-1}(X_i)F(X_i) \tag{1.18}$$

Where,

$$J = [\partial f i^2 / \partial x_i \partial x_j]_{ij} \tag{1.19a}$$

$$X = [x_1 \quad x_2 \quad x_3 \dots x_N] \tag{1.19b}$$

$$F(X_i) = [f_1 \quad f_2 \quad f_3 \dots f_N]^T \tag{1.20}$$

Stopping conditions

(1) For each $f(x_{i+1})$, check $f(x_{i+1}) \le$ given value, (say 10^{-3})

(2) Or check for each x: $\mathrm{abs}(x_{i+1} - x_i) \le \mathrm{abs}(x_i \times 10^{-3})$

(3) Or check the iteration not to exceed the given maximum iteration (say, $N_{\max} = N * 100$)

If stopping condition (1) or (2) are not met, X_i is replaced by X_{i+1} from Eq. (1.18) and then get the next improved value from Eq. (1.21).

$$X_{i+2} = X_{i+1} - J^{-1}(X_{i+1})F(X_{i+1}) \tag{1.21}$$

Repeat the above step till the stopping condition (1) or (2) is met.

Example 3 By the Newton Raphson (NR) method, obtain the numerical solution for the simultaneous nonlinear algebraic equations Eq. (1.22, Eq. 1.23 and Eq. 1.24).

$$f_1(x_1, x_2, x_3) = 0 \Rightarrow -0.5x_1 + (1/3)(x_2^3 + x_3^3) + 0.3 = 0 \tag{1.22}$$

$$f_2(x_1, x_2, x_3) = 0 \Rightarrow 0.5x_1 x_2 - 0.5x_2 + 0.5x_3^5 = 0 \tag{1.23}$$

$$f_3(x_1, x_2, x_3) = 0 \Rightarrow (1/3)x_2^3 - (1/4)x_1 + 0.04 = 0 \tag{1.24}$$

The Jacobian matrix is obtained as Eq. (1.25).

$$J(X) = \begin{bmatrix} -0.5 & x_2^2 & x_3^2 \\ 0.5x_2 & 0.5(x_1 - 1) & 2.5x_3^4 \\ -0.25 & x_2^2 & 0 \end{bmatrix} \tag{1.25}$$

The iteration equations are given by Eq. (1.26).

$$X_{i+1} = X_i - J^{-1}(X_i)F(X_i) \tag{1.26}$$

With the initial guess as in Eq. (1.27).

$$X_0 = [0 \quad 0 \quad 0]^T \tag{1.27}$$

The NR iteration steps are executed. The results for 4 iterations are given in Eq. (1.28):

$$
\begin{aligned}
i &= 1, & X &= [0.60 \quad 0 \quad 0.16]^T \\
i &= 2, & X &= [0.6027307 \quad 0.0001049 \quad 0.16]^T \\
i &= 3, & X &= [0.6027307 \quad 0.0001056 \quad 0.16]^T \\
i &= 4, & X &= [0.6027307 \quad 0.0001056 \quad 0.16]^T
\end{aligned}
\tag{1.28}
$$

1.7 Conversion of Nth Order ODE to State Variable Form

There are several ways of converting an ODE model to a state variable form. If the variables representing the energy storage in the system are chosen as the state variable, the model must then be manipulated algebraically to produce the standard state variable form. The required algebra is not always obvious. However, we note here that the model in Eq. (1.29).

$$\frac{d^n y}{dt^n} = f\left[t, y, \frac{dy}{dt}, \ldots, \frac{d^{n-1}}{dt^{n-1}}, v(t)\right] \tag{1.29}$$

The initial conditions are:

$$\text{At } t = 0, \quad y = a_1; \quad dy/dt = a_2, \ldots, dy^{n-1}/dt^{n-1} = a_n$$

with the input $v(t)$, can always be put into state variable form by the following choice of state variables in Eq. (1.30).

$$
\begin{aligned}
x_1 &= y \\
x_2 &= \frac{dy}{dt} \\
&\vdots \\
x_n &= \frac{d^{n-1}}{dt^{n-1}}
\end{aligned}
\tag{1.30}
$$

By differentiating each equation in Eq. (1.30), the resulting state model is obtained as Eq. (1.31).

$$
\begin{aligned}
x_1' &= x_2 \\
x_2' &= x_3 \\
&\;\;\vdots \\
x_{n-1}' &= x_n \\
x_n' &= f\,[t, x_1, x_2, \ldots, x_n, v(t)]
\end{aligned}
\tag{1.31}
$$

at $t = 0$; $x_1 = a_1$; $x_2 = a_2$; $x_3 = a_3$; ...; $x_n = a_n$

1.8   Coupled Higher Order Models

Equation (1.29) is somewhat general in that it can be nonlinear, but it does not cover all cases. For example, consider the coupled higher order model in Eq. (1.32).

$$
\begin{aligned}
y' &= f_1(t, y, y', z, z') \\
z' &= f_2(t, y, y', z, z')
\end{aligned}
\tag{1.32}
$$

With the appropriate initial conditions. Choose the state variables as in Eq. (1.33).

$$
\begin{aligned}
x_1 &= y \\
x_2 &= y' \\
x_3 &= z \\
x_4 &= z'
\end{aligned}
\tag{1.33}
$$

By differentiating Eq. (1.33) we get state model as Eq. (1.34).

$$
\begin{aligned}
x_1' &= x_2 \\
x_2' &= f_1(t, x_1, x_2, x_3, x_4) \\
x_3' &= x_4 \\
x_4' &= f_2(t, x_1, x_2, x_3, x_4)
\end{aligned}
\tag{1.34}
$$

It can be stated that control theory started and flourished using transfer function methods. Then the state variables approach was developed and for many years it was synonymous with modern control.

The selection of state variables is not a unique process. Various set of state variables can be used. Some are easier to derive, whereas others are easier to work with once they are obtained. The starting information about a system is derived from an experimentally obtained transfer function. There is often a good reason to select states which have physical significance. These can potentially be measured, perhaps by adding additional instrumentation.

The state variable approach originally was able to define and explain more completely many system characteristics and attributes. Currently, most of the advantages and insights gained by the use of state variable methods in the early years have been found to have counterparts in the new and expanded input–output transfer function methods of multivariable systems. Both the two approaches are now being used. They require somewhat different mathematical tools, but they complement each other in various ways. Some of the newest aspects of transfer function methods appear in this manner.

1.9 Reported Transfer Function Matrix Models from Experiments

1.9.1 Distillation Column for Binary Ethanol–Water Separation

Ogunnaike et al. (1983) determined a model relating outputs to inputs through pulse testing of the column and fitting simple First Order Plus Time Delay (FOPTD) transfer functions to most elements of the process model, and in some cases some slightly more complicated forms. The distillation column studied was a 19 plate, 12-inch diameter copper column having variable feed and side stream draw-off locations. Temperatures were measured on each tray as well as in the overhead, reflux, feed, and product lines. Compositions were determined through various online sensors (densitometry, refractometry, etc.). Data acquisition and computer control were carried out by interfacing to a PDP 11/55 minicomputer. This study was concerned with the binary ethanol–water system, but later results included ternary mixtures.

For the binary ethanol–water system, the column model takes the form Eq. (1.35).

$$
G = \begin{bmatrix}
\dfrac{0.66e^{-2.6s}}{6.7s+1} & \dfrac{-0.61e^{-3.5s}}{8.64s+1} & \dfrac{-0.0049e^{-1.0s}}{9.06s+1} \\[2ex]
\dfrac{1.11e^{-6.5s}}{3.25s+1} & \dfrac{-2.3e^{-3s}}{5.0s+1} & \dfrac{-0.01e^{-1.2s}}{7.09s+1} \\[2ex]
\dfrac{-34.68e^{-9.2s}}{8.15s+1} & \dfrac{46.2e^{-9.4s}}{10.9s+1} & \dfrac{0.85e^{-0.83s}}{6.60s+1}
\end{bmatrix}
\tag{1.35}
$$

Where the outputs are:

y_1 = overhead ethanol mole fraction

y_2 = side stream ethanol mole fraction

y_3 = tray #19 temperature, °C (corresponding to bottom composition)

The inputs are:

u_1 = reflux flow rate, gpm (m^3/s)

u_2 = side stream product flow rate, gpm (m^3/s)

u_3 = reboiler stream pressure, psig (kPa)

Ogunnaike et al. (1983) reported experimental data from two pulse- tests and the resulting transfer function model fits of the data. Parameters were estimated by regular least squares in the frequency domain and by direct minimization of the sum of squares surface on the data in the time domain using a standard optimization routine.

1.9.2   Wood and Berry Column

Wood and Berry (1972) reported the transfer functions models characterizing the distillation column (for ethanol and water systems) dynamics established by pulse testing. This study was performed using a 9-inch diameter, 8 tray column equipped with a total condenser and a basket type reboiler. The trays, at a spacing of 12 inch, were fitted with four 23 × 18 inch bubble caps arranged in a square pattern. A continuous in-line capacitance cell was developed and used for analysis of the overhead composition. Bottoms composition was analysed by means of a Beckman Series C gas chromatograph modified so that sample injection and analysis of the chromatogram could be performed by the computer.

Parameters of assumed FOPTD transfer functions were determined from the transient data. The time delays were established visually and the gains and time constants determined by a least squares fit employing direct search technique. The simplified dynamic model was expressed as in Eq. (1.36).

$$G_p(s) = \begin{bmatrix} \dfrac{12.8e^{-s}}{16.7s+1} & \dfrac{-18.9e^{-3s}}{21s+1} \\ \dfrac{6.6e^{-7s}}{10.9s+1} & \dfrac{-19.4e^{-3s}}{14.4s+1} \end{bmatrix} \tag{1.36}$$

1.9.3 Pilot Scale Acetone and Isopropyl Alcohol Distillation Column

Mutalib (2014) conducted an experimental study of the system identification for a pilot scale project for separating acetone and isopropyl alcohol in a distillation column. He reported experiment data on step testing Reflux (R) and separately on step testing Steam Flow Rate (S) on the composition of distillate (X_D) and composition in the bottom plate (X_B). The compositions of the distillate and bottom products were analysed by using a refractometer. The overall mathematical model is given in Eq. (1.37) and Eq. (1.38).

$$[x_D(s) \quad x_B(s)]^T = G_P(s)[R(s) \quad S(s)]^T \tag{1.37}$$

Where, $G_P(s)$ is given by Eq. (1.38).

$$G_p(s) = \begin{bmatrix} \dfrac{0.187}{1.29s+1} & \dfrac{-0.0086}{0.73s+1} \\ \dfrac{0.0031}{0.96s+1} & \dfrac{-0.00024}{0.84s+1} \end{bmatrix} \tag{1.38}$$

1.9.4 Babji and Saraf Distillation Column

Babji and Saraf (1991) reported an experimental study of distillation column by transfer function matrix by step test experiments on ethanol–water mixture. The pilot plant column consisted of 16 stage sieve tray column of 22.8 cm diameter and 10 metre height made of 316 stainless steel. Temperature in 6th and 16th plates were measured and from these values compositions were calculated if pressure was held constant. The compositions were considered as output variables to be controlled, and reflux flow and steam pressure were considered manipulated variables.

Step change were given in reflux flow rate, and separately in reboiler steam pressure. A random search optimization was used to find the model parameters from each of the step response data. It was found that the FOPTD model was able to fit the input and output data satisfactorily. The transfer function model given in Eq. (1.39) and Eq. (1.40) were obtained.

$$[T_6(s) \quad T_{16}(s)]^T = G_P(s)[R \quad S_T]^T \tag{1.39}$$

$$G_p(s) = \begin{bmatrix} \dfrac{-0.12e^{-0.5s}}{1.9s+1} & \dfrac{0.82e^{-0.9s}}{4.6s+1} \\ \dfrac{-0.11e^{-0.9s}}{3.7s+1} & \dfrac{1.05e^{-1.1s}}{5.1s+1} \end{bmatrix} \tag{1.40}$$

1.9.5 Crude Oil Distillation Process

The crude oil distillation unit lies at the front end of a refinery. This unit performs the initial distillation of the crude oil into several boiling range fractions. The crude is pumped in from storage tanks and, after desalination, is preheated against crude tower products and overhead streams. The crude is then partially vaporized in two parallel fuel gas-fired heaters. The vapour and liquid from the heaters enter the flash zone at the bottom of the crude column. Muske et al. (1991) have considered the crude distillation unit at Cosmo Oil's Sakai Refinery. They have given the transfer function for three general crudes and average crude. Crude 2 is considered in this work and its transfer function is given in Eq. (1.41). In this example, controlled variables are naphtha/kerosene cut point (y_1), kerosene/light gas oil (LGO) cut point (y_2), LGO/HGO cut point (y_3) and measured over flash (y_4). Manipulated variables are top temperature (u_1), kerosene yield (u_2), LGO yield (u_3), heavy gas oil (HGO) yield (u_4) and heater outlet temperature (u_5).

$$G(s) = \begin{bmatrix} \dfrac{3.8(16s+1)}{140s^2+14s+1} & \dfrac{2.9e^{-6s}}{10s+1} & 0 & 0 & \dfrac{-0.73-16s+1)e^{-4s}}{150s^2+20s+1} \\ \dfrac{3.9(4.5s+1)}{96s^2+17s+1} & \dfrac{6.3}{20s+1} & 0 & 0 & \dfrac{16se^{-2s}}{(5s+1)(14s+1} \\ \dfrac{3.8(16s+1)}{140s^2+14s+1} & \dfrac{6.1(12s+1)e^{-s}}{337s^2+34s+1} & \dfrac{3.4e^{-2s}}{6.9s+1} & 0 & \dfrac{22se^{-2s}}{(5s+1)(10s+1} \\ \dfrac{-0.73-16s+1)e^{-4s}}{150s^2+20s+1} & \dfrac{-1.53(3.1s+1)}{5.1s^2+7.1s+1} & \dfrac{-1.3(7.6s+1)}{4.7s^2+7.1s+1} & \dfrac{-0.6e^{-s}}{2s+1} & \dfrac{0.32(-9.1s+1)e^{-s}}{12s^2+15s+1} \end{bmatrix} \tag{1.41}$$

Here each element $G(i, j)$ is given by y_i/u_j; $i = 1, 4$ and $j = 1, 4$

1.9.6 Two Coupled Pilot Plant Distillation Column

Levien and Morari (1987) reported an experimental evaluation of methods to analyse process resilience based on Internal Model Control (IMC). A pilot plant consisting of two coupled distillation columns was designed for control studies using an LSI 11/23 microcomputer. Two alternative sets of inputs were chosen and the two full 7 (all outputs) × 4 (all inputs) transfer matrix models of the pilot plant were identified. From these models, a large number of potential 3 × 3 control structures were first screened for different levels of achievable control. In this book, the above system having non-minimum phase is considered with 2 outputs and 3 inputs. Here the outputs are a mole fraction of ethanol in distillate (y_1) and a mole fraction of water in bottoms (y_2), and manipulated variables are distillate flow rate (u_1), steam flow rate (u_2), and product fraction from the side column (u_3). The system transfer function is given as in Eq. (1.42).

$$G(s) = \begin{bmatrix} \dfrac{0.052e^{-8s}}{19.8s + 1} & \dfrac{-0.03(1 - 15.8s)}{108s^2 + 63s + 1} & \dfrac{0.012(1 - 47s)}{181s^2 + 29s + 1} \\ \dfrac{0.0725}{890s^2 + 64s + 1} & \dfrac{-0.0029(1 - 560s)}{293s^2 + 51s + 1} & \dfrac{0.0078}{42.3s + 1} \end{bmatrix} \tag{1.42}$$

Here each element $G(i, j)$ is given by y_i/u_j; $i = 1, 3$ and $j = 1, 3$

1.9.7 A Reactor Control Problem

This example is an industrial-scale polymerization reactor control problem (Chien et al., 1999). In order to improve the consistency of the quality of the product, a control improvement project was initiated. The first phase of the project was to do automatic reactor condition control. The manipulated variables of this control tier were the set points of various reactor feed flow control loops. The future higher tier product quality control loops will be setting the set points of the reactor condition control loops.

For proprietary reasons, we will not describe in detail this reactor control system. The process model for the reactor composition control loops is obtained from model identification given in Eq. (1.43).

$$G(s) = \begin{bmatrix} \dfrac{22.89e^{-0.2s}}{4.572s + 1} & \dfrac{-11.64e^{-0.4s}}{1.807s + 1} \\ \dfrac{4.689e^{-0.2s}}{2.174s + 1} & \dfrac{5.8e^{-0.4s}}{1.801s + 1} \end{bmatrix} \tag{1.43}$$

The time scales are in hours, so the dynamic response of the process is quite slow. The two controlled variables are two measurements representing the reactor condition, and the two manipulated variables are the set points of two reactor feed flow loops with load disturbance as the purge flow of the reactor.

1.9.8   Transfer Function Model of a Packed Distillation Column

Taiwo (2005) reported the experimental transfer function matrix of a packed distillation column. The packed distillation column used for the experiments was a part of the pilot plant in the Department of Chemical Engineering at the University Manchester Institute of Science and Technology (UMIST). This continuous fractionating column separates a mixture of methanol and iso-propanol into 97 wt% pure product methanol as a distillate. The column was constructed from mild steel 23 cm in diameter and packed with 13 mm ceramic Raschig rings to a total height of 8.52 m. The products taken off at the top and bottom of the column were returned to the appropriate feed tank for recycling (). The reboiler was heated by steam at 4.5 bar and the total condenser was cooled by water. Both the reflux and the steam flow rates were controlled by electro-pneumatic valve positioners and monitored by orifice plates fitted with differential pressure cells. Temperatures were measured using platinum resistance thermometers which, in the case of the top of the column, was connected to a continuous transmitter (accuracy $\pm 0.1°C$) and at the bottom was connected to a Kent six-point recorder, which scanned each temperature at 30 second intervals (accuracy $\pm 0.2°C$).

A least squares fit employing Rosenbrock's (1969) direct search technique was used to determine the parameters of simple models which describe plant responses. The model adopted for the column is given by Eq. (1.44).

$$[T_1(s) \quad T_2(s)]^T = G_P(s)[R(s) \quad S(s)]^T \tag{1.44}$$

Where,

$$G_P(s) = \begin{bmatrix} \dfrac{-0.86e^{-s}}{(35.4s+1)} & \dfrac{0.6}{(18.9s+1)(2.6s+1)} \\ \dfrac{-1.5e^{-4s}}{(74s+1)} & \dfrac{1.22}{(30.6s+1)} \end{bmatrix}$$

The time constants and time delays are in minutes and the gains have the units Khr/kg. A good agreement was reported between plant outputs and the transfer assumed model.

1.9.9   Transfer Function Model of a Distillation Column

Binder et al. (1984) have reported the transfer function matrix model of a distillation column. The experimental system considered is mixture of water–methanol. The output variables are top composition (X_D) and flow rate (L_D). The manipulated variables are heating power (Q_B) and reflux flow rate (R). The obtained model is given by Eq. (1.45).

$$[X_D(s) \quad L_D(s)]^T = G_P(s)[Q_B(s) \quad R(s)]^T \tag{1.45}$$

Where, T is the transpose of the vector and,

$$G_p(s) = \begin{bmatrix} \dfrac{-1.9e^{-15s}}{(67s+1)} & \dfrac{1.2e^{-30s}}{(40s+1)} \\ \dfrac{1.8e^{-15s}}{(25s+1)} & \dfrac{-10.95}{(20s+1)} \end{bmatrix}$$

1.9.10 Transfer Function Model of an Industrial Distillation Column

Tyreus (1979) reported experimental work on 30 tray, 2 m diameter industrial distillation column separating a partly vaporized multi-component feed into two streams. The transfer function matrix is reported as Eq. (1.46).

$$G(s) = \begin{bmatrix} \dfrac{1.318e^{-2.58s}}{(20s+1)} & \dfrac{0.333e^{-4s}}{s} & \dfrac{-0.11e^{-12s}}{s} \\ \dfrac{-0.038(182s+1)e^{-0.59s}}{(27s+1)(10s+1)(6.5s+1)} & \dfrac{0.36}{s} & \dfrac{-0.12e^{-38s}}{s(14s+1)(8s+1)} \\ \dfrac{0.313e^{-7.75s}}{s(29s+1)(17s+1)} & \dfrac{2.222}{s(10s+1)(6.7s+1)} & \dfrac{1.563e^{-27.33s}}{s(17s+1)} \end{bmatrix} \tag{1.46}$$

Where,

$u_1 =$ Distillate flow

$u_2 =$ Steam flow to the reboiler

$u_3 =$ Reflux flow rate

$y_1 =$ Pressure in the reflux drum

$y_2 =$ Temperature on the bottom plate

$y_3 =$ Pressure in the condenser

Summary

To conclude, a brief introduction of Classical Control Theory and the Modern Control Theory has been given in this chapter. The modelling of systems by state space model and by transfer function matrix model have been reviewed. Some of the reported experimental transfer function matrices have been given. In the next chapters, the identification of transfer function models for single input and single output systems and multi input and multi output systems and design of controllers will be considered in detail.

Problems

1. Luyben and Vinante (1972) reported the transfer function matrix model for a distillation column by conducting experiments as:

$$G_p(s) = \begin{bmatrix} \dfrac{-2.16e^{-s}}{(8s+1)} & \dfrac{1.26e^{-0.3s}}{(9.5s+1)} \\ \dfrac{-2.75e^{-1.8s}}{(9.5s+1)} & \dfrac{4.28e^{-0.35s}}{(9.2s+1)} \end{bmatrix}$$

   Prepare a write-up on the binary system used and list the output variables and the manipulated variables and the type of experiments used by them.

2. Schwanke et al. (1977) reported the transfer function matrix model for a distillation column by conducting experiments as:

$$G(s) = \begin{bmatrix} \dfrac{-10.8(3.08s+1)}{(2.13s^2+2.04s+1)} & \dfrac{0.52(3.125s+1)}{(1.78s^2+1.87s+1)} \\ \dfrac{-28.14e^{-0.65s}}{(1.9s^2+2.21s+1)} & \dfrac{1.84}{(1.87s^2+2.19s+1)} \end{bmatrix}$$

   Prepare a write-up on the binary system used and list the output variables, the manipulated variables, and the type of experiments used by them.

3. The transfer function matrix of a reformer (ammonia production) is reported as:

$$G(s) = \begin{bmatrix} \dfrac{-0.562e^{-4s}}{(8s+1)} & \dfrac{-0.01e^{-s}}{(3s+1)} & \dfrac{0.378e^{-4s}}{(8s+1)} \\ \dfrac{0.0135e^{-19s}}{(5.6s+1)} & 0 & \dfrac{-0.0159e^{-21s}}{(10s+1)} \\ \dfrac{-0.002e^{-10s}}{(7s+1)} & \dfrac{-0.00125e^{-4s}}{(3s+1)} & \dfrac{0.12e^{-7.5s}}{(10s+1)} \end{bmatrix}$$

   Where,

   $u_1 =$ The process natural gas flow rate

   $u_2 =$ The process air flow rate

u_3 = The fuel natural gas flow rate

y_1 = Primary reformer coil outlet temp

y_2 = Secondary reformer methane leak

y_3 = The H_2/N_2 ratio

Prepare a write-up on the system used and list the output variables, the manipulated variables, and the type of experiments used by them.

4. Johansson (2000) reported transfer function model for a Quadruple-tank Process as:

$$G(s) = \begin{bmatrix} \dfrac{1.5}{(63s+1)} & \dfrac{2.5}{(63s+1)(39s+1)} \\ \dfrac{2.5}{(91s+1)(56s+1)} & \dfrac{1.6}{(91s+1)} \end{bmatrix}$$

Where,

u_1 = Input voltage to pump 1

u_2 = Input voltage to pump 2

y_1 = Voltage from level measurement device 1

y_2 = Voltage from level measurement device 2

Prepare a write-up on the system used and list the output variables, the manipulated variables, and the type of experiments used by them.

5. Wang et al. (2008) reported transfer function model for a depropanizer column as:

$$G(s) = \begin{bmatrix} \dfrac{-0.2698e^{-27.5s}}{(97.5s+1)} & \dfrac{1.978e^{-53.5s}}{(118.5s+1)} & \dfrac{0.07724e^{-56s}}{(96s+1)} \\ \dfrac{0.4881e^{-117s}}{(56s+1)} & \dfrac{5.26e^{-26.5s}}{(58.5s+1)} & \dfrac{0.19996e^{-35s}}{(58.5s+1)} \\ \dfrac{0.6e^{-16.5s}}{(40.5s+1)} & \dfrac{5.5e^{-15.5s}}{(19.5s+1)} & \dfrac{-0.5e^{-17s}}{(18s+1)} \end{bmatrix}$$

Where,

u_1 = Column top reflux flow

u_2 = Column bottom steam flow

u_3 = Column overhead pressure

y_1 = Top butane concentration

y_2 = Bottom propane concentration

y_3 = Column DP flooding

Prepare a write-up on the system used and list the output variables, the manipulated variables, and the type of experiments used by them.

2

Identification and Control of SISO Systems

In this chapter, basics of open loop and closed loop identification methods of transfer function models of Single Input Single Output (SISO) systems are reviewed. Open loop identification methods are proposed to identify the parameters of First Order Plus Time Delay (FOPTD) model and Critically Damped Second Order Plus Time Delay (CSOPTD) model. The methods under closed loop identification include the reaction curve method and the optimization method. A review of these methods is given in this chapter. Methods of designing PI controllers for transfer function models for SISO systems are brought out.

2.1 Identification of SISO Systems

Process identification is the method of obtaining a model which can be used to predict the behaviour of the process output for a given process input. Transfer function model identification is important for the design of controllers. Identification can be carried out in an open loop manner or in a closed loop manner. Excellent books are available for process identification (Ljung, 1998; Soderstrom and Stoica, 1989; Wang et al., 2008; Sung et al., 2009). Pintelon and Schoukens (2001) have given an excellent description of the methods of system identification from the data represented in frequency domain. Clark (2005) plotted the normalized step response curve in time domain for second order systems and third order systems

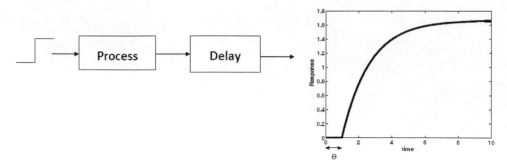

FIGURE 2.1   Open loop identification

for different damping ratios. These graphs can be used for the estimation of model parameters. Keesman (2011) has given an excellent account of the method available for system identification based on the discrete time models.

2.1.1   Open Loop Identification of Stable Systems

In open loop identification of the system, for a given change in the input variable, the change in output response is noted. The block diagram of the open loop identification is shown in Fig. 2.1. Many works have been reported on open loop identification methods. Identification of a FOPTD model is adequate for design of Proportional Integral Derivative (PID) controllers. Sundaresan and Krishnaswamy (1978) (SK method) presented a reaction curve method for process identification of FOPTD models. Sundaresan et al. (1978) pointed out the error sensitivity involved in the inflection point in the graphical method and suggested a new method of parameter estimation for non-oscillatory, oscillatory, and systems with delay, which involves utilizing the first moment of the step response curve. An open loop identification for second order parameters from step and impulse response data was given by Huang and Clements (1982). In this method, they developed correlations for parameter estimation from step response data using one point (inflection point), two point or three points.

2.1.2   Identification of CSOPTD Systems

The transfer function model is required to design a PID controller. Usually, a FOPTD model or a SOPTD model is sufficient to design a PID controller. The Critically Damped Second Order Plus Time Delay Model (CSOPTD) has only three parameters (Stephanopoulos, 1984) as given in Eq. (2.1),

$$\Delta y(s)/\Delta u(s) = k_P \exp(-Ls)/(\tau s + 1)^2 \tag{2.1}$$

when compared to four parameters of an over-damped SOPTD model as given in Eq. (2.2).

$$\Delta y(s)/\Delta u(s) = k_P \exp(-Ls)/(\tau_e^2 s^2 + 2\tau_e \zeta s + 1) \tag{2.2}$$

The time delay (L) arises due to measurement delay and/or due to the approximation of a higher order system by a simple FOPTD or SOPTD model. Excellent reviews on the identification of transfer function are given by Liu et. al. (2013), Liu and Gao (2012), and Okiy (2015). Simi and Chidambaram (2015) have showed that a significant improvement is obtained in the performance of the control system based on a critically damped SOPTD model (CSOPTD) over a FOPTD model.

Open loop identification methods are available to obtain an over-damped SOPTD model for higher order systems (Huang and Clements, 1982; Huang and Huang, 1993; Rangaiah and Krishnaswamy, 1994). Huang and Huang (1993) have proposed a method to estimate the parameters of SOPTD by utilizing four points of the fractional step response data using correlation equations. In Eq. (2.3), k_P is the ratio of the change in the steady state value of output and the change in the steady state value of input:

$$k_P = \Delta y_\infty / \Delta u_\infty \tag{2.3}$$

Hence, the model to be identified from the fractional step response is given by Eq. (2.4).

$$\Delta y(s)/\Delta u(s) = \exp(-Ls)/(\tau_e^2 s^2 + 2\tau_e \zeta s + 1) \tag{2.4}$$

Rangaiah and Krishnaswamy (1994) have utilized three points of the fractional step response curve for an over-damped system. The model is assumed to intersect with the system response at three points, that is, (y_1, t_1), (y_2, t_2), and (y_3, t_3). Rangaiah and Krishnaswamy (1994) have considered for each set of τ, ζ, and L, and solved numerically the derivative equations of IAE (integral absolute error between the actual system and the model) with respect to τ, ζ, and L, to get the values of y_1, y_2, and y_3 and hence to obtain t_1, t_2 and t_3. Rangaiah and Krishnaswamy (1994) have varied the damping coefficient ($\zeta = 1.0$ to 3.0 with $\tau_e = 1$ and $L = 0.4$) and from the fractional step response plot noted the three values of fractional responses and the corresponding times (t_1, t_2, and t_3). The average values for y_1, y_2, and y_3 over all the simulation studies are considered as 0.14, 0.55 and 0.91. The correlation (polynomial of order 5) between ζ and α, defined by $(t_3 - t_2)/(t_2 - t_1)$, is proposed. Two more correlations (polynomial of order 4) are obtained for $[(t_3 - t_1)/\tau]$ versus ζ, and $[(t_2 - L)/\tau]$ versus ζ. A simple set of equations is to be interpreted from the correlations by letting $\zeta = 1$ to calculate the modal parameters of the critically damped SOPTD model as in Eq. (2.5) and Eq. (2.6).

$$(t_2 - t_1)/\tau = 1.189 \tag{2.5}$$

$$(t_2 - L)/\tau = 1.843 \tag{2.6}$$

Where, t_1 and t_2 are noted from the fractional step responses respectively at $y_1 = 0.14$ and $y_2 = 0.54$. For the CSOPTD model, only two parameters to be identified; k_P is already calculated from Eq. (2.3) and hence (y_1, t_1) and (y_2, t_2) are only required. As seen above, the derivation for the model parameters of CSOPTD is complicated in the work of Rangaiah and Krishnaswamy (1994) [known as RK method]. Narasimha Reddy and Chidambaram (NC) (2020) presented a simple method (NC method) for identifying the model parameters of the CSOPTD model. This method (NC) is discussed in the next section.

2.2   Narasimha Reddy and Chidambaram (NC) Method

2.2.1   Identification of CSOPTD Model by NC method

In the NC method, a simple derivation is given to calculate the model parameters of a CSOPTD system. This method makes use of the unique property of such systems (Harriott, 1964), that the time required to reach the fractional step response of 0.73, is given by Eq. (2.7).

$$[(t^* - L)/2\tau] \approx 1.3 \tag{2.7}$$

One more equation is to be formulated to solve for the two parameters τ and L. The fractional step response of the CSOPTD system (for a step change of magnitude 'a') is given by Harriott (1964) as in Eq. (2.8).

$$y/(k_P a) = 1 - \{1 + [(t - L)/\tau] \exp[-(t - L)/\tau]\} \tag{2.8}$$

Equation (2.8) can be recast as Eq. (2.9a).

$$f = (1 + \theta) \exp(-\theta) \tag{2.9a}$$

Where,

$$f = 1 - [y/(k_P a)] \qquad (2.9b)$$

$$\theta = (t - L)/\tau \qquad (2.10)$$

In Eq. (2.9a), by substituting various values of θ, we can get the corresponding value of f. Fig. 2.2 shows f versus θ behaviour. For easy interpretation, Table 2.1 gives the values of θ for various values of f. There is no need to solve numerically the nonlinear algebraic equation Eq. (2.9a). For the given $y/(k_P a) = 0.35$ as the fractional step response (that is, $f = 0.65$), we get the value of θ (denoted by θ^+) from Fig. 2.2 or from Table 2.1 and the time t^+ is noted from the fractional step response of the system.

Given the value of f, the value of θ can be noted.

From Eq. (2.10), we have Eq. (2.11).

$$(t^+ - L)/\tau = \theta^+ \qquad (2.11)$$

When the fractional step response is 0.73, we have noted the time (t^*) from the fractional step response and $\theta*$ from Fig. 2.2 or Table 2.1 as 2.58. Hence, we have

TABLE 2.1 f versus θ

f	0.09	0.2	0.27	0.35	0.4	0.45	0.5	0.55	0.65	0.7	0.8	0.86	0.91	1.0
θ	3.89	2.99	2.58	2.218	1.99	1.84	1.67	1.52	1.23	1.09	0.79	0.65	0.53	0

$(1 - f)$ is the fractional step response and $f = (1 + \theta)\exp(-\theta)$ and $\theta = (t - L)/\tau$

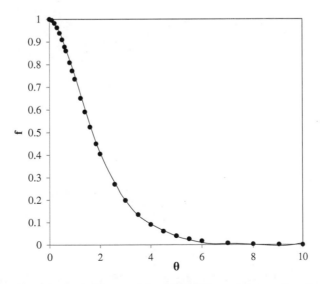

FIGURE 2.2   Graph of f versus θ for CSOPTD model: $f = (1 + \theta)\exp(-\theta)$

Eq. (2.12).

$$(t^* - L)/(2\tau) = 1.29 \qquad (2.12)$$

Harriott (1964) reported an approximate value of 1.3 for $(t^* - L)/(2\tau)$. Solving the two equations Eq. (2.11) and Eq. (2.12), we get the values of L and τ as given in Eq. (2.13a) and Eq. (2.13b).

$$L = (2.58t^+ - t^*\theta^+)/(2.58 - \theta^+) \qquad (2.13a)$$

$$\tau = (t^* - t^+)/(2.58 - \theta^+) \qquad (2.13b)$$

We can also get the values of L and τ from any two data. The values of fractional step response and its corresponding time (f_1, t_1) and (f_2, t_2) are to be noted from the original system and hence we can calculate the corresponding values of θ_1 and θ_2 from Table 2.1. Then solving the two equations Eq. (2.14a) and Eq. (2.14b), we get the model parameters (L, τ) of CSOPTD.

$$(t_1 - L)/\tau = \theta_1 \qquad (2.14a)$$

$$(t_2 - L)/\tau = \theta_2 \qquad (2.14b)$$

We get the equations Eq. (2.15a) and Eq. (2.15b) for L and τ from equations Eq. (2.14a) and Eq. (2.14b).

$$\tau = (t_1 - t_2)/(\theta_1 - \theta_2) \qquad (2.15a)$$

$$L = (t_2\theta_1 - t_1\theta_2)/(\theta_1 - \theta_2) \qquad (2.15b)$$

2.2.2   Identification of FOPTD Model

For SOPTD system, Eq. (2.12) is rewritten as Eq. (2.16).

$$(t^* - L)/(\tau_1 + \tau_2) = 1.29 \qquad (2.16)$$

Since for FOPTD system $(\tau_2 = 0)$, we have Eq. (2.17).

$$(t^* - L)/\tau = 1.29 \qquad (2.17)$$

Where, t^* is the time noted from the system response to reach the fractional step response of 0.73.

The fractional step response of a FOPTD system is given by Eq. (2.18).

$$y/(k_P a) = 1 - \exp[-(t - L)/\tau] \tag{2.18}$$

The time taken (t^+) to reach 0.35 (approximately half of 0.73) of fractional step response is given by Eq. (2.19).

$$(t^+ - L)/\tau = -\ln(0.65) \tag{2.19}$$

$$\text{i.e.,} \quad (t^+ - L)/\tau = 0.43 \tag{2.20}$$

By solving Eq. (2.17) and Eq. (2.20), we get the parameters τ and L of the FOPTD model as in Eq. (2.21) and Eq. (2.22).

$$L = (3t^+ - t^*)/2 \tag{2.21}$$

$$\tau = (t^* - C)/(1.29) \tag{2.22}$$

Where, $C = (3t^+ - t^*)/2$. In case, we consider FOPTD model, as in Eq. (2.18), to get the time to reach 0.73 fractional step response, we get 1.31 instead of 1.29 which we got in Eq. (2.17). By using 1.31 rather than 1.29 in Eq. (2.22), the model parameters of FOPTD and the ISE values are calculated. There are no significant changes in the ISE values which are obtained in this manner.

2.2.3   Simulation Examples

Let us consider the five simulation examples for the present work (Simi and Chidambaram, 2015) as in equations Eq. (2.23) to Eq. (2.27).

$$\text{Example 1 } G_P(s) = \exp(-10s)/(20s + 1)^2 \tag{2.23}$$

$$\text{Example 2 } G_P(s) = \exp(-2s)/[(20s + 1)(10s + 1)(s + 1)] \tag{2.24}$$

$$\text{Example 3 } G_P(s) = (1.5s + 1)\exp(-2s)/[(10s + 1)(20s + 1)] \tag{2.25}$$

$$\text{Example 4 } G_P(s) = 1/(s + 1)^8 \tag{2.26}$$

$$\text{Example 5 } G_P(s) = \exp(-4s)/(63.291s^2 + 25.597s + 1) \tag{2.27}$$

Table 2.2 gives the identified parameters of a CSOPTD model. Table 2.3 gives the comparisons of IAE and ISE values, where the error is defined as the difference between the step response of the actual system and that of the CSOPTD model.

In Table 2.2, t_1 and t_2 are the time required to reach respectively 0.14 and 0.55 fractional responses for the CSOPTD model by Rangaiah and Krishnaswamy (RK) method and for the present method, t_1 and t_2 correspond to fractional responses of 0.35 and 0.73.

TABLE 2.2   Parameters of the identified CSOPTD model

| | Values of t_1, t_2 | | | | Model Parameters | | | |
| | CSOPTD (RK) | | CSOPTD (Present) | | Present | | RK | |
Example	t_1	t_2	t_1	t_2	τ	θ	τ	θ
1. Eq. (2.23)	34.7	61.73	23.1	46.9	20.02	10.07	20.01	10
2. Eq. (2.24)	20.9	41.55	12.35	30.05	15.29	2.085	14.88	2.61
3. Eq. (2.25)	18.4	39	9.92	27.53	15.26	−0.368 [0]	14.81	0.23
4. Eq. (2.26)	6.65	9.48	5.01	8.02	2.098	4.072	2.53	3.36
5. Eq. (2.27)	16.74	36.84	9.978	25.18	14.88	−1.57[0]	12.78	1.62

TABLE 2.3   Comparison of IAE and ISE values for present and RK methods for CSOPTD model

| | ISE | | IAE | |
Example	Present Method	RK Method	Present Method	RK Method
1. Eq. (2.23)	0.4099	0.4015	4.9288	4.8909
2. Eq. (2.24)	0.0016	0.0041	0.3525	0.5343
3. Eq. (2.25)	0.0036	0.0046	0.5025	0.5978
4. Eq. (2.26)	0.0082	0.0182	0.2899	0.5284
5. Eq. (2.27)	0.0494	0.0763	1.7410	2.4546

Error is defined as the difference between the system and the model

2.2.4   Comparison with SK Method

The model parameters of FOPTD are calculated and the values are given in Table 2.4. for the present method and by the SK method (Sundaresan and Krishnaswamy, 1978). For the present method Eq. (2.21) and Eq. (2.22) are used to identify the model parameters of the FOPTD system. Eq. (2.28) and Eq. (2.29) are used for the SK method.

$$L = 1.3t_1 - 0.29t_2 \qquad (2.28)$$

$$\tau = 0.67(t_2 - t_1) \qquad (2.29)$$

Where t_1 and t_2 are the time taken to reach 0.35 and 0.85 of the fractional step responses.

In Table 2.4, t_1 and t_2 are the times required to reach, respectively, 0.35 and 0.85 fractional responses for the FOPTD model by the SK method. For the Present method, t_1 and t_2 correspond to fractional responses of 0.35 and 0.73.

TABLE 2.4   Parameters of the identified FOPTD model

| | Values of t_1, t_2 | | | | Model Parameters | | | |
| | FOPTD (SK) | | FOPTD (Present) | | Present | | | SK |
Example	t_1	t_2	t_1	t_2	τ	θ	τ	θ
1. Eq. (2.23)	34.7	77.45	34.7	61.73	31.46	21.15	28.64	22.65
2. Eq. (2.24)	20.9	54.00	20.9	41.55	24.03	10.55	22.17	11.51
3. Eq. (2.25)	18.4	51.46	18.4	39.00	23.97	8.07	22.15	8.99
4. Eq. (2.26)	6.65	10.90	6.65	9.48	3.30	5.23	2.84	5.49
5. Eq. (2.27)	16.74	51.46	16.74	36.84	23.39	6.66	23.26	6.83

TABLE 2.5   ISE values for the present method and SK method for FOPTD model

| | ISE | | |
Example	Present Method	Present Method [Eq. (2.21) and Eq. (2.22)]	SK Method
1.	0.6385	0.6604	0.7842
2.	0.0485	0.0401	0.0675
3.	0.0458	0.0466	0.0631
4.	0.0303	0.0298	0.0359
5.	0.0038	0.0041	0.0050

Error is defined as the difference in value between the system and the model

By identifying a FOPTD model for examples 3 and 5, we do not get negative values for the time delay. For all the examples 1 to 5, the FOPTD model is identified and the model parameters are given in Table 2.4. The comparisons of the fractional step responses are given for the Example 4 in Fig. 2.3 and Fig. 2.4. The present simple method of identifying the FOPTD system gives a similar fit as the SK method. The present study also shows that identifying CSOPTD model gives better fit than the FOPTD model.

2.3   Closed Loop Identification of Stable Systems

In the closed loop identification method, the response of the system for a given change in the set point is noted and the closed loop model is identified. From this, the open loop model is identified. Closed loop identification is preferred over open loop identification as they are insensitive to noise and disturbance. Open loop identification cannot be applied to unstable systems. The block diagram of the closed loop identification is shown in Fig. 2.5.

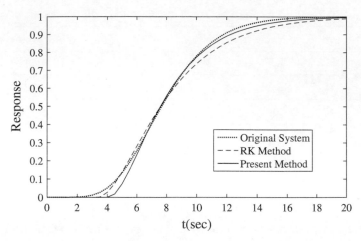

FIGURE 2.3 Comparisons of step response of the original system with that of CSOPTD model (Example 4)

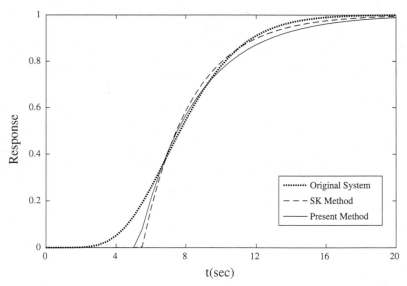

FIGURE 2.4 Comparisons of step response of the original system with that of FOPTD model (Example 4)

Closed loop identification methods can be classified into three types (Ljung, 1998):

(a) Direct Identification: In this the process input (additional measurement) and output data are used. Identification is carried out in the same way as open loop process identification, neglecting any feedback.

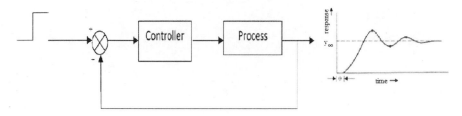

FIGURE 2.5   Closed loop identification

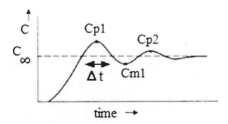

FIGURE 2.6   Closed loop response of an under-damped process

(b) Indirect Identification: The closed loop system is first identified by using the known reference signal to the process set point and the measured output data. By using the controller settings, the process model parameters are estimated.

(c) Joint Input–Output Identification: The input and output are jointly regarded as the output of the system, driven by the reference signal and noise. From this joint model, the knowledge of the system and the controller is obtained.

Liu et al. (2013) have given an excellent survey of the closed loop identification methods using step or relay methods developed in the past three decades. In the early stages of closed loop identification, many works were reported using a P controller. Yuwana and Seborg (1982) proposed a method for tuning controllers online based on a single experimental test performed during closed loop operation under P controller. A step change is given in the set point and a FOPTD process model is back-calculated from the response data. The set point response is assumed to be oscillatory as shown in Fig. 2.6. From the response data, the first and the second peak values (c_{p1} and c_{p2} respectively), the first minimum (cm_1), time period (T), and steady state value (c_1) are noted and the model parameters are estimated. In this method they have approximated the time delay term to a first order Pade approximation. However, this method was not recommended for processes with large time delays. The expressions for the effective time constant (τ_e) and the damping coefficient (ζ) are obtained as in Eq. (2.30) to Eq. (2.36).

$$v_1 = \frac{c_\infty - c_{m1}}{c_{p1} - c_\infty} \tag{2.30}$$

$$v_2 = \frac{c_{p2} - c_\infty}{c_{p1} - c_\infty} \tag{2.31}$$

$$\zeta_1 = \frac{-\ln v_1}{[\pi^2 + \ln(v_1)^2]^{0.5}} \tag{2.32}$$

$$\zeta_2 = \frac{-\ln v_2}{[4\pi^2 + \ln(v_1)^2]^{0.5}} \tag{2.33}$$

$$\zeta = \zeta_1 + \zeta_2 \tag{2.34}$$

$$T = \frac{2\pi\tau_e}{(1 - \zeta^2)^{0.5}} \tag{2.35}$$

$$\tau_e = \frac{T(1 - \zeta^2)^{0.5}}{2\pi} \tag{2.36}$$

Jutan and Rodriguez (1983) have extended the Yuwana and Seborg (1982) method by using a separate but similar method by approximating the time delay term in the process model to a second order term. They have shown that the extended version produced a better process model and a better tuning parameter. Bogere and Ozgen (1989) have extended the method of Yuwana and Seborg (1982) to second order dead time process. They have used a three-term modified Taylor approximation to the exponential term in the closed loop characteristic equation. Lee (1989) modified Yuwana and Seborg (1982) for processes with large time delay. For determining the process parameters, the dominant poles of the closed loop system are matched with a second order process response. Chen (1989) presented the method of determining the ultimate gain and frequency of the open loop system directly from step change in set point under P control mode and from this the first order model parameters are determined.

Jutan (1989) compared Yuwana and Seborg (1982), Jutan and Rodriguez (1983), and Lee (1989) and found that Lee (1989) gave the best tuning. Taiwo (1993) compared Yuwana and Seborg (1982), Jutan and Rodriguez (1983), Lee (1989), and Bogere and Ozgen (1989) and found that Chen (1989) and Lee (1989) methods are robust in the estimation of parameters of FOPTD model and Jutan and Rodriguez (1983) method gave satisfactory results.

2.4 Closed Loop Identification Method under PI Control

2.4.1 Ananth and Chidambaram Method

Ananth and Chidambaram (2000) have proposed a closed loop identification method of stable system under PI control action. For the purpose of designing controllers, a stable FOPTD model is assumed to represent the system given in Eq. (2.37).

$$G_P = y(s)/u(s) = k_P \exp(-Ls)/(\tau s + 1) \tag{2.37}$$

The PID controller in Eq. (2.38) is used to stabilize the system.

$$u = k_C[e + (1/\tau_I) \int edt - \tau_D dy/dt] \tag{2.38}$$

where $e = y_r - y$.

The closed loop transfer function model is given by Eq. (2.39).

$$y(s)/y_r(s) = [k_C k_P (1+\tau_I s) \exp(-Ls)]/[(\tau s+1)\tau_I s + k_C k_P (1+\tau_I s+\tau_I \tau_D s^2) \exp(-Ls)] \tag{2.39}$$

In Eq. (2.39), it is assumed that the derivative is taken only for the process output and not for the set point value. This is to avoid the excessive overshoot under step input change in the set point. Eq. (2.39) can be approximately written as Eq. (2.40).

$$y(s)/y_r(s) = \exp(-Ls)/[\tau_e s^2 + 2\tau_e \zeta s + 1] \tag{2.40}$$

For a given step change in the set point, the closed loop response is obtained (Fig. 2.6). The response is to be approximated to that given by Eq. (2.40) and the parameters τ_e and ζ are obtained from the equations given by Yuwana and Seborg (1982). From the values of τ_e and ζ, the dominant closed loop poles ($\alpha \pm j\beta$) of Eq. (2.39) can be obtained as Eq. (2.41) and Eq. (2.42).

$$\alpha = -\zeta/\tau_e \tag{2.41}$$

$$\beta = (1 - \zeta^2)^{0.5}/\tau_e \tag{2.42}$$

We can substitute the dominant poles of the closed loop in the characteristic equation of Eq. (2.40). On equating to 0, the real and imaginary parts of the resultant equation, we get Eq. (2.43) to Eq. (2.51).

$$A_1\tau + A_2 k_P = A_3 \tag{2.43}$$

$$B_1\tau + B_2 k_P = B_3 \tag{2.44}$$

Where,

$$A_1 = (\alpha^2 - \beta^2) \tag{2.45}$$

$$A_2 = k_C E_1\{[\tau_D(\alpha^2 - \beta^2) + \alpha + (1/\tau_I)]\cos(L\beta) + [2\alpha\beta\tau_D + \beta]\sin(L\beta)\} \tag{2.46}$$

$$A_3 = -\alpha \tag{2.47}$$

$$E_1 = \exp(-L\alpha) \tag{2.48}$$

$$B_1 = 2\alpha\beta \tag{2.49}$$

$$B_2 = k_C E_1\{[2\alpha\beta\tau_D + \beta]\cos(L\beta) - [\tau_D(\alpha^2 - \beta^2) + \alpha + (1/\tau_I 0]\sin(L\beta)\} \tag{2.50}$$

$$B_3 = -\beta \tag{2.51}$$

The values of gain can be calculated from the steady state value of y and u as $k_P = (y/u)_{t=\infty}$. The nonlinear equations Eq. (2.43) and Eq. (2.44) can be solved for τ and L. The initial value for L can be considered from the delay noted in the closed loop response and the initial value for τ can be considered as $t_s/8$ where t_s is the settling time of the closed loop response.

Mamat and Fleming (1995) proposed an improved method for the online identification of FOPTD parameters under PI mode. They used a step input of magnitude A. The closed loop response is approximated by a second order plus time delay transfer functions. From the closed loop response, first peak (C_{p1}), second peak (C_{p2}), time taken for the first peak and second peak (t_{p1} and t_{p2}) respectively, and the final steady state value (C_{ss}) are noted. Using the time domain solution of the closed loop transfer, the equations for closed loop gain (K), damping coefficient (ζ), effective time constant (τ_e), and closed loop time delay (d) are calculated. The relationship between the closed loop system, controller, and open loop system in terms of amplitude ratio and phase angle at phase cross over frequency ω_c was developed. The two nonlinear equations are solved for the open loop time constant (τ) and time delay (d). Open loop gain (K_P) is obtained by using the final steady state value, controller settings and the characteristic area.

$$\tau_p = \frac{\sqrt{(K_cK_p)^2(1 + T_i^2\omega_c^2) - M^2T_i^2\omega_c^2}}{MT_i\omega_c^2} \tag{2.52}$$

$$d_p = \frac{1}{\omega_c}\left[\tan^{-1}(\omega_cT_i) + \tan^{-1}\left(\frac{1}{\tau\omega_c}\right)\right] \tag{2.53}$$

$$K_p = \frac{T_i}{K_cS_c}C_{ss} \tag{2.54}$$

Hwang (1995) developed a time domain approach to identify the process as a SOPTD model via an under-damped transient. In their work, they reported that the identification test can be conducted for any test input such as step, pulse, or impulse, using any controller mode (P, PI, or PID). From the resulting under-damped response, the first peak value C_P and the first minimum value C_m are noted. The closed loop response for a step input of magnitude A is derived as:

For P control:

$$C(t) = \frac{Akk_p}{1 + kk_p} - M_1e^{\sigma(t-d)}\sin([\omega(t-d) + \phi]) \tag{2.55}$$

For PI control:

$$C(t) = A - M_1e^{\sigma(t-d)}\sin([\omega(t-d) + \phi]) + M_2e^{\sigma(t-d)} \tag{2.56}$$

Where, $\sigma \pm j\omega$ are the PI or PID control dominant complex conjugate poles. They used a Pade 2/2 (Stephanopoulos, 1984) approximation for time delay and M_1, M_2, and φ were derived in terms of k_p, σ, ω, and d by matching the resultant equation with the first peak and minimum data.

$$\frac{dC(t_p)}{dt} = 0 \tag{2.57}$$

$$\frac{dC(t_m)}{dt} = 0 \tag{2.58}$$

$$C(t_p) = C_p \tag{2.59}$$

$$C(t_m) = C_m \tag{2.60}$$

By using appropriate root finding techniques, they solved these equations for d and ω.

Based on Yuwana and Seborg (1982) method, Srividya and Chidambaram (1997) reported a method for online tuning of controllers for an integrator plus time delay (where $g_p = k_p/s$) system. They compared the results obtained with that of relay feedback identification method and found that the new method of identification gives robust model parameters. PI controller parameters for the identified model was calculated from the reported tuning formula. Simulation application was done on an isothermal continuous co-polymerization reactor.

Park et al. (1998) proposed a closed loop identification method using least square in the frequency domain for identifying SOPTD parameters using P controller. Suganda et al. (1998) reported a method for the online identification of SOPTD model parameters under PI control. Here the closed loop response on introducing a step change was modelled to be a second order delay process. For identifying the model, they used Mamat and Fleming (1995) method for under-damped response and estimated the FOPTD parameters.

Melo and Friedly (1992) presented a method for the online identification of all open loop transfer function for a MIMO process from the closed loop data obtained for a step change in the set point under PID control mode. A step change was given in each controller set point keeping other set points unchanged. This was repeated for all the set points and the output responses were obtained. Thus, using the set point matrix (X_s) and the output matrix (X) the process transfer function matrix (G) is obtained from the equation as Eq. (2.61).

$$G_p = YY_r^{-1} \left(I - YY_r^{-1} \right)^{-1} G_c^{-1} \tag{2.61}$$

Where, G_c is the feedback controller matrix and I is the unitary matrix. The frequency response from the measured time response can be obtained by using Laplace or Fourier transform. The non-parametric frequency response model obtained was used directly for PID control algorithm design. Simulation studies were performed on 2×2, 3×3, and 4×4 systems.

2.5   Optimization Method

Pramod and Chidambaram (2000, 2001) have identified FOPTD model for both stable and unstable systems using the optimization method. PI/PID controller is used to obtain either step or pulse response of the system. The method is as follows:

The response (y_{pc}) of the closed loop system using a PI or PID controller is obtained for a known step magnitude in the step point. The model for the open loop process is assumed to be a FOPTD system. The initial guess values for the model parameters are assumed. Using this model and the same controller settings, the closed loop response for the same step magnitude is obtained by simulation.

Let this response be y_{mc}. The final model parameters are obtained by minimizing the sum of squared difference between y_{pc} and y_{mc} using least square optimization method. Since all the data points are used rather than dominant data points (as in the case of reaction curve method), the robustness of the method to measurement noise will be enhanced. In this method, optimization routine *leastsq* of Matlab is used, which implements Leveberg–Marquadrt algorithm.

The major criticism in using any optimization method is the selection of the initial guess values of the parameters. The following method is used for initial guess values:

(1) The guess value of time delay is equal to the closed loop delay (usually open loop delay will be less than closed loop delay).

(2) From the settling time t_s (time required to reach 98% of the final value and remain within that limit) of the closed loop response, the dominant time constant of the system is considered as $t_s/5$. Usually, the open loop time constant is lower than the close loop system. Hence, open loop time constant is assumed to be $t_s/8$.

(3) Guess values for k_P:

    (a) For stable system: For higher order systems, the minimum of the maximum value of $k_p k_c$ is 1. Hence the minimum value of the design value for $k_p k_c$ is 0.5, and hence the guess value of k_p is $0.5/k_c$.

    (b) For unstable system: The guess value of k_P is obtained from the formula of proportional controller gain (De Paor and O'Malley, 1989) as $k_p k_c = (\tau/\tau_D)^{0.5}$. Hence, $k_P = (\tau/\tau_D)^{0.5}/k_c$

Pramod and Chidambaram (2000) have used constrained optimization (*constr* of Matlab using a constraint $\tau_d > 0.01$) for unstable systems. Instead of a step response, a pulse response can also be considered. If we would like to identify a SOPTD model, then the initial guess values for the model parameters k_P, L and τ_e are similar to k_p, L and τ of the FOPTD model parameters guess values. The value for the damping coefficient (ζ) is assumed to be 1.0.

2.5.1 Simulation Examples

The higher order transfer function models assumed here for stable systems are the same as considered by Park et al. (1997) and Lee (1989).

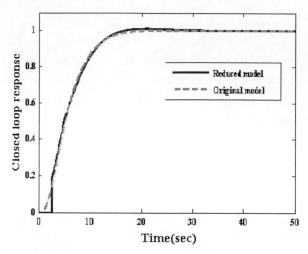

FIGURE 2.7   Closed loop response of the original process and the reduced FOPTD model $k_c = 0.5$, $\tau_I = 3$ and $\tau_D = 5$ (Example 6)

Example 6

Let the process be given by $1/(s+1)^5$.

The PID controller settings used for identification are given by Park et al. (1998) as $k_c = 0.5$, $\tau_I = 3$ and $\tau_D = 1.0$. The closed loop response (y_{pc}) for a unit step change in the set point using Simulink is obtained. From the response, the initial delay is noted as 1 and the settling time constant (for 98%) is noted as 15. Hence the time constant (τ) for the model is considered as $(15/8) = 1.875$. The gain (k_p) for the model is taken as $0.5/k_c$ (= 1.0). With these guess values and using 300 data points in the time interval of 0.1 to 30, the optimization method (unconstrained) is carried out. The final converged values for the parameters are $k_p = 1.008$, $L = 2.6$ and $\tau = 2.688$. The step response of the closed loop system using the model parameters is also shown in Fig. 2.7, which shows a good fit. The initial guess values of the model parameters are also changed by +20% of each of these values. In another simulation run, the value of the guess foe k_p is assumed as 2.0. In all these cases, the optimization method converged to the same values of the model parameters as obtained earlier. The identification is also carried out using the closed loop response obtained earlier as well as using different controller settings (details in Table 2.6). The converged model parameters are the same as those obtained earlier. The maximum computational time taken is 32 s on a Pentium computer.

Example 7

The process is assumed as $\exp(-s)/[(9s^2 + 2.4s + 1)]$

TABLE 2.6   Model parameters by optimization method for the system $1/(s+1)^5$

S. No.	PID settings			Guess values of			Converged			No of iterations
k_c	τ_I	τ_D	k_p	τ	L	k_p	τ	L		
1	1.0	3.0	1.0	0.5	1.875	0.8	1.020	3.001	2.6	109
2	0.5	3.0	1.0	1.0	1.875	0.8	1.008	2.68	2.6	148
3	2.0	8.0	1.0	0.25	3.75	0.8	0.994	4.08	2.57	93
4	0.25	4.0	2.0	2.0	1.875	0.8	1.002	2.63	2.52	74

Time taken for the optimization method for S. No. 2 is 32 s on a Pentium computer.

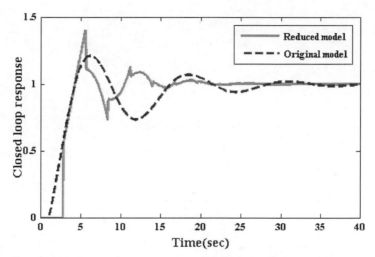

FIGURE 2.8   Closed loop response of original model and FOPTD model (Example 7) $k_c = 1.596$, $\tau_I = 5.25$ and $\tau_D = 1.3125$

The controller settings are $k_c = 1.596$, $\tau_I = 5.25$ and $\tau_D = 1.313$. Using the closed loop response (Fig. 2.8), the initial guesses for the model parameters are obtained as $k_p = 0.63$, $\tau = 930/80 = 3.75$, $L = 1$. The model parameters obtained from the optimization are obtained as $k_p = 1.154$, $\tau = 4.44$ and $L = 2.8$. The closed loop servo responses of the process and the FOPTD model with the original PID settings are compared in Fig. 2.8, which shows that the FOPTD model is not adequate. Hence, there is a need for identifying the SOPTD model. The initial guesses for the model parameters are $\zeta = 1.0$, $k_p = 0.63$, $\tau_e = (30/8) = 3.75$ and $L = 1.0$. The optimization method gives the converged values as $\zeta = 0.4$, $k_p = 1.0$, $\tau_e = 3$ and $L = 1.0$. These values are exactly the same as that of the process. Thus, for this problem, SOPTD model is preferable to FOPTD model. The optimization method of identifying unstable SOPTD with a 0 proposed by Dhanya Ram et al. (2014) is presented in Appendix A.

2.6   Design of PI Controller

2.6.1   FOPTD System by IMC Method

From Internal Model Control (IMC) method, controller G_C is obtained in Eq. (2.62).

$$G_C = Q/[1 - QG_P] \tag{2.62}$$

Where, Q is the inverse of the invertible part (of G_P) multiplied by an IMC filter (f).
    For FOPTD system, we have:

$$Q = [(\tau s + 1)/k_P]f \tag{2.63}$$

Where,

$$f = 1/(\tau_c s + 1)^r \tag{2.64}$$

with $r = 1$.
    Using the approximation $[\exp(-Ls) = 1 - Ls]$, we get G_C as a PI controller:

$$k_c = \tau/[k_P(\tau_c s + 1)] \tag{2.65}$$

$$\tau_I = \tau \tag{2.66}$$

If we use first order Pade's approximation for $\exp(-Ls) = (1 - 0.5Ls)/(1 + 0.5Ls)$,
we get a PID controller:

$$k_{cC} = (1/k_P)[(\tau + 0.5Ls)/(\tau_c + 0.5Ls)] \tag{2.67}$$

$$\tau_I = (\tau + 0.5L) \tag{2.68}$$

$$\tau_D = (0.5Ls)/(\tau + 0.5L) \tag{2.69}$$

The tuning parameter τ_C is selected as:

$$(\tau_C/L) > 0.8 \text{ and } \tau_C > 0.1\tau \text{ (Rivera et al., 1986)} \tag{2.70}$$

$$\tau > \tau_C > L \text{ (Chien and Fruehauf, 1990)} \tag{2.71}$$

$$\tau_C = L \text{ (Skogestad, 2003)} \tag{2.72}$$

2.6.2 SOPTD System with a Zero

$$G_P = k(\tau_3 s + 1)\exp(-Ls)/[(\tau_1 s + 1)(\tau_2 s + 1)] \qquad (2.73)$$

With the approximation $\exp(-Ls) = (1 - Ls)$, we get a PID controller with

$$k_C = (\tau_1 + \tau_2 - \tau_3)/[k(\tau_3 + L)] \qquad (2.74)$$

$$\tau_I = (\tau_1 + \tau_2 - \tau_3) \qquad (2.75)$$

$$\tau_D = [\tau_1 \tau_2 - (\tau_1 + \tau_2 - \tau_3)\tau_3]/(\tau_1 + \tau_2 - \tau_3) \qquad (2.76)$$

2.6.3 PID Settings by Skogestad Method

For FOPTD systems, $[G_P(s) = k\exp(-\theta s)/(\tau s + 1)]$, Skogestad (2003) recommended the PI settings as:

$$k_c = \tau/[k(\tau_c + \theta)] \qquad (2.77)$$

$$\tau_I = \min[\tau, 4(\tau_c + \theta)] \qquad (2.78)$$

For SOPTD systems, Skogestad (2003) recommended for PID settings as:
   For $\tau_1 \leq (8\theta)$

$$k_c = 0.5(\tau_1 + \tau_2)/(k_P \theta) \qquad (2.79)$$

$$\tau_I = (\tau_1 + \tau_2) \qquad (2.80)$$

$$\tau_D = \tau_1/[1 + (\tau_1/\tau_2)] \qquad (2.81)$$

For $\tau_1 \geq (8\theta)$

$$k_c = [0.5\tau_1/(k_P \theta)](1 + m) \qquad (2.82)$$

$$\tau_I = (8\theta + \tau_2) \qquad (2.83)$$

$$\tau_D = \tau_2/[1 + m] \qquad (2.84)$$

Where,

$$m = \tau_2/(8\theta) \qquad (2.85)$$

2.6.4   PI and PID Settings by Stability Analysis Method

Two equations are formulated using phase angle criterion and amplitude criterion for the process transfer function model $G_P G_V G_M$ along with a P controller. By solving these equations analytically for some cases or numerically, we get $k_{c,\max}$ and ω_C. Then by using the continuous cycling method (Ziegler and Nichols, 1942), we can calculate the PI settings as:

$$k_c = 0.45 k_{c,\max} \tag{2.86}$$

$$\tau_I = P_u/1.2 \tag{2.87}$$

$$P_u = 2\pi/\omega_c \tag{2.88}$$

For PID Controller

$$k_c = 0.6 k_{c,\max} \tag{2.89}$$

$$\tau_I = P_u/2 \tag{2.90}$$

$$\tau_D = P_u/8 \tag{2.91}$$

2.7   Unstable Systems

A system described by a transfer function model having at least one pole in the Right Half-Plane (RHP) is known as an unstable system. Such systems move away from the steady state even for small perturbation of the system parameters or operating conditions. Linearization of the nonlinear mathematical model equations of such systems around the operating point gives an unstable transfer function model. The response of such transfer function models for a small perturbation may be ever increasing, but its nonlinear model (i.e., real systems), the system steady state moves to other but stable steady state. In case the system has a single steady state, but which is unstable, then the actual response is oscillatory around this unstable steady state point. A time delay is introduced into the transfer description of such systems due to the measurement delay or actuator delay or by the approximation of higher order dynamics using a lower order plus time delay system. Many real systems exhibit multiple steady states due to certain nonlinearity of the system. Some of the steady states may be unstable.

In a FOPTD system, if the time constant is large, then the time constant in the denominator of the transfer function can be clubbed with process gain and the resulting model is called Integrating Plus Time Delay System. This essentially consists of only two parameters (k_P and τ_D) and the model is very simple for identification.

It may be necessary to operate the system in an unstable steady state for economic and/or safety reasons. But the performance of the control system is limited for an unstable transfer function model. The performance specifications like overshoot and settling time are large for unstable systems compared to stable systems. There exists a minimum value of controller gain for unstable systems below which a closed loop system cannot be stabilized. There also exists a maximum value of controller gain above which a closed loop system cannot be stabilized. The design value of the controller gain is the average value of these two limits. For unstable FOPTD systems, PI control can stabilize the system for $\tau_D/\tau \leq 0.7$ and the PID controller can stabilize the system for $\tau_D/\tau \leq 1.2$.

The PI settings for an unstable FOPTD system are given by Chidambaram and Padmasree (2006) as given in Eq. (2.92).

$$
\begin{aligned}
k_C k_P &= 49.53 \exp(-21.644\varepsilon) && \text{for } \varepsilon < 0.1 \\
k_C k_P &= 08668\varepsilon^{-0.829} && \text{for } 0.1 \leq \varepsilon \leq 0.7 \\
\tau_I/\tau &= 0.1523 \exp(7.9425\varepsilon) && \text{for } 0 \leq \varepsilon \leq 0.7
\end{aligned}
\tag{2.92}
$$

For PID settings with a first order filter,

$$
\begin{aligned}
k_c k_P &= 4282\varepsilon^2 - 1344.6\varepsilon + 101 && \text{for } \varepsilon < 0.1 \\
k_C k_P &= 1.1161\varepsilon^{-0.9427} && \text{for } 0.2 \leq \varepsilon \leq 1.0 \\
\tau_I/\tau &= 36.842\varepsilon^2 - 10.3\varepsilon + 0.8288 && \text{for } 0 \leq \varepsilon \leq 0.8 \\
\tau_I/\tau &= 76.24\varepsilon^{6.77} && \text{for } 0.8 \leq \varepsilon \leq 1.0 \\
\tau_D/\tau &= 0.5\varepsilon && \text{for } 0 \leq \varepsilon \leq 1.0
\end{aligned}
\tag{2.93}
$$

The first order filter constant (α) is given by Eq. (2.94).

$$
\alpha/\tau = 0.1233\varepsilon + 0.0033
\tag{2.94}
$$

In this chapter, open loop identification methods are proposed to identify the parameters of FOPTD and CSOPTD models. A review of methods under closed loop identification method is given for reaction curve method and optimization method. Methods of designing PI controllers for transfer function models for SISO systems are brought out. An overview of a simple method of designing PI controllers for multivariable systems based on Steady State Gain Matrix (SSGM) and also based on transfer function matrix are also reported.

2.8   Estimation Methods

Estimation methods minimize certain appropriately defined error criteria as a means to optimally fit the system data. The error criteria can be defined in three ways. Let us consider a discrete model with 'a' and 'b' as model parameters. The system output and input variables are y_k and u_k at a sampling instant k.

$$y_k = ay_{k-1} + bu_{k-1} \tag{2.95}$$

(1) The error can be defined as a deviation of the parameter estimates from the true values (parameter error) as in Eq. (2.96).

$$e_k = \theta_k - \theta_{m,k} \tag{2.96}$$

where the subscript m denotes model. This method is known as parameter error method. Since true parameter values are unknown, this method may be difficult to follow.

(2) The error can be defined as difference between the output of the system and that of the model:

$$e_k = y_k - y_{m,k} \tag{2.97}$$

Where (for example),

$$y_k = ay_{k-1} + bu_{k-1} \tag{2.98}$$

and

$$y_{m,k} = a_m \, y_{m,k-1} + b_m u_{k-1} \tag{2.99}$$

where the subscript m denotes model. This method is known as output error method. The same value of input (u_k) is used in the process and in the model. In this method, the solution of the model equation for the given sequence of u should be known.

The model parameters usually appear nonlinearly in the resulting solution of the model equation. Hence the estimation of the parameters will require an optimization method for the nonlinear algebraic equations which require the initial guess for the parameter values. The possibility of local minima rather than global minima may also arise.

For example, consider the model equation in Eq. (2.100).

$$y_k = ay_{k-1} + u_{k-1} \tag{2.100}$$

with given initial condition on y_{-1}. For a unit step change in u_k, the solution of the model equation can be shown by taking the z-transform, followed by inverse z-transform as in Eq. (2.101).

$$y_{m,k} = a^{k+1}y_{m,k-1} - [1/(a-1)] - [a^{k+1}/(1-a)] \tag{2.101}$$

For the model Eq. (2.101), let us formulate a least squares problem as in Eq. (2.102).

$$e_k = y_k - y_{m,k} \tag{2.102}$$

$$\sum_{k=1}^{N} e_k^2 = \sum_{k=1}^{N} \left[y_k - \left\{ a^{k+1}y_{m,k-1} - [(1/a-1)] - \left[\frac{(a^{k+1})}{(1-a)} \right] \right\} \right]^2 \tag{2.103}$$

Here N is the number of data points. The optimization problem is stated as the selection of the model parameters so as to minimize the sum of the error squared over all the experimental data (N) points. If the error is formulated based on output error, then we find that the model parameter 'a' is present as a nonlinear form. This is an optimization problem and it requires a good initial guess value for the model parameter.

For the model Eq. (2.95), let us formulate a least squares problem.

$$e_k = y_k - y_{m,k} \tag{2.104}$$

$$\sum_{k=1}^{N} e_k^2 = \sum_{k=1}^{N} [y_k - a_m y_{m,k-1} - b_m u_{k-1}]^2 \tag{2.105}$$

Select the model parameters a_m and b_m so as to minimize the sum of the error squared over all the experimental data (N) points. This is an optimization problem since $y_{m,k-1}$ depends on a_m and b_m through Eq. (2.101). This method requires good initial guess values for the model parameters a_m and b_m.

(3) The error can be defined as a discrepancy between the model equation and the measured input and output data ($e = y_k - y_{m,k}$, for the example case we have $y_k = ay_{k-1} + bu_{k-1}$ and

$y_{m,k} = ay_{k-1} + bu_{k-1}$). This method is known as the equation error method or the prediction error method. Here, there is no solution of the model equation as equation error required and the model parameters appear linearly in the error equation. We substitute the experimental data sequences $[y_k, u_k]$ in the model equations and find the expression for the resulting error of the equation. Hence, the method is called identification based on error equation.

The prediction error method (equation error method) compares the behaviour of the model. The model-based prediction with experimental data and tries to adjust the model parameters to obtain a better fit. As stated earlier, the output error method thus relies on the accuracy of the future output modeling. The output error identification is a nonlinear estimation problem, whereas the equation error identification is a linear estimation problem.

2.8.1   Least Square Estimation of ARX Model

The Autoregressive with Exogenous Input (ARX) model is given by Eq. (2.106).

$$y_k = -\sum_{i=1}^{n} a_i y_{k-i} + \sum_{i=1}^{n} b_i u_{k-i} + e_k \qquad (2.106)$$

Here n is the number of model parameters in a_i. The same number of model parameters is assumed in b_i also. Let us define x_k as in Eq. (2.107).

$$\mathbf{x}_k = [-y_{(k-1)}, -y_{(k-2)}, \ldots, -y_{(k-n)}, u_{(k-1)}, \ldots, u_{(k-n)}]^T \qquad (2.107)$$

The size of x is $Nx(2n)$.

$$\boldsymbol{\theta} = [a_1, a_2, \ldots, a_n, b_1, b_2, \ldots, b_n]^T \qquad (2.108)$$

We can rewrite Eq. (2.106) as Eq. (2.109).

$$\mathbf{y} = \mathbf{x}\boldsymbol{\theta} + \mathbf{e} \qquad (2.109)$$

Where,

$$\mathbf{y} = [y_{(n+1)}, y_{(n+2)}, \ldots, y_{(n+N)}]^T \qquad (2.110)$$

$$\mathbf{e} = [e_{(n+1)}, e_{(n+2)}, \ldots, e_{(n+N)}]^T \qquad (2.111)$$

$$\mathbf{x} = [\mathbf{x}_{(n+1)}^T \mathbf{x}_{(n+2)}^T \cdots \mathbf{x}_{(n+N)}^T]^T \qquad (2.112)$$

and

$$\mathbf{x}_{(n+1)}^T = [-y_n, \ldots, -y_1, u_{(n+1)}, \ldots u_1] \qquad (2.113)$$

Using the data set $\{y_k, u_k\}$ for $k = 1, 2, \ldots, (N + n)$, we can evaluate matrix $\mathbf{x}$ and vector $\mathbf{y}$. By the least square method, we can estimate the parameters vector θ. The details are as per Eq. (2.114) onwards.

We can write

$$\mathbf{e} = \mathbf{y} - \mathbf{x}\boldsymbol{\theta} \tag{2.114}$$

Where,

$$\mathbf{e} = [e_1, e_2, \ldots, e_n] \tag{2.115}$$

The parameter vector θ is to be calculated in order to minimize the scalar objective function.

$$J = \sum_{i=1}^{n} e_i^2 \tag{2.116}$$

$$= e^T e \tag{2.117}$$

$$= (y - x\theta)^T (\mathbf{y} - \mathbf{x}\boldsymbol{\theta}) \tag{2.118}$$

$$= \mathbf{y}^T \mathbf{y} - \boldsymbol{\theta}^T \mathbf{x}^T \mathbf{y} - \mathbf{y}^T \mathbf{x}\boldsymbol{\theta} + \boldsymbol{\theta}^T \mathbf{x}^T \mathbf{x}\boldsymbol{\theta} \tag{2.119}$$

For J to be minimized, we require the necessary and sufficient conditions as in Eq. (2.120).

$$\partial J / \partial \boldsymbol{\theta} = 0 \tag{2.120}$$

$\partial^2 J / \partial \boldsymbol{\theta}^2 > 0$ (i.e., positive value)

On differentiation of J with respect to θ in Eq. (2.119), we get Eq. (2.121).

$$\partial J / \partial \boldsymbol{\theta} = -2x^T y + 2x^T x\boldsymbol{\theta} = 0 \tag{2.121}$$

Which gives,

$$\boldsymbol{\theta} = (\mathbf{x}^T \mathbf{x})^{-1} \mathbf{x}^T \mathbf{y} \tag{2.122}$$

We can check that,

$$\partial^2 J / \partial \boldsymbol{\theta}^2 = 2\mathbf{x}^T \mathbf{x}$$

$$> 0 \text{ (i.e., positive value)} \tag{2.123}$$

Let us formulate, by output error method, a least square problem for the ARX model Eq. (2.106).

$$e_k = y_k - y_{m,k} \tag{2.124}$$

$$\sum_{k=1}^{N} e_k^2 = \sum_{k=1}^{N} \left\{ y_k - \left[-\sum_{i=1}^{n} a_{m,i} y_{m,(k-i)} + \sum_{i=1}^{n} b_{m,i} u_{(k-i)} \right] \right\} \tag{2.125}$$

Here, y_k is the experimental data from the system at sampling instant k and $y_{m,k}$ is the model output from Eq. (2.126).

$$y_{m,k} = -\sum_{i=1}^{n} a_{m,i} y_{m,(k-i)} + \sum_{i=1}^{n} b_{m,i} u_{(k-i)} \qquad (2.126)$$

Here k is the sampling instant and m denotes the model. Select the model parameters $a_{m,i}$ and $b_{m,i}$ so as to minimize the sum of the error squared over all the experimental data points (N). Here $y_{m,(k-1)}$ depends on the model parameters $a_{m,i}$ and $b_{m,i}$. This is an optimization problem.

Summary

A simple derivation is given to calculate the parameters τ and L of a critically damped SOPTD model from the fractional step response. Five simulation examples are given to show the simplicity of the proposed method. The method makes use of a unique property of a Critically Damped Second Order Plus Time Delay system, i.e., the fractional response of 0.73 needs a time of 2.58 τ. The present method is simple when compared to the Rangaiah and Krishnaswamy (RK) method. The proposed method gives a better fit and gives a lesser value of ISE. The method is also modified to identify a FOPTD model. The present method gives better fit than the SK method. The model identification by closed loop method using the optimization method is reviewd. The tuning formulae for PI and PID controllers settings for FOPTD and SOPTD systems are presented.

Problems

1. Using the open loop identification method, estimate the parameters of the CSOPTD and FOPTD models:

   $y(s)/u(s) = 2\exp(-s)/[(9s^2 + 6s + 1)]$

   Compare the open loop step responses.

2. Using the open loop identification method, estimate the parameters of the CSOPTD and FOPTD models:

   $y(s)/u(s) = 2\exp(-s)/[(9s^2 + 6s + 1)]$

   Compare the open loop step responses.

3. Using the open loop identification method, estimate the parameters of the CSOPTD and FOPTD models:

$y(s)/u(s) = 3\exp(2s)/(9s^3 + 16s^2 + 7s + 1)$

Compare the open loop step responses.

4. Carry out closed loop optimization method to identify the parameters for the FOPTD model:

$G_P(s) = (1.5s + 1)\exp(-2s)/[(10s + 1)(20s + 1)]$

with $k_c = 3.0$, $\tau_I = 15$, $\tau_D = 4$

(a) Compare the open loop step responses and

(b) Compare the closed loop step responses with the controller designed by the IMC method.

5. Carry out closed loop optimization method to identify the parameters for the FOPTD model:

$G_P(s) = 1/(s + 1)^8$

with $k_c = 0.7$, $\tau_I = 5$, $\tau_D = 1.2$

(a) Compare the open loop step responses and

(b) Compare the closed loop step responses with the controller designed by the IMC method.

3

Introduction to Linear Multivariable Systems

In this chapter, interactions in multivariable systems, interaction measure, and suitable variables pairing are described. The design methods of decentralized controllers using a detuning factor to reduce the interactions are discussed. The methods of designing decouplers followed by the methods of designing single loop controllers are also discussed. The use of relay tuning to design decentralized Proportional and Integral (PI) controllers is brought out. The methods of designing simple centralized PI controllers are reviewed. The performance and robust stability analysis of the closed loop system are discussed. –

3.1 Stable Square Systems

Consider Fig. 3.1, where the connection of inputs and outputs are shown for TITO multivariable systems. For the step change in u_1, record the change in y_1 (as y_1 versus t); this is called *response* and the change on y_2 is called *interaction* of u_1 on y_2. Similarly, for the step change in u_2, the change in y_2 is called the response and that on y_1 is called interaction. The control of output y_1 by changing u_1 and control of output y_2 by u_2 by the two controllers as shown in Fig. 3.2 is called decentralized

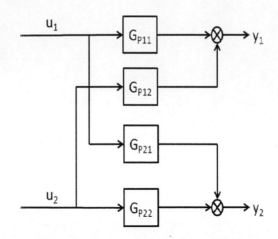

FIGURE 3.1   Open loop multivariable systems

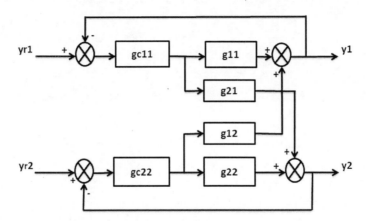

FIGURE 3.2   Decentralized control system of a TITO process g_{ij} is the process transfer function model

control system. Here we have Eq. (3.1).

$$[u_1 \quad u_2] = G_c E \tag{3.1}$$

$$G_c = \begin{bmatrix} g_{c11} & 0 \\ 0 & g_{c22} \end{bmatrix} \tag{3.2}$$

It is a diagonal matrix:

$$E = [e_1 \quad e_2]^{\mathrm{T}} \tag{3.3}$$

For n input and n output systems, there are n controllers for a decentralized control system. If the interactions are significant, then the centralized control system is required as shown in Fig. 3.3.

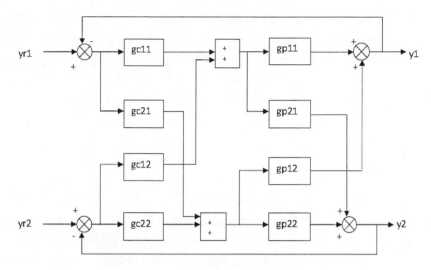

FIGURE 3.3   Centralized control system

Here, $[u_1 \quad u_2] = G_c E$

Where,

$$G_C = \begin{bmatrix} g_{c11} & g_{c12} \\ g_{c21} & g_{c22} \end{bmatrix} \tag{3.4}$$

is a full matrix. And the error vector is defined the same as earlier. For n input and n output systems, there are $n \times n$ controllers which are to be used for the centralized control system.

Decentralized control system is popular in industries due to the following reasons:

(1) Decentralized controllers are easy to implement.

(2) They are easy for the operator to understand.

(3) The operator can easily retune the controllers to take into account changing process conditions.

(4) Tolerance for failures in measuring devices or in the final control elements is more easily incorporated into the design of decentralized controllers.

(5) The operator can easily bring the control system into service during process start up and can take it gradually out of service during shutdown.

Since we know how to design a single loop control system, we would first like to form n single loop systems from the given n input and n output systems. Then single loop controller can be designed for each loop. Given the n input variables and n output variables, the question is, then, which input is to be paired with which output to form the loop. Even for the stable multivariable systems, the method of designing controllers is complicated due to the interactions among the loops. Methods of designing decentralized PI controllers for stable MIMO system are reviewed by Luyben and Luyben (1997) and Maciejowski (1989). More rigorous methods of designing multivariable PI controllers are reviewed by Wang et al. (2008) and Wang and Nie (2012). Simple methods of designing centralized PI controllers for stable systems are reviewed by Tanttu and Lieslehto (1991) and Katebi (2012). Davison (1976) proposed a simple method for PI controller tuning for multivariable systems if the inverse of the SSGM matrix of the process exists.

For a process with N controlled outputs and N manipulated variables, there are $N!$ ways to form the control loops (Stephanopoulos, 1984). To select the best control configuration the interactions among the loops for all $N!$ loop configurations are considered and the one in which the interactions are minimal is considered. Many methodologies are available for the interaction measurement. The Relative Gain Array (RGA) proposed by Bristol (1966) provides a way to select input–output variables pairing in order to minimize the number of interactions among the resulting loops. The RGA analysis is based on SSGM.

3.2   Relative Gains

The process gain relating output (y_i) and manipulated variable (u_j) is defined by $\partial y_i / \partial u_j$. This gain value would be evaluated experimentally or from the model equations of the system. The gain when all other loops are open (i.e., all other manipulated variables are constant) and the gain of the system with all other loops closed (when all other y are constant), is known. Then, the relative gain λ_{ij} for the selected pair of variables y_i and u_j is defined by Eq. (3.5) and Eq. (3.6).

$$\lambda_{ij} = (\partial y_i / \partial u_j)_u / (\partial y_i / \partial u_j)_y \qquad (3.5)$$

$$= (\partial y_i / \partial u_j)_u \times (\partial y_j / \partial u_i)_y \qquad (3.6)$$

Consider a set of gains $(\partial y_i / \partial u_j)$ for all i, j arranged in a matrix denoted by $\mathbf{K}$. If this matrix is inverted and transposed, its elements become the inverse of the gains for all other loops closed, that is, $(\partial y_j / \partial u_i)_y$.

Let the transpose of the inverse of matrix $\mathbf{K}$ be denoted by $\mathbf{R}$. Then the relative gain λ_{ij} can be calculated by multiplying each corresponding element from the matrices **K** and **R** as in Eq. (3.7).

$$\lambda_{ij} = (ij^{\text{th}} \text{ element of } \mathbf{K} \text{ matrix}) \times (ij^{\text{th}} \text{element from } \mathbf{R} \text{ matrix}) \qquad (3.7)$$

It is the corresponding element by element multiplication and not the total matrix multiplication. These values of λ_{ij} for all i and j are arranged in a matrix (Λ) known as Relative Gain Array (RGA). The array has very useful properties of having the numbers in every row summing to 1, and also numbers in every column summing to 1. Another useful property of this matrix Λ is that it is unaffected by scaling.

The significance of the relative gain numbers can be easily explained. If a process gain is the same with or without all other loops are closed, then those other loops have no effect on the first loop and hence there is no interaction. For this situation we have $\lambda_{11} = 1$. The deviation of λ_{ij} from 1 in either direction shows increasing interaction. If any λ_{ij} is negative then it is very clear that, by interaction, loop ij becomes unstable. Hence, we should not pair that input and output which has negative relative gain. A very large positive value of $\lambda \gg 1$ indicates that other λ in that row or in that column have large negative values in order to make the sum of elements in a row (or column) equal to 1. Thus, larger value of λ_{ij} indicates that the interaction is severe and a decentralized controller is not recommended. For such a situation, a full multivariable (centralized) control scheme is required.

Consider the process gain matrix K as in Eq. (3.8).

$$K = \begin{bmatrix} 12.8 & -18.9 \\ 6.6 & -19.4 \end{bmatrix} \qquad (3.8)$$

In addition, R, the transpose of inverse of K_P is given by Eq. (3.9).

$$R = (-1/123.58) \begin{bmatrix} -19.4 & -6.6 \\ 18.9 & 12.8 \end{bmatrix} \qquad (3.9)$$

Multiply the ij element of matrix K_P with the ij^{th} element of R. Hence, the relative gain array is given by Eq. (3.10).

$$\Lambda = \begin{bmatrix} 2.01 & -1.01 \\ -1.01 & 2.01 \end{bmatrix} \qquad (3.10)$$

The sum of each row of the relative gain matrix (Λ) is found to be 1 and similarly the sum of each column is found to be 1. We should avoid those pair having a negative relative gain. Hence, for this example, the recommended pairing are:

(a) y_1 and u_1

(b) y_2 and u_2

3.3   Single Loop and Overall Stability

Decentralized controllers are generally designed with a failure tolerance. The controlled and manipulated variables are paired such that each single loop and the overall control loop system are stable. The Niederlinski Index NI (Niederlinski, 1971; Grosdidier et al., 1985) is evaluated for this purpose. NI is defined by Eq. (3.11).

$$NI = \det[K_P]/\prod K_{P,ii} \qquad (3.11)$$

Where, det $[K_P]$ is the determinant of K_P. For an integrally controlled system with a stable individual loop, $NI < 0$ indicates that the closed loop system will be unstable. The 3×3 and higher sized system with $NI > 0$ may be stable or unstable. Therefore, the value of NI can be used to estimate unworkable pairings which obviously failed to stabilize an individual loop and the overall system at the same time. For systems which are already non-interacting (i.e., off-diagonal elements in Steady State Gain Array (SSGA) are 0), NI will be 1. The basic steps in deriving Eq. (3.11) is as follows (Grosdidier et al., 1985):

For diagonal controller matrix with each PI or PID mode, the closed loop characteristic equation can be obtained. On application of the necessary condition for instability by Routh Hurwitz criteria (Stephanopoulos, 1984), we get the condition in Eq. (3.12).

$$\prod k_{c,ii}\det[K_P] < 0 \qquad (3.12)$$

The product of controller gain and the process gain should be greater than 0 for the stability of the individual loop, i.e.,

$$k_{c,ii}K_{P,ii} > 0 \qquad (3.13)$$

Therefore,

$$\prod k_{c,ii}K_{P,ii} > 0 \qquad (3.14)$$

Dividing Eq. (3.12) by Eq. (3.14), we get Eq. (3.11). Chien and Arkun (1990) have established the relation between RGA and NI. The significance of RGA is that it is explicitly related to system integrity (stability under loop failure) instead of stability directly. It is assumed that K, the SSGM, has been arranged so that the diagonal elements correspond to the proposed pairing. Zhu and Jutan (1996) have proposed the following rules: The input–output variables in a decentralized control system should be paired such that the corresponding NI is positive and close to unity. In addition, to ensure the system integrity, the corresponding RGA elements should be positive.

3.4 Design of Decentralized SISO Controllers

After the selection of suitable input–output variables for pairing using the corresponding transfer function model, PI controller settings can be calculated using the IMC method or any other method discussed earlier. These settings should be detuned by decreasing the proportional gain, increasing the integral time, and reducing the derivative time suitably to take into account the interactions among the SISO loops.

$$(K_{C,i})_{\text{des}} = K_{c,i}/F \qquad (3.15a)$$

$$(\tau_{Ii})_{\text{des}} = \tau_{I,i}F \qquad (3.15b)$$

$$(\tau_{Di})_{\text{des}} = \tau_{D,i}/F \qquad (3.15c)$$

It must be noted that the detuning factor (F) may be the same for all the loops. The value of F varies from 1 to 4 depending on the interactions (i.e., based on the relative gain elements). The value of F is selected by trial and error by checking the responses and interactions of the multivariable control systems by a simulation study. In the next section, we will discuss a method to select the value of F systematically.

Example 1 Consider a TITO example as given in Eq. (3.16).

$$\begin{bmatrix} \dfrac{12.8e^{-s}}{16.7s+1} & \dfrac{-18.9e^{-3s}}{21s+1} \\ \dfrac{6.6e^{-7s}}{10.9s+1} & \dfrac{-19.4e^{-3s}}{14.4s+1} \end{bmatrix} \qquad (3.16)$$

The PI controller based on the diagonal element is designed by closed loop ZN method (based on $k_{c,\text{max}}$ and ω_c) as represented in Eq. (3.17).

$$G_c(s) = K_c + \frac{K_I}{s} \qquad (3.17)$$

Let us calculate the controller settings for the SISO loops (1,1) and (2,2).

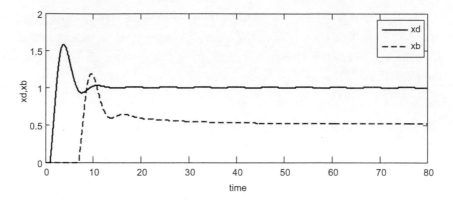

FIGURE 3.4a   MIMO systems. Solid line: $u_1 - y_1$ loop closed; dashed line: $u_2 - y_2$ loop open

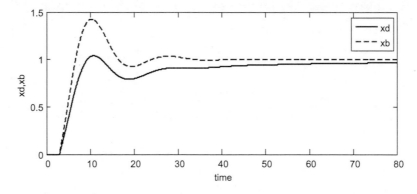

FIGURE 3.4b   MIMO systems. Solid line: $F_R - x_d$ loop open; dashed line: $Q_r - a_b$ loop closed

(a) Loop (1,1)

Ultimate Gain, $K_{cu} = 2$; Ultimate Period, $P_u = 4$
   PI Controller Setting s: Gain, $K_c = 0.9$; $T_I = 3.33$

(b) Loop (2,2)

Ultimate Gain, $K_{cu} = -0.422$; Ultimate Period, $P_u = 11.12$
   PI Controller Setting: Gain, $K_c = -0.189$; $T_I = 9.26$

The closed loop performance for a step change in the set point $y_{1,r}$ is obtained by the Simulink package. The other loop is kept open. Fig. 3.4a shows the response of the first loop and the interaction on the second loop. Similarly, consider the first loop is open and the second loop is closed. The set point response and interaction are shown in Fig. 3.4b. Now consider both the loops are closed. With the same PI settings, the closed performances for a step change in y_{r1} are

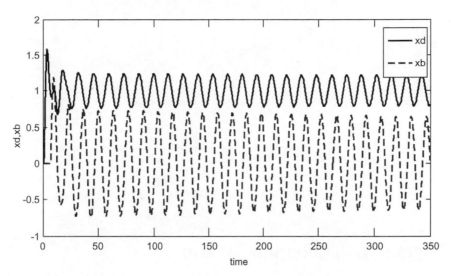

FIGURE 3.4c Set point response when both the loops are closed. Set point change in x_d TITO systems

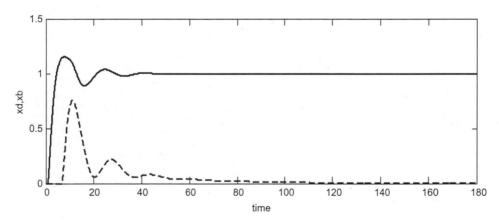

FIGURE 3.4d Set point response for the detuned controllers ($F = 2$) when both the loops are closed. Set point change in x_d

evaluated. Fig. 3.4c shows that the response and the interaction become unstable (continuously sustained oscillations). Thus, the PI controller tuning based on single loop systems does not work on the TITO systems. There is a need to detune the settings. The PI settings are detuned ($F = 2$; decreasing the controller gain and increasing the integral time). With the detuned PI settings, the performances are found to be good as shown in Fig. 3.4d and Fig. 3.4e.

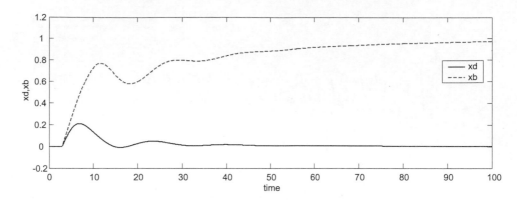

FIGURE 3.4e   Detuned controllers ($F = 2$). Set point response for step change in x_b (dashed line)

3.5   Biggest Log–Modulus Tuning

Luyben (1986) proposed the Biggest Log–Modulus Tuning (BLT) method. The method achieves the conservative objectives of arriving at reasonable controller settings with only a small amount of computational effort. The method is an extension of SISO Nyquist stability criterion in multivariable systems. The essential steps of the algorithm for the method are as follows:

Step 1: Determine the ultimate gain K_u and the ultimate frequency ω_u, of each diagonal process transfer function $G_P(s)$ by the classical SISO method.

Step 2: Calculate the corresponding Ziegler–Nichols setting (K_{cZN}, τ_{IZN}, and τ_{DZN}) for each loop.

Step 3: Assume a detuning factor F. Typical values are 2 to 5.

Step 4: The gains and the integral and derivative times are detuned as in equations Eq. (3.18a), Eq. (3.18b), and Eq. (3.18c).

$$(K_{C,i})_{\text{des}} = K_{cZN,i}/F \tag{3.18a}$$

$$(\tau_{Ii})_{\text{des}} = \tau_{IZN,i}F \tag{3.18b}$$

$$(\tau_{Di})_{\text{des}} = \tau_{DZN,i}/F \tag{3.18c}$$

It must be noted that the value of the detuning factor (F) remains the same for all the loops.

Step 5: Compute the closed loop log modulus L_{cm} using the designed controllers for a specified frequency range, as in Eq. (3.19).

$$L_{cm} = 20\log[W/(1+W)] \tag{3.19}$$

Where,

$$W = -1 + \det[I + GK(s)] \tag{3.20}$$

Step 6: Compute $L_{cm}^{\max}$ (the biggest log–modulus) from the data of L_{cm} versus frequency.

Step 7: Check if $L_{cm}^{\max} = 2N$, where N is the order of the process matrix (N for N input and N output system). If:

$$L_{cm}^{\max} = 2N, \tag{3.21}$$

then stop, otherwise return to Step 3.

The larger the detuning factor F, the more stable the system will be, but the servo responses and the regulatory responses will become more sluggish. The BLT method yields a value of F which gives a reasonable compromise between stability and performance in a multivariable system. The criterion $L_{cm}^{\max} = 2N$ is obtained by simulation studies on several systems (Luyben, 1986). The criterion suggests that the higher the order of the system, the more under-damped the closed loop system must be to achieve reasonable responses. The above method has been found valid by simulation for stable multivariable systems.

3.6 Pairing Criteria for Unstable Systems

The problem of loop interaction can be minimized by a proper choice of input–output pairing. The degree of interaction is quantified using the relative gain array analysis (RGA) which helps in choosing the manipulated–controlled variable pairings that best suit the control problem. For stable plants, the pairing is selected corresponding to the positive values of NI (Niederlinski Index) and RGA.

$$NI = \frac{\det[K_P]}{\prod k_{P,ii}} \tag{3.22}$$

$$RGA(\Lambda) = K_P \otimes K_P^{-T} = \begin{bmatrix} \lambda_{11} & \lambda_{12} \\ \lambda_{21} & \lambda_{22} \end{bmatrix} \tag{3.23}$$

where the special symbol indicates the element by element multiplication rather than matrix multiplication. The pairing criteria for unstable systems (Hovd and Skogestad, 1993) will differ when the number of open loop unstable poles of $G_P(s)$

is different from $G_P*(s) = \text{diag}[g_{P,ii}(s)]$. Hence, the input and output variables are to be paired in the following way: For an $n \times n$ plant with one unstable pole which appears in all elements of $G_P(s)$, the pairing should be such that NI is positive if n is odd and NI is negative if n is even.

For the special case of $n \times n$ plant with P unstable poles, which appear in all the elements of $G(s)$, pairing should be such that NI is positive if $(n-1)P$ is even and negative if $(n-1)P$ is odd. Pair on positive RGA elements if P is even and on negative elements if P is odd. For systems with mild interactions, detuning the controller settings, as discussed in the previous section, will be desirable. However, if the interactions are significant, we need to design a suitable decoupler to reduce the interactions of the combined process and the decoupler. Consider the example given by Govindakannan and Chidambaram (1997, 2000), containing all elements unstable, with equal poles as in Eq. (3.24).

$$G_P(s) = \begin{bmatrix} \dfrac{-1.6667e^{-s}}{-1.6667s+1} & \dfrac{-1e^{-s}}{-1.6667s+1} \\ \dfrac{-0.8333e^{-s}}{-1.6667s+1} & \dfrac{-1.6667e^{-s}}{-1.6667s+1} \end{bmatrix} \tag{3.24}$$

Since the number of open loop unstable poles of $G(s)$ are different from $G(s) = \text{diag}[g_{ii}(s)]$, the pairing criteria for this system will differ from that of a stable system. Hence, the pairing is carried out according to Section 3.6:

$$K = \begin{bmatrix} -1.6667 & -1 \\ -0.8333 & -1.6667 \end{bmatrix} \quad \Lambda = \begin{bmatrix} 1.4283 & -0.4283 \\ -0.4283 & 1.4283 \end{bmatrix} \tag{3.25}$$

The NI for this 2×2 system is 0.5833 and hence the columns are interchanged to get the correct pairing. Therefore, the newly paired system is given by Eq. (3.26).

$$G(s) = \begin{bmatrix} \dfrac{-1e^{-s}}{-1.6667s+1} & \dfrac{-1.6667e^{-s}}{-1.6667s+1} \\ \dfrac{-1.6667e^{-s}}{-1.6667s+1} & \dfrac{-0.8333e^{-s}}{-1.6667s+1} \end{bmatrix} \tag{3.26}$$

3.7 Decoupling Control

A possible strategy to achieve a non-interacting closed loop system is decoupling. Consider the open loop system with a decoupler as in Fig. 3.5. The objective is to

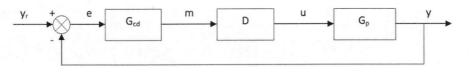

FIGURE 3.5 Decoupler and controller for multivariable systems
D: decoupler matrix; G_P: process matrix; G_{cd}: controller matrix for the decoupled system

find a matrix $D(s)$ such that

$$Y(s) = G_P(s)\, D(s)\, U(s) \tag{3.27}$$

$$= Q(s)\, U(s) \tag{3.28}$$

where the matrix $Q(s)$ has some desirable properties. Depending on the extent of decoupling desired, we could specify $Q(s)$. For a TITO system (Fig. 3.5) we represent as in Eq. (3.29).

$$G_P(s) = \begin{bmatrix} g_{P11} & g_{P12} \\ g_{P21} & g_{P22} \end{bmatrix} \tag{3.29}$$

For ideal (perfect) decoupling, we specify the matrix $Q(s)$ as in Eq. (3.30).

$$Q(s) = \begin{bmatrix} g_{P11} & 0 \\ 0 & g_{P22} \end{bmatrix} \tag{3.30}$$

This eliminates the off-diagonal (interaction) terms completely and results in the decoupling structure of the decoupling matrix as in Eq. (3.31). It can be checked that GPD denoted by P is a diagonal matrix. Similarly, it can be checked that GC calculated by DGcd is a full matrix.

$$D(s) = \begin{bmatrix} d_{11} & d_{12} \\ d_{21} & d_{22} \end{bmatrix} \tag{3.31}$$

Where,

$$d_{11} = g_{P11}g_{P22}/T; \ d_{12} = -g_{P12}g_{P22}/T \tag{3.32}$$

$$d_{21} = -g_{P21}g_{P11}/T; \ d_{22} = g_{P22}g_{P11}/T \tag{3.33}$$

$$T = g_{P11}g_{P22} - g_{P12}g_{P21} \tag{3.34}$$

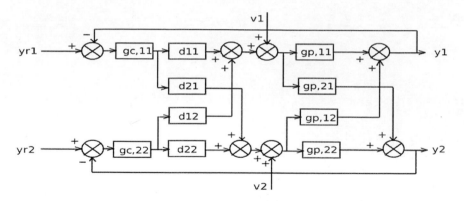

FIGURE 3.6a   A TITO process with an ideal decoupler

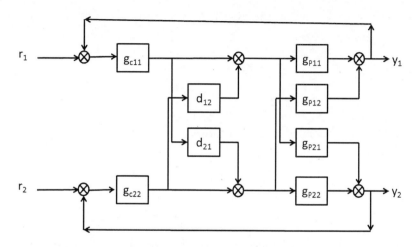

FIGURE 3.6b   Simplified decoupled control system of a TITO process

For perfect decoupling, the decouplers are complicated ones (transfer function matrix). Let us consider the simplified decouplers where $d_{11} = 1$ and $d_{22} = 1$ as shown in Fig. 3.6b. We can derive the conditions in Eq. (3.35a) and Eq. (3.35b) to cancel the interaction from the first loop to the second loop.

$$d_{12} = -g_{P12}/g_{P11} \tag{3.35a}$$

and

$$d_{21} = -g_{P21}/g_{P22} \tag{3.35b}$$

This results in the decoupled system given in Eq. (3.36).

$$Q(s) = \begin{bmatrix} q_{11} & 0 \\ 0 & q_{22} \end{bmatrix} \tag{3.36}$$

Where,

$$q_{11} = g_{P11}/(1 - h) \tag{3.37a}$$

$$q_{22} = g_{P22}/(1 - h) \tag{3.37b}$$

Where,

$$h = g_{P12}g_{p21}/(g_{P11}g_{P22}) \tag{3.38}$$

The transfer functions q_{11} and q_{22} should be reduced into a FOPTD model or a SOPTD model to design a PI controller. For this, the step response of q_{11} is to be obtained and using SK method (Sundaresan and Krishnaswamy, 1978), we can get an approximated FOPTD model. Similarly, a suitable FOPTD model can be obtained for q_{22}. The decoupler is to be considered a part of the overall control system as in Eq. (3.39).

$$G_{cd} = G_{c,PI}D(s) \tag{3.39}$$

where $G_{c,PI}$ is a diagonal matrix with the conventional PI or PID controller designed for the coupled system, Q.

The resulting combined diagonal PI controller matrix and the decoupler matrix $[G_C^D]$ is usually a full matrix where each element can then be approximated into a PI controller with a lead-lag filter. We can also design a centralized PI controller system (Fig. 2.7) directly from the transfer function matrix of the process. These methods will be discussed in the next section.

3.8 Gain and Phase Margin Method

The Gain and Phase Margins (GPMs) are typical loop specifications associated with the frequency response (Franklin et al., 1989). The GPMs have always served as important measures of robustness. It is known from classical control that phase margin is related to the damping of the system and can therefore also serve as a performance measure (Franklin et al., 1989). The controller design methods to satisfy the GPM criteria are not new, and are widely used. Simple formulae are given by Maghade and Patre (2012) to design the PI/PID controller to meet user-defined gain margin and phase margin specifications. Using definitions of GPM, the set of Eq. (3.40) to Eq. (3.43) can be written.

$$\arg[l_{ii}(j\omega_{pii})k_{ii}(j\omega_{pii})] = -\pi \tag{3.40}$$

$$A_{mii} = \frac{1}{|l_{ii}(j\omega_{pii})k_{ii}(j\omega_{pii})|} \tag{3.41}$$

$$l_{ii}(j\omega_{gii})k_{ii}(j\omega_{gii}) = 1 \tag{3.42}$$

$$\varphi_{mii} = \arg[l_{ii}(j\omega_{gii})k_{ii}(j\omega_{gii})] + \pi \tag{3.43}$$

where A_{mii} and Φ_{mii} are GM and PM respectively. Also, ω_{gii} and ω_{pii} are gain and phase crossover frequencies. The PI controller parameters are given for the FOPTD model in Eq. (3.44).

$$k_{c,ii} = \frac{\omega_{pii}T_{ii}}{A_{mii}k_{p,ii}}; \quad T_{I,ii} = \left(2\omega_{pii} - \frac{4\omega_{pii}^2 l_{ii}}{\pi} + \frac{1}{T_{ii}}\right)^{-1} \tag{3.44}$$

Where,

$$\omega_{pii} = \left(\frac{A_{mii}\varphi_{mii} + \frac{1}{2}\pi(A_{mii} - 1)}{(A_{mii}^2 - 1)l_{ii}}\right) \tag{3.45}$$

PID controller parameters with a first order filter will be as in Eq. (3.46).

$$k_{c,ii} = \frac{\omega_{pii}T_{ii}}{A_{mii}k_{p,ii}}; \quad T_{Iii} = \left(2\omega_{pii} - \frac{4\omega_{pii}^2 l_{ii}}{\pi} + \frac{1}{T_{ii}}\right)^{-1};$$

$$T_{Dii} = T_{Iii}; \quad T_{Fii} = T_{Iii} \tag{3.46}$$

3.9   Relay Tuning Method

The continuous cycling Ziegler–Nichols (ZN) method is the recommended method to design a PID controller for a SISO system (Yu, 2006). The ZN tuning process requires trial-and-error experiments to determine the critical frequency and the critical gain of the controller. The relay auto-tuning method (Astrom and Hagglund, 1984, 1995) is an automated technique to obtain the ultimate gain and the frequency of oscillation of the controller. The method is easy to implement where the PID controller is replaced by a relay. The appropriate switching sign of the steady state gain of the system is required for existence of the stable limit cycles. For the positive steady state gain system, relay is defined as $U = U_{\max}$ for $e \geq 0$ and $U = U_{\min}$ for $e < 0$. Whereas for the negative steady state gain process, it is defined as $U = U_{\max}$ for $e \leq 0$ and $U = U_{\min}$ for $e > 0$. In case of a change in the process parameters, the system can switch over to the relay, calculate the parameters of the PID controller and update the settings. Yu (2006), Wang et al. (2003), Chidambaram and Vivek (2014), and Chidambaram and Nikita (2018) give good reviews about the relay tuning of PID controllers.

The implementation of the ZN method is difficult when applied to MIMO systems due to the presence of interactions. The fundamental step of the method is the determination of critical points of the plant (critical frequency and critical gain, together are known as the critical points). The main deviation from SISO case is that MIMO systems exhibit infinite number of critical points that can lead to stable limit cycles. The set of critical points are referred as the stability limit of the system. With the popularity of the SISO–PID auto-tuning in industries and in commercial control systems, the relay tuning method is to be extended to MIMO–PID systems (Palmor et al., 1995). The relay auto-tuning method is one of the most trusted methods to determine the critical points. It has the several advantages over the independent loop tuning method (Skogestad and Morari, 1989) and the sequential loop tuning method (Hovd and Skogestad, 1993). By varying the magnitudes of the relay (i.e., relay heights) different critical points can be identified. One of the main advantages of relay auto-tuning is the decoupling effect. The limit cycles obtained are used to design the controllers. It is a single experiment and does not require the detuning of the controller parameters.

For MIMO systems, selection of the appropriate ratios of relay heights is important. Palmor et al. (1995) proposed an algorithm to determine the Desired Critical Point (DCP) for the TITO system which was later generalised by Halevi et al. (1997) for MIMO systems. The algorithm also identifies the steady state gain of the process in a closed loop. The DCP obtained is used to calculate the PID settings using the relay auto-tune formula proposed by Astrom and Hagglund (1995) for the SISO system given in Eq. (3.47).

$$K_{u,i} = \frac{4h_i}{\pi a_i} \tag{3.47}$$

where, h_i is the relay height for i^{th} loop and a_i is the output amplitude of limit cycle of ith loop. The expression in Eq. (3.47) for the ultimate gain of the controller $(K_{u,i})$ gives only an approximate value, since the process is assumed to have the low pass filter characteristics. The conventional method of analysing the relay tuning method assumes only the principal harmonics. However, most of the processes encountered do not follow this assumption. Yu (2006) demonstrated that, for SISO FOPTD processes, different delay to time constant (D/τ) ratio gives different shapes in the relay feedback tests varying from triangle to rectangle forms. The limit cycles obtained during the relay feedback test deviates largely from the sinusoidal shape for higher D/τ ratios. Similar results are obtained by Friman and Waller (1994). Higher order harmonics plays an important role in the shape of the output limit cycles. Chidambaram and Sathe (2014) proposed the modified relay auto-tuning method to improve the controller performance by considering the higher order harmonics during the evaluation of the ultimate values for stable SISO systems.

There is a need for carrying out the relay auto-tuning method, particularly for MIMO systems. One should keep in mind that incorporating the higher order harmonics changes only the controller gains of the system and does not affect derivative time and reset time (Yu, 2006; Astrom and Hagglund, 1984; Palmor et al. 1995). The results appear to be the detuned values of the controller gains. Without the relay tuning method, the calculation of the controller settings may take considerable time and may not be simple. Nikita and Chidambaram (2018) extended the method of relay auto-tuning of multivariable stable systems (Palmor et al. 1995) taking into account the higher order harmonics. The details of this method are given here.

3.10   Relay Tuning of MIMO Systems

The conventional method of analysis of relay tuning of multivariable systems that is reported by Palmor et al. (1995) is briefly reviewed here. Appendix A gives the essential steps of the method. The decentralized control system for a TITO system is shown in Fig. 3.7. The linear plant is represented in Eq. (3.48) by transfer function matrix $G_p(s)$, with $y(t)$ as the output vector, $u(t)$ as the manipulated variable vector, and $y_r(t)$ as the reference signal vector.

$$G_p(s) = \begin{bmatrix} g_{p11}(s) & g_{p12}(s) \\ g_{p21}(s) & g_{p22}(s) \end{bmatrix} \tag{3.48}$$

The transfer function matrix of decentralized PID controller matrix $G_C(s)$ is given in Eq. (3.49).

$$G_C(s) = \begin{bmatrix} c_1(s) & 0 \\ 0 & c_2(s) \end{bmatrix} \tag{3.49}$$

The SISO system has only one critical point, which brings the system to the verge of instability. However, as stated earlier, in a TITO system many critical points exist that can lead to generation of limit cycle, collectively known as the stability limits

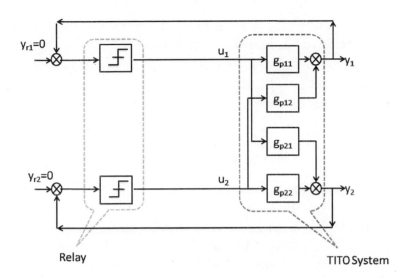

FIGURE 3.7   TITO decentralized relay system

(Palmor et al., 1995). The knowledge of the critical points is necessary for the tuning procedure of controllers. During the identification of the points, both the controllers are replaced by the relays. Relays consist of sinusoidal waves of odd multiples of fundamental frequency ω (Yu, 2006). The magnitude of the relay is defined by the relay height.

The limit cycles generated from both the loops have a common time period, $P_{u,i}$ with different amplitude, a_i $(i = 1, 2)$. The proof for this is given by Atherton (1975) and Zhuang and Atherton (1994). Wang et al. (1997) also reported this behaviour while identifying the transfer function matrix model for the TITO system. All the pairs of relay magnitude (h_1, h_2) correspond to same critical point provided relay ratio h_1/h_2 is constant (Palmor et al, 1995). Also, by changing the relay ratio, critical point moves along the stability limit. Once the critical points are known, then the controllers can be designed similar to SISO systems using the ZN formulae.

3.11   Improved Relay Analysis

The assumption considered during the derivation of the ultimate gain, that the higher order harmonics of the relay output is filtered by the system, is not always applicable for all the systems (Yu, 2006; Chidambaram and Vivek, 2014). If the system shows low pass filter characteristics, then this assumption is valid. In most cases, as the steady state relay output does not take the sinusoidal waveform, the presence of higher order harmonics is indicated. The expression to determine the ultimate gain value needs to be modified to incorporate the higher order harmonic terms. The Fourier series expansion of amplitude of the limit cycle is suitably modified to take into account the effect of higher order harmonics (Appendix B). Chidambaram and Sathe (2014) proposed a method to incorporate the higher order harmonics for stable SISO systems.

It is reported for SISO systems that whenever the relay oscillations obtained have waveform closer to a rectangular wavefrom, then the result for smaller $\tau\omega$ is used for determining the improved ultimate gain value, whereas for waveforms closer to triangular waveform the result for larger $\tau\omega$ is used for calculating the same (Appendix B gives the derivation of the equation). If the relay oscillations are close to sine waveform, then the standard equation is considered. It should be noted that 'a' in Eq. (3.47) is not the amplitude which we observe from the limit cycles; it is to be calculated by Eq. (3.50a) and Eq. (3.50b).

For smaller $\tau\omega$ (similar to a rectangular waveform) Eq. (3.50a)is applicable.

$$a = \frac{y(t^*)}{\left[1 - \frac{1}{3} + \frac{1}{5} - \frac{1}{7} + \frac{1}{9} - \cdots \frac{1}{N}\right]} \tag{3.50a}$$

For larger $\tau\omega$ (similar to a triangular waveform) Eq. (3.50b) is applicable.

$$a = \frac{y(t^*)}{\left[1 + \frac{1}{3^2} + \frac{1}{5^2} + \frac{1}{7^2} + + \cdots \frac{1}{N^2} \cdots\right]} \tag{3.50b}$$

where $y(t^*)$ is to be noted from the relay response curve at the time $t^* = 0.5\pi/\omega$ (t^* is noted from where the stable sustained limit cycles are obtained). The number of terms to be incorporated depends on the degree of deviation of the limit cycle waveform from the sinusoidal behaviour and the initial transient dynamics. As stated earlier, due to the decoupling effect of the simultaneous relay auto-tuning method, the improved relay auto-tuning of single loop system can be applied directly to each of the diagonal loops. Here, the method is applied by simulation on three different TITO systems with varying degree of interactions. Adequate numbers of higher order harmonic terms to get the improved estimates of the ultimate values are selected, depending on the waveform. Also, it should be noted that by incorporating all the higher order harmonic terms, it may lead to error in evaluating the ultimate values. Therefore, an appropriate evaluation is required for determining the ultimate gains. In the present work, the case study from the work of Palmor et al. (1995) is considered and the effect of incorporating the higher order harmonic terms is analysed. Here, the method is applied to two more case studies to highlight the importance of the higher order harmonic terms.

3.12 Simulation Examples

Example 2 The methanol–water distillation column model studied by Wood and Berry (1973) is considered. The transfer function matrix is given by Eq. (3.51).

$$G_p = \begin{bmatrix} \dfrac{12.8e^{-4s}}{16.7s+1} & \dfrac{-18.9e^{-3s}}{21s+1} \\ \dfrac{6.6e^{-7s}}{10.9s+1} & \dfrac{-19.4e^{-3s}}{14.4s+1} \end{bmatrix} \tag{3.51}$$

The relative gain array of the system is calculated as in Eq. (3.52).

$$\text{RGA} = \Lambda = K_p. * [K_p^{-1}]' \tag{3.52}$$

$$\text{RGA} = \Lambda = \begin{bmatrix} 2.0094 & -1.0094 \\ -1.0094 & 2.0094 \end{bmatrix} \tag{3.53}$$

The RGA shows that the system is highly interactive. To design the decentralized PID controller for the process, it is required to determine the value of critical points for the system. For determination of the critical points, the PID controllers are replaced by relays. The appropriate relay heights are to be specified. It is observed that the limit cycles obtained have common frequency, different amplitudes but do not take sinusoidal shape completely, implying the presence of higher order harmonics. Based on initial dynamics of the response of system to relay input, higher order harmonics terms are considered for evaluating the critical points. The system responds to different relay heights by generating the limit cycle for each relay, but the optimum performance is obtained at defined, desired critical point only. The height ratio of relays used for the system is taken from the studies reported by Palmor et al. (1995). The results are reviewed in Table 3.1. Fig. 3.8 shows the system oscillations (both dynamics and steady state) of the relay test for the system.

The plant model considered provides a limited filtering. This is clearly reflected in the relay response curve. The relay oscillations obtained in Fig. 3.8 takes the waveform closer to that of a triangular waveform, so the results for value of larger $\tau\omega$ are used for determination of critical points. The number of the higher order harmonics terms, to be considered for evaluating the improved ultimate gains of the controllers, depends on the initial dynamics of the response of the system to the relay input and on the extent of deviation of the relay response curve from

TABLE 3.1 Results using the relay response (Example 2)

h_1	h_2	a_1	a_2	P_u	ω_u	t^*	$y(t^*)_1$	$y(t^*)_2$
0.059	−0.08	0.3901	0.3904	13.89	0.452	3.752	0.2721	0.3441

TABLE 3.2 PID controller setting (Example 2)

	K_u (CM)	K_u (IM)	K_c (CM)	K_c (IM)	τ_i	τ_d
Controller 1	0.1925	0.3068	0.1155	0.1840 ($N=3$)	6.9504	1.7376
Controller 2	−0.261	−0.3290	−0.1566	−0.1974 ($N=3$)	6.9504	1.7376

CM = conventional method; IM = improved method

TABLE 3.3 Table 3.3 Time domain performance analysis (Example 2)

	RESPONSE (11)		INTERACTION (12)		INTERACTION (21)		RESPONSE (22)	
	ISE	ITAE	ISE	ITAE	ISE	ITAE	ISE	ITAE
CM	7.77	121.4	0.698	69.97	4.793	152.1	4.65	55.6
IM	6.27	96.99	0.916	73.74	3.35	107	4.36	50.7

CM = Conventional method; IM = Improved method

the sinusoidal waveform. $N = 3$ in the summation term in Eq. (3.50), indicates that one higher order harmonics term is included while determining the ultimate gains to the controllers. The values of ultimate gains are calculated by both the conventional method and by the improved method. The controller is designed using the ZN method. The values of the ultimate gains and the controller setting are given in Table 3.2. The improved process responses and the reduced interactions are obtained as shown in Fig. 3.9. Although the relay ratio specified by Palmor et al. (1995) was optimum and the settings resulting from the critical point gave good performances, but still, on including the higher order dynamics terms, a large improvement can be seen in the process response. A comparison of the design methods, quantified based on the time domain performance, showed better results for improved method over the conventional method as shown in Table 3.3. It can be seen that a reduction of 30% in the ISE values and 29% in the ITAE value is obtained for interaction in case of step change in the second reference signal. Also, approximately 20% reduction in ISE or ITAE values is obtained in the responses in the case of step change in the first reference signal.

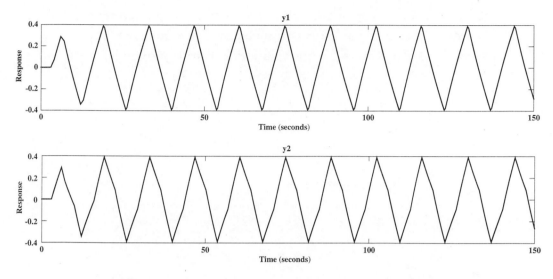

FIGURE 3.8   Relay response curve of the process (Example 2)

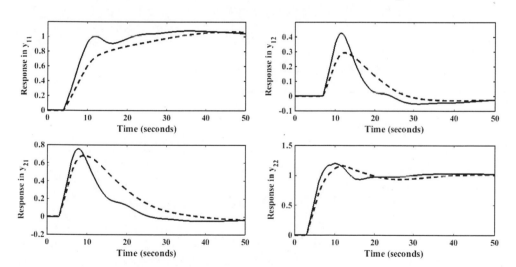

FIGURE 3.9   Process response and interaction curve for unit step change (Example 2). Solid line: improved method; dashed lines: conventional method

3.13   Design of Simple Multivariable PI Controllers

An excellent review of the selection method of control configuration for linear MIMO system is given by Sujatha and Panda (2013). Some of the available tools for the interaction measurement include Niederlinski Index (NI) (Niederlinski, 1971), Singular Value Decomposition (SVD) (Morari, 1982), Morari Resiliency Index (MRI) (Yu and Luyben, 1986), Dynamic Relative Gain Array (DRGA) (Tung and Edgar, 1981). For systems with significant interactions, a centralized PI control system is preferred rather than a decentralized control system.

Davison (1976) proposed a multivariable PI controller, given by Eq. (3.54) and Eq. (3.55).

$$K_c = \delta[G(s=0)]^{-1} \tag{3.54}$$

$$K_I = \epsilon[G(s=0)]^{-1} \tag{3.55}$$

where δ and ε are the tuning parameters.

Maciejowski (1989) suggested a method for designing multivariable PI controllers. The matrix K_c is given by Eq. (3.56).

$$K_c = \delta[G(s=i\omega_b)]^{-1} \tag{3.56}$$

where ω_b is the desired bandwidth of the system. K_I is same as given by Davison (1976) method. Usually $[G(s=i\omega_b)]^{-1}$ is complex, hence a real approximation of it is needed for designing.

Ogunnaike et al. (1983) reported an experimental study of the multivariable control of a pilot plant distillation column having side streams and overhead and bottom products. A multivariable controller consisting of multivariable time delay compensator coupled with PI controller was developed. An improved disturbance rejection was observed.

3.14   Design of Multivariable PID Controllers

Tanttu and Lieslehto (1991) and Lieslehto et al. (1993) developed a heuristic multivariable controller tuning method. Using the IMC method (Morari and Zafiriou, 1989), PID controllers for each of the scalar transfer function of the process are designed first. The multivariable PID controllers can then be designed as given in Eq. (3.57), Eq. (3.58), and Eq. (3.59).

$$K_c = \begin{bmatrix} 1/k_{c,11} & 1/k_{c,12} & \cdots & 1/k_{c,1n} \\ 1/k_{c,21} & 1/k_{c,22} & & 1/k_{c,2n} \\ \vdots & \vdots & \ddots & \vdots \\ 1/k_{c,n1} & 1/k_{c,n2} & \cdots & 1/k_{c,nn} \end{bmatrix}^{-1} \tag{3.57}$$

$$K_I = \begin{bmatrix} 1/k_{i,11} & 1/k_{i,12} & \cdots & 1/k_{i,1n} \\ 1/k_{i,21} & 1/k_{i,22} & & 1/k_{i,2n} \\ \vdots & \vdots & \ddots & \vdots \\ 1/k_{i,n1} & 1/k_{i,n2} & \cdots & 1/k_{i,nn} \end{bmatrix}^{-1} \tag{3.58}$$

$$K_D = \begin{bmatrix} 1/k_{d,11} & 1/k_{d,12} & \cdots & 1/k_{d,1n} \\ 1/k_{d,21} & 1/k_{d,22} & & 1/k_{d,2n} \\ \vdots & \vdots & \ddots & \vdots \\ 1/k_{d,n1} & 1/k_{d,n2} & \cdots & 1/k_{d,nn} \end{bmatrix}^{-1} \tag{3.59}$$

Each SISO systems can have different tuning parameters, but better results are obtained when all the tuning parameters have the same value. In order to get a better insight of the method, let G be a $m \times m$ transfer function matrix. Assume all elements of G to be minimum phase transfer functions. An IMC controller can be suggested for the process as in Eq. (3.60).

$$Q = G^{-1}F \tag{3.60}$$

Where,

$$F = [1/(\lambda s + 1)]I \tag{3.61}$$

The classic controller C and IMC controller Q can be related as given in Eq. (3.62).

$$= Q(I - GQ)^{-1} \tag{3.62}$$

Substituting the equation for Q in Eq. (3.62), we get:

$$C = G^{-1}F(I - GG^{-1}F)^{-1} \tag{3.63}$$

$$C = G^{-1}\frac{1}{(\lambda s + 1)}I \left(I - GG^{-1}\frac{1}{(\lambda s + 1)}I \right)^{-1} \tag{3.64}$$

$$C = \frac{1}{\lambda s}G^{-1} \tag{3.65}$$

Assume that the controller is designed using K_C, K_I, K_D and the matrix is given by the transfer function mentioned in Eq. (3.57), Eq. (3.58), and Eq. (3.59). The elements (g_{ij}) of the transfer function matrix are considered to be the minimum phase and the same fine-tuning parameter is applied for all controllers. Then each of the scalar elements of the controller is given by Eq. (3.66).

$$c_{ij} = \frac{1}{\lambda s}\frac{1}{g_{ij}} \tag{3.66}$$

Now the multivariable controller C can be computed as in Eq. (3.67).

$$C = \begin{bmatrix} 1/c_{11} & 1/c_{12} & \cdots & 1/c_{1n} \\ 1/c_{21} & 1/c_{22} & & 1/c_{2n} \\ \vdots & \vdots & \ddots & \vdots \\ 1/c_{n1} & 1/c_{n2} & \cdots & 1/c_{nn} \end{bmatrix}^{-1} \tag{3.67}$$

If the process models are simple, then the scalar controllers are in PID form given in Eq. (3.68).

$$c_{ij} = k_{c_{ij}} + k_{i_{ij}}\frac{1}{s} + k_{d_{ij}}s \tag{3.68}$$

If the controller form in Eq. (3.68) is substituted in Eq. (3.67), a complex multivariable control structure is obtained. Hence an approximation is made based on the assumption that, at all frequencies, one of the terms in Eq. (3.68) is dominating. For example, at low frequency Eq. (3.68) is approximated as $cij \approx k_{I,ij}(1/s)$. Substituting this approximation into Eq. (3.67) gives the I-part of the multivariable controller. Thus K_C, K_I and K_D of Eq. (3.57), Eq. (3.58), and Eq. (3.59) give a controller that approximates the multivariable controller C of Eq. (3.68).

Vu and Lee (2010a) demonstrated an analytical method for the design of a multi loop proportional integral controller based on direct synthesis method. A MIMO process with time delay was considered. Based on the desired closed loop response and relative gain, ideal multi loop PI controllers are designed. The ideal multi loop PI controllers are approximated using Maclaurin series expansion and standard multi loop PI controllers are obtained. Simulation studies on a 2×2 system and a 3×3 system illustrated the superior performance of the proposed controller.

3.15   Performance and Robust Stability Analysis

A control system is robust if it is sensitive to differences between the actual system and the model of the system used to design the controllers. These differences are known as model or plant mismatches or simply model uncertainties. In practice, model uncertainty arises because the theoretical process model G is often not an accurate representation of the real plant G_P. This is because the model parameters are not known exactly due to the assumptions made in the models of the processes and due to the linearization made during the derivation of transfer function model. Moreover, the behaviour of the plant itself changes with time, which is not captured in the model. Many dynamic perturbations that may occur in different parts of a system can, however, be lumped into one single perturbation block Δ, for instance, some un-modelled, high frequency dynamics. A number of robustness analysis methods are listed in Table 3.4. The controller robustness property can be carried out by the maximum singular value method.

Using maximum singular value: The closed loop system is stable if and only if,

$$\bar{\sigma}[H(j\omega)] < \frac{1}{\bar{\sigma}[\Delta(j\omega)]} \quad \forall \omega \tag{3.69}$$

where $\bar{\sigma}$ is the maximum singular value and H is the complementary sensitive function of the entire closed loop system. By plotting the frequency plot of $\bar{\sigma}$ for the different methods, we can compare the robust stability of the methods. The controller is robust if the value $\bar{\sigma}[H(j\omega)]$ is smaller in all the frequency ranges.

One of the problems in using $\bar{\sigma}[H(j\omega)]$ in robust stability analysis is that the stability criterion may be too conservative.

First, for a process multiplicative input uncertainty as $G(s)[I+\Delta_I(s)]$, the closed loop system is stable (Maciejowski 1989) if,

$$\|\Delta_I(j\omega)\| < 1/\bar{\sigma} \tag{3.70}$$

where $\bar{\sigma}$ is the maximum singular value of $[I + G_C(j\omega)G(j\omega)]^{-1}G_C(j\omega)G(j\omega)$

It can be derived from the characteristic equation of:

$$
\begin{aligned}
T(s) &= \det\{I + G(s)[I + \Delta_I(s)]G_C(s)\} \\
&= \det[I + G(s)G_C(s) + G(s)\Delta_I(s)G_C(s)] \\
&= \det[I + G(s)G_C(s)]\det\{I + [I + G(s)G_C(s)]^{-1}G(s)\Delta_I(s)G_C(s)\} \\
&= \det[I + G(s)G_C(s)]\det\{I + G_C(s)[I + G(s)G_C(s)]^{-1}G(s)\Delta_I(s)\} \\
&= \det[I + G(s)G_C(s)]\det\{I + [I + G_C(s)G(s)]^{-1}G_C(s)G(s)\Delta_I(s)\} \tag{3.71}
\end{aligned}
$$

We can also use multiplicative output uncertainty $[I+\Delta_O(s)]G(s)$. The closed loop system is stable if,

$$\|\Delta_O(j\omega)\| < 1/\bar{\sigma} \tag{3.72}$$

where $\bar{\sigma}$ is maximum singular value of $[I + G(j\omega)G_C(j\omega)]^{-1}G(j\omega)G_C(j\omega)$.

TABLE 3.4   Various robust stability analysis methods reported in literature

S. No.	Robust stability method	Stability criterion	Remarks
1	Maximum singular value	System is stable if maximum singular value of complementary sensitive function is smaller in all frequencies.	Stability description is general and is too conservative. Singular values are affected by similarity transformations. Singular values may be used for both structured and unstructured uncertainties.
2	Inverse of maximum singular value	Stability of the system depends on the area under the curve. More the area, more the stability of the system.	This method is easy to use to compare different controller system stabilities. It may be used for any type of uncertainties.
3	Internal stability	Nyquist stability criterion is used for analysing stability of each element in internal stability matrix.	Difficult to analyse the system's internal stability for uncertainties. Nyquist stability test is difficult for MIMO systems.
4	Spectral radius	System is stable if spectral radius is below unity,	It can be used for all types of uncertainties. Often used for comparing different system stabilities.
5	Structured Singular Value (SSV)	System is stable if SSV is less than one.	SSV is a conservative test for internal parameter variations. It can be used only for structured uncertainties.

The frequency plot obtained for the right-hand side of Eq. (3.71) or Eq. (3.72) indicates stability bounds of the closed loop system. The area under the curve represents stability of the system. More area under the curve indicates higher stability of the system. By using this plot, it is easy to compare the stability of different controllers. The control system that gives the maximum area under the curve is more stable.

Many dynamic perturbations that may occur in different parts of a system can, however, be lumped into one single perturbation block Δ for instance, some un-modelled, high frequency dynamics. This uncertainty representation is referred to as an unstructured uncertainty. The unstructured dynamics uncertainty in a control system is commonly described by Eq. (3.73a) and Eq. (3.73b) (Gu et al., 2003).

3.15.1 Input Multiplicative Uncertainty

$$G_p(s) = G_o(s)[I + \Delta_I(s)] \tag{3.73a}$$

3.15.2 Output Multiplicative Uncertainty

$$G_p(s) = [I + \Delta_o(s)]G_o(s) \tag{3.73b}$$

Where,

$G_p(s)$ = actual perturbed system dynamics; $G_o(s)$ = nominal model.

The additive uncertainty representation gives an account of absolute error between the actual dynamics and the nominal models, while the multiplicative uncertainty representation shows relative errors. The process multiplicative input uncertainties, shown in Fig. 3.10, can be loosely interpreted as the process input actuator uncertainties; the process multiplicative output uncertainties, shown in Fig. 3.11, can be practically viewed as the process output measurement

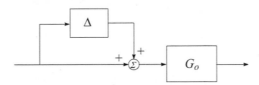

FIGURE 3.10   Input multiplicative perturbation configuration

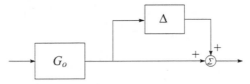

FIGURE 3.11   Output multiplicative perturbation configuration. $G_p(s)$ actual perturbed system dynamics; $G_o(s)$ nominal model

uncertainties. The comparison of the controller robustness property can be carried out by using different robust analysis methods.

3.15.3   Using Maximum Singular Value

The comparison of the controller robustness property can be carried out by assuming a plant with an input uncertainty to describe modelling errors. The closed loop system is stable if and only if,

$$\bar{\sigma}[H(j\omega)] < \frac{1}{\bar{\sigma}[\Delta(j\omega)]} \quad \forall \omega \tag{3.74}$$

where $\bar{\sigma}(.)$ is the maximum singular value and H is the complementary sensitive function of the entire closed loop system. By plotting the frequency plot of $\bar{\sigma}[H(j\omega)]$ for the different methods, we can compare the robust stability of these methods. The controller is robust if the value $\bar{\sigma}[H(j\omega)]$ is smaller in all the frequency ranges.

First, for a process multiplicative input uncertainty represented as $G(s)[I + \Delta_I(s)]$, the closed loop system is stable if (Maciejowski 1989),

$$\|\Delta_I(j\omega)\| < 1/\bar{\sigma} \tag{3.75}$$

where $\bar{\sigma}$ is the maximum singular value of $[I + G_C(j\omega)G(j\omega)]^{-1}G_C(j\omega)G(j\omega)$.

This can be derived from the characteristic equation of,

$$
\begin{aligned}
T(s) &= \det\{I + G(s)[I + \Delta_I(s)]G_C(s)\} \\
&= \det[I + G(s)G_C(s) + G(s)\Delta_I(s)G_C(s)] \\
&= \det[I + G(s)G_C(s)]det\{I + [I + G(s)G_C(s)]^{-1}G(s)\Delta_I(s)G_C(s)\} \\
&= \det[I + G(s)G_C(s)]det\{I + G_C(s)[I + G(s)G_C(s)]^{-1}G(s)\Delta_I(s)\} \\
&= \det[I + G(s)G_C(s)]det\{I + [I + G_C(s)G(s)]^{-1}G_C(s)G(s)\Delta_I(s)\} \tag{3.76}
\end{aligned}
$$

Since,

The product of the singular values of A equals the value of its determinant,

$$\sigma_1\sigma_2\sigma_3\cdots\sigma_n = \det(A) \tag{3.77}$$

from the small gain theorem,

$$\|T(j\omega)\| < 1 \tag{3.78}$$

The inverse of the maximum singular value of $[I+G_C(s)G(s)]^{-1}G_C(s)G(s)\Delta_I(s) < 1$ is represented as,

$$\Delta_I(s) < 1/\sigma([I + G_C(s)G(s)]^{-1}G_C(s)G(s)) \tag{3.79}$$

From Eq. (3.74) and Eq. (3.75), Eq. (3.79) can be written as,

$$\|\Delta_I(s)\| < 1/\bar{\sigma} \tag{3.80}$$

For comparing the robustness of any control system, by plotting the inverse maximum singular values over the frequency, the robust control system gives maximum area under the curve.

We can also use multiplicative output uncertainty $[I + \Delta_o(s)]G(s)$. The closed loop system is stable if,

$$\|\Delta_O(j\omega)\| < 1/\bar{\sigma} \tag{3.81}$$

where $\bar{\sigma}$ is the maximum singular value of $[I + G(j\omega)G_c(j\omega)]^{-1}G(j\omega)G_c(j\omega)$.

The frequency plot obtained for the right-hand side part of either Eq. (3.80), or, Eq. (3.81) indicates stability bounds of the closed loop system. The area under the curve represents stability of the system. More area under the curve indicates higher stability of the system. By using this plot, it is easy to compare the stability of different controllers. The control system that gives the maximum area under the curve is more stable. Since Δ_I is unknown, sensitivity to relative error is analysed by Eq. (3.82).

$$\bar{\sigma}\,\|\Delta_I\| < \frac{1}{\bar{\sigma}(T)} \tag{3.82}$$

Summary

In this chapter, open loop reaction curve method, closed loop reaction curve method, and optimization method of identifying transfer function models were presented. Methods of designing PI controllers for transfer function models for SISO systems were brought out. A simple method of designing PI controllers for multivariable systems methods based on SSGM and based on transfer function matrix are also reviewed.

Problems

1. Marchetti et al. (2002) present the application study of multivariable sequential DCS relay auto tuning method to a pilot distillation system under industrial DCS monitoring and control environment. The performance of alternate tuning strategies is compared for both set point changes and disturbance rejection. Give a write-up about the experimental facility and the tuning performance of the relay tuning method.

2. Consider the system whose transfer function matrix is given as:

$$G_P = \begin{bmatrix} \dfrac{22.89e^{-2s}}{4.572s+1} & \dfrac{-11.64e^{-0.4s}}{1.807s+1} \\ \dfrac{4.687e^{-0.2s}}{2.174s+1} & \dfrac{5.80e^{-0.9s}}{1.801s+1} \end{bmatrix}$$

Find the RGA and select suitable input–output pairing. Design PI controllers for the diagonal elements of the paired system. Selecting a suitable detuning factor and calculate the decentralized PI settings.

3. For Problem 1, consider relay heights of 0.025 and 1.0, perform a relay tuning simulation and obtain the $k_{c,\max}$ and ω_c for each loop.

4. Consider the system whose transfer function matrix is given as:

$$G_p = \begin{bmatrix} \dfrac{2.5e^{-7s}}{15s+1} & \dfrac{5e^{-3s}}{4s+1} \\ \dfrac{1e^{-4s}}{5s+1} & \dfrac{-4e^{-10s}}{20s+1} \end{bmatrix}$$

Find the RGA and select suitable input–output pairing. Design PI controllers for the diagonal elements of the paired system. Selecting a suitable detuning factor and calculate the decentralized PI settings.

5. For Problem 3, consider relay heights of 0.2 and −0.1, perform a relay tuning simulation and obtain the $k_{c,\max}$ and ω_c for each loop.

6. Consider the system whose transfer function matrix is given as:

$$G(s) = \begin{bmatrix} \dfrac{-0.2698e^{-27.5s}}{(97.5s+1)} & \dfrac{1.978e^{-53.5s}}{(118.5s+1)} & \dfrac{0.07724e^{-56s}}{(96s+1)} \\ \dfrac{0.4881e^{-117s}}{(56s+1)} & \dfrac{5.26e^{-26.5s}}{(58.5s+1)} & \dfrac{0.19996e^{-35s}}{(58.5s+1)} \\ \dfrac{0.6e^{-16.5s}}{(40.5s+1)} & \dfrac{5.5e^{-15.5s}}{(19.5s+1)} & \dfrac{-0.5e^{-17s}}{(18s+1)} \end{bmatrix}$$

Find the RGA and select suitable input–output pairing. Design PI controllers for the diagonal elements of the paired system. Selecting a suitable detuning factor and calculate the decentralized PI settings.

7. For problem 4, after selecting suitable input and output pairing, select suitable relay heights and perform a relay tuning simulation. Calculate the $k_{c,\max}$ and ω_c for each loop. Select the design settings of PID controllers for the decentralized controllers.

8. Chidambaram and Nikita (2018) discuss the method designing centralized PI controllers by relay tuning method. Read the paper and give a summary of the method.

9. For the packed distillation column model given by Taiwo (2015):

$$G_p(s) = \begin{bmatrix} \frac{-0.86e^{-s}}{35.4s+1} & \frac{0.6}{(18.9s+1)(2.6s+1)} \\ \frac{-1.5e^{-4s}}{(74s+1)} & \frac{1.22}{(30.6s+1)} \end{bmatrix}$$

Design centralized PI controllers by the Davison method (Section 3.13). Evaluate the servo responses of the controllers by using Matlab and Simulink.

10. For the distillation column model given by Babji and Saraf (1991):

$$G_p(s) = \begin{bmatrix} \frac{-0.12e^{-0.5s}}{1.9s+1} & \frac{0.82e^{-0.9s}}{4.6s+1} \\ \frac{-0.11e^{-0.9s}}{3.7s+1} & \frac{1.05e^{-1.1s}}{5.1s+1} \end{bmatrix}$$

design centralized PI controllers by the Davison method (Section 3.13). Evaluate the servo responses of the controllers by using Matlab and Simulink.

11. For the distillation column model given by Mutalib (2014):

$$G_p(s) = \begin{bmatrix} \frac{0.187}{1.29s+1} & \frac{-0.0086}{0.73s+1} \\ \frac{0.0031}{0.96s+1} & \frac{-0.00024}{0.84s+1} \end{bmatrix}$$

design centralized PI controllers by the Davison method (Section 3.13) and by Tanttu and Lieslehto (Section 3.13) method. Evaluate the servo responses of the controllers by using Matlab and Simulink.

4

CRC Method for Identifying TITO Systems

In this chapter, a Closed Loop Reaction (CRC) curve method for the identification of stable TITO system is discussed. The system under consideration is controlled by decentralized PI or PID controllers. The responses and interactions are modelled by the extension of Yuwana and Seborg (1982) method and the transfer function matrix for the closed loop system is obtained. Using the relation between the open loop and the closed loop transfer function matrices, the open loop transfer function matrix is obtained in Laplace (s) domain. The responses of the obtained and the actual transfer functions matrix with the original controller settings are compared. Better results are obtained if the parameters of the identified transfer function models are used as initial guess values for any optimization method and the transfer function model is identified. The higher order models are approximated to a FOPTD model. Since the method is proposed for stable systems, many of the higher order stable systems can be approximated to a FOPTD system.

4.1   Identification Method

4.1.1   Identification of Individual Responses

Consider a stable 2×2 transfer function matrix (G_P) as in Eq. (4.1a).

$$G_p = \begin{bmatrix} G_{p11} & G_{p12} \\ G_{p21} & G_{p22} \end{bmatrix}$$

(4.1a)

Where,

$$G_{pij} = \frac{k_{pij}e^{-\theta_{ij}s}}{\tau_{ij}s + 1}$$

(4.1b)

Let the controller settings be defined by the matrix given in Eq. (4.2).

$$G_c = \begin{bmatrix} G_{c11} & 0 \\ 0 & G_{c22} \end{bmatrix} \text{ where } G_{cij} = k_{cij}\left(1 + \frac{1}{\tau_{Iij}s} + \tau_{Dij}s\right)$$

(4.2)

The PI/PID controllers are selected so as to get a closed loop under-damped response. Initial controller settings are taken, based on the knowledge of the gain alone, using Davison method (Section 3.13).

Consider Fig. 4.1a. A step change of known magnitude is given to the set point y_{r1} and the other set point is kept unchanged. The main response obtained is y_{11} and the interaction response is y_{21}. Similarly for Fig. 4.1b. A step change of the

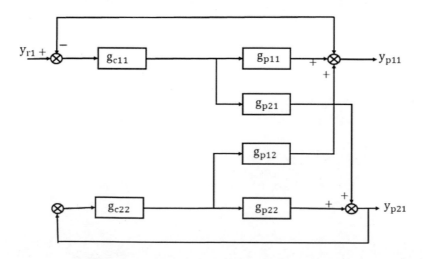

FIGURE 4.1a   TITO process with set point given to y_{r1}

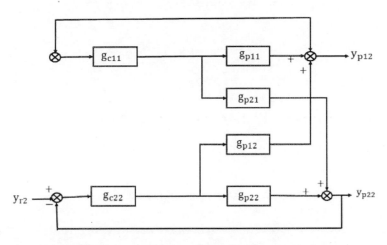

FIGURE 4.1b TITO process with set point given to y_{r2}

same magnitude is given as y_{r2} and the other set point is kept unchanged. Let the set point matrix be given as in Eq. (4.3).

$$Y_r = \begin{bmatrix} y_{r1} & 0 \\ 0 & y_{r2} \end{bmatrix} \qquad (4.3)$$

The main response is denoted as y_{22}, the interaction as y_{21}, and the output matrix is given in Eq. (4.4).

$$Y = \begin{bmatrix} y_{11} & y_{12} \\ y_{21} & y_{22} \end{bmatrix} \qquad (4.4)$$

The transfer function of the closed loop responses is assumed to be an under-damped SOPTD system. From the closed loop response curve for each case, the values y_{p1}, y_{p2}, y_{m1} and T are noted (Fig. 2.6). Using the formulas given by Yuwana and Seborg (1982), the value of τ_e and ζ are noted for each case. The closed loop time delay is noted from the responses. Due to the presence of the integral actions, the closed loop process gain k_{pc} for the main responses (y_{11} and y_{22}) are taken as 1. For the interaction responses (y_{21} and y_{12}), k_{pc} is determined by solving the Laplace inverse of y_{21} and y_{12} respectively. Now the responses can be approximated in the form given in Eq. (4.5), Eq. (4.6) and Eq. (4.7).

Main response:

$$y_{ii}(s) = \frac{k_{pcii}e^{-\theta_{ii}s}}{s(\tau_{eii}^2 s^2 + 2\zeta_{ii}\tau_{eii}s + 1)} \qquad (4.5)$$

Interaction response:

$$y_{ij}(s) = \frac{k_{pcij}e^{-\theta_{ij}s}}{(\tau_{eij}{}^2 s^2 + 2\zeta_{ij}\tau_{eij}s + 1)} \qquad (4.6)$$

4.1.2   Identification of Transfer Function Model

Melo and Friedly (1992) solved the equation for the transfer function of a closed loop response as given in Eq. (4.7).

$$G_p = YY_r^{-1}(I - YY_r^{-1})^{-1}G_c^{-1} \qquad (4.7)$$

Substituting G_p, y, y_r, I, G_c in the matrix and solving the equations, the values for G_{p11}, G_{p21}, G_{p12}, and G_{p22} are obtained.

$$G_{p11} = \frac{(y_{12}y_{21} - y_{11}y_{22})G_{c22}s^2 + (y_{11}G_{c22} - y_{12}G_{c21})s}{[(y_{11}y_{22} - y_{12}y_{21})s^2 - (y_{11} + y_{22})s + 1][G_{c11}G_{c22} - G_{c21}G_{c12}]} \qquad (4.8)$$

$$G_{p21} = \frac{(y_{11}y_{22} - y_{12}y_{21})G_{c21}s^2 + (y_{21}G_{c22} - y_{22}G_{c21})s}{[(y_{11}y_{22} - y_{12}y_{21})s^2 - (y_{11} + y_{22})s + 1][G_{c11}G_{c22} - G_{c21}G_{c12}]} \qquad (4.9)$$

$$G_{p12} = \frac{(y_{11}y_{22} - y_{12}y_{21})G_{c12}s^2 + (y_{12}G_{c11} - y_{11}G_{c12})s}{[(y_{11}y_{22} - y_{12}y_{21})s^2 - (y_{11} + y_{22})s + 1][G_{c11}G_{c22} - G_{c21}G_{c12}]} \qquad (4.10)$$

$$G_{p22} = \frac{(y_{12}y_{12} - y_{11}y_{22})G_{c11}s^2 + (y_{22}G_{c11} - y_{21}G_{c12})s}{[(y_{11}y_{22} - y_{12}y_{21})s^2 - (y_{11} + y_{22})s + 1][G_{c11}G_{c22} - G_{c21}G_{c12}]} \qquad (4.11)$$

Substituting the equations for y_{11}, y_{12}, y_{21}, y_{22}, G_{c11}, G_{c21}, G_{c12}, and G_{c22}, the values of G_{p11}, G_{p21}, G_{p12}, and G_{p22} are obtained as a function of s. A graph is plotted between G_p vs s for each case. By equating G_{p11} to a FOPTD model, i.e., $G_{p11} = \frac{k_{p11}e^{-\theta_{11}s}}{\tau_{11}s+1}$, suitable curve fitting method is applied to solve for k_{p11}, τ_{11}, and θ_{11}. The procedure is repeated for G_{p21}, G_{p12}, and G_{p22} in order to obtain k_{p21}, τ_{21}, and θ_{21}; k_{p12}, τ_{12}, and θ_{12}; and k_{p22}, τ_{22}, and θ_{22}. Thus, the transfer function matrix is identified as in Eq. (4.12).

$$G_p(s) = \begin{bmatrix} G_{p11} & G_{p12} \\ G_{p21} & G_{p22} \end{bmatrix} = \begin{bmatrix} \dfrac{k_{p11}\,e^{-\theta_{11}s}}{\tau_{11}s + 1} & \dfrac{k_{p12}\,e^{-\theta_{12}s}}{\tau_{12}s + 1} \\[2ex] \dfrac{k_{p21}\,e^{-\theta_{21}s}}{\tau_{21}s + 1} & \dfrac{k_{p22}\,e^{-\theta_{22}s}}{\tau_{22}s + 1} \end{bmatrix} \qquad (4.12)$$

The main and the interaction responses of the obtained transfer function matrix are compared with the actual process using the same controller settings. A better result can be obtained if these parameters are used as the initial guess and the transfer function model is identified through an optimization method.

4.2 Simulation Examples

Example 1 Consider the transfer function given by Chien et al. (1999) given in Eq. (4.13).

$$G_p(s) = \begin{bmatrix} \dfrac{22.89\,e^{-0.2\,s}}{4.572s+1} & \dfrac{-11.64\,e^{-0.4\,s}}{1.807s+1} \\[2ex] \dfrac{4.689\,e^{-0.2\,s}}{2.174s+1} & \dfrac{5.8\,e^{-0.4\,s}}{1.801s+1} \end{bmatrix} \tag{4.13}$$

The controller settings are designed using Davison (1976) method so as to get an under-damped response.

$$G_c(s) = \begin{bmatrix} 0.263 + 0.1852/s & 0 \\ 0 & 0.163 + 0.09209/s \end{bmatrix} \tag{4.14}$$

The simulation is carried out for 15 s with a sample time of 0.01. The transfer function of the main response is assumed as a SOPTD model given in Eq. (4.15).

$$G_{pii}(s) = \frac{k_{pcii}\,e^{-\theta_{ii}s}}{\left(\tau_{eii}^{2}s^2 + 2\zeta_{ii}\tau_{eii}s + 1\right)} \tag{4.15}$$

The analytical method proposed by Yuwana and Seborg (1982) is used to identify the model equations. For this, the first and the second peak values, the first minimum valley, and the period of oscillation are noted. From this the effective time constant (τ_e) and zeta (ζ) are obtained. Closed loop delay is taken to be the same as the open loop delay (θ). Process gain (k_p) is found from the final steady state value. In order to identify y_{22}, this procedure is repeated. The value of k_{p11} and k_{p22} are taken as 1. Table 4.1 shows the observed peak values, the minimum valley, calculated period of oscillation, time constant, zeta, and time delay for the main and interaction responses. The response can be written in the form of Eq. (4.15). The transfer function of the interaction is assumed as 's' times of a SOPTD model.

$$G_{pij}(s) = \frac{k_{pcij}s\,e^{-\theta_{ij}s}}{\left(\tau_{eij}^{2}s^2 + 2\zeta_{ij}\tau_{eij}s + 1\right)} \tag{4.16}$$

To obtain the value of k_{p21} and k_{p12}, the Laplace inverse of Eq. (4.6) is taken. Substituting the values of τ_e, ζ, y, t, in each case of interaction response, the values

TABLE 4.1   Main and interaction responses (Example 1)

Parameter	y_{11}	y_{21}	y_{12}	y_{22}
y_{p1}	1.369	0.331	−0.5486	1.1162
y_{p2}	1.034	0.0261	–	–
y_{m1}	0.8913	−0.124	0.254	0.9475
T	4.7450	4.7570	2.244	4.330
τ_e	0.7049	0.7132	0.6938	0.6681
ζ	0.3587	0.3354	0.2381	0.2452
θ	0.2	0.2	0.4	0.4

of k_{p21} and k_{p12} are obtained. The obtained responses can be written as in Eq. (4.17).

$$y_{11}(s) = \frac{1}{s(0.4969s^2 + 0.5057s + 1)}e^{-0.2s} \qquad (4.17)$$

A better matching of the response is obtained when the y_{11} is of the form of a SOPTD model with numerator dynamics as in Eq. (4.18).

$$y_{ii}(s) = \frac{k_{pii}(ps + 1)e^{-\theta_{ii}s}}{s(\tau_{eii}^2 s^2 + 2\zeta_{ii}\tau_{eii}s + 1)} \qquad (4.18)$$

The value of p can be obtained from the Laplace inverse of Eq. (4.18):

$$y(s) = \frac{(1 + ps)e^{-\theta_{cii}s}}{s(\tau_{eii}^2 s^2 + 2\zeta\tau_{eii}s + 1)} \qquad (4.19a)$$

Laplace inverse of Eq. (4.19) is given by Eq. (4.19b).

$$y(t) = \frac{pq}{\tau_e}\left[\sin(b)\exp\left(\frac{-\zeta t'}{\tau_e}\right)\right] + \left[1 - \exp\left(\frac{-\zeta t'}{\tau_e}\right)[q\sin(b) + \cos(b)]\right] \qquad (4.19b)$$

$$y(s) = \frac{k_p e^{-\theta s}}{(\tau_e^2 s^2 + 2\zeta\tau_e s + 1)} \qquad (4.19c)$$

Laplace inverse of (4.19c) is given by Eq. (4.19d).

$$y(t) = \frac{k_p q}{\tau_e}\left[\exp\left(\frac{-\zeta t'}{\tau_e}\right)\sin(b)\right] \qquad (4.19d)$$

Where,

$$t' = t - \theta; \ b = \frac{t'}{\tau_e}(1 - \zeta^2); q = (1 - \zeta^2)^{-0.5} \qquad (4.19e)$$

Substituting the values of k_{p11}, τ_e, ζ, y, and t, the value of p is obtained as 0.4448.

Hence we have,

$$y_{11}(s) = \frac{(1 + 0.4448s)}{s(0.4969s^2 + 0.5057s + 1)}e^{-0.2s} \tag{4.20}$$

For the response y_{22}, the first peak is at 1.1162, but the first valley is at 0.895. This posed a problem in response identification as the obtained damping coefficient is much less. Hence, an approximate model is obtained by drawing a smooth curve from the first peak, taking around the midpoint of the first valley and the steady state value 1, and reaching the second peak. Thus, the responses are fitted as given in Eq. (4.21), Eq. (4.22), and Eq. (4.33).

$$y_{22}(s) = \frac{1}{s(0.4464s^2 + 0.3276s + 1)}e^{-0.4s} \tag{4.21}$$

$$y_{21}(s) = \frac{0.3658}{(0.5087s^2 + 0.9896s + 1)}e^{-0.2s} \tag{4.22}$$

$$y_{12}(s) = \frac{-0.5372}{(0.4814s^2 + 0.3304s + 1)}e^{-0.4s} \tag{4.23}$$

The fitted response and the actual responses are shown in Fig. 4.2. The four equations from Eq. (4.20) to Eq. (4.23) are substituted in equations Eq. (4.8) to Eq. (4.11) and a graph is plotted between each G_p vs s in each case. Fig. 4.3 shows the G_p vs s graph of the identified responses. Each of the G_p is fitted to a FOPTD model, and the values of k_p, τ, and θ are found in each case. For this, 10 values of G_p for 10 values of s are taken in each case and are solved by using *fmincon* in Matlab. The obtained values of k_p, τ, and θ in each transfer function element are shown in Table 4.2. Fig. 4.4 shows the response curve for the identified model and the actual model using the same controller settings. From Fig. 4.4, it can be seen that the responses are matching satisfactorily. The slight mismatch may be due to the approximation of the main and the interaction responses to a second order delay system. Even in SISO systems, an accurate matching of the actual and the identified model is not reported. MIMO system identification encounters more problems because of the interactions from all the transfer functions.

However, the model fitting can be further improved by using an optimization method such that the model parameters (gain, time constant, time delay) are selected to minimize the sum of the squared errors between the model and the actual process responses.

$$\text{Minimize } \phi = \sum_{i=1}^{2}\sum_{j=1}^{2}\sum_{k=1}^{n}[y_{mij}(t_k) - y_{ij}(t_k)]^2\Delta t \tag{4.24}$$

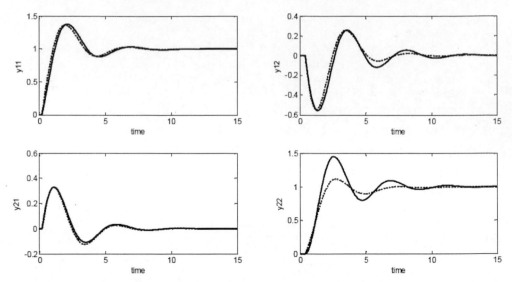

FIGURE 4.2   Comparison plot of fitted response and actual response (Example 1). Solid line: fitted response; dashed line: actual response

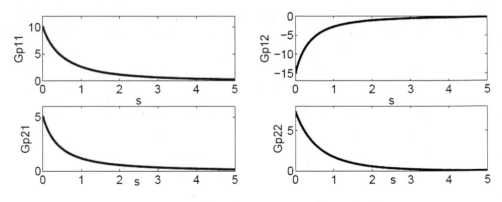

FIGURE 4.3   Plot of Gp vs s curve (Example 1)

TABLE 4.2   Transfer function parameters obtained from Gp vs s curve (Example 1)

	k_p	τ	θ
G_{p11}	10.2890	1.9820	0.2635
G_{p21}	5.1600	2.6449	0.1773
G_{p12}	−15.1857	2.7003	0.3978
G_{p22}	7.3514	1.2367	0.6447

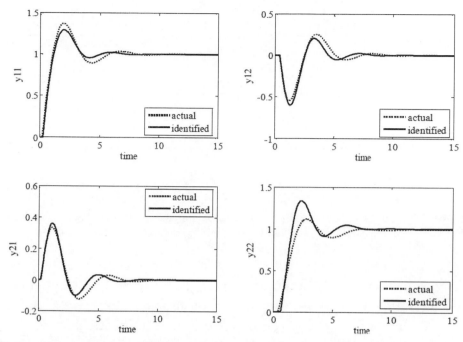

FIGURE 4.4 Comparison plot of the closed loop response of identified vs actual transfer function model for the same controller settings (Example 1)

TABLE 4.3 Converged parameters from optimization using the parameters in Table 3.2 as initial guess (Example 1)

	k_p	τ	θ
G_{p11}	22.8890	4.5719	0.2002
G_{p21}	4.6888	2.1749	0.2003
G_{p12}	−11.6401	1.8075	0.4001
G_{p22}	5.8003	1.7998	0.4002

Where, ϕ = objective function

$$i = 1, 2; j = 1, 2$$

n is the number of the data points in the process and the model responses.

The optimization problem is solved by using the Matlab routine *lsqnonlin* with the trust-region reflective algorithm. For the case of limits for optimization, the lower bound of k_p (for positive values) is taken as $1/20$ times the initial guess and the upper bound as 20 times the initial guess. For negative values of k_p the lower bound of k_p is fixed as 20 times the initial guess and the upper bound as $1/20$ times the initial guess value. For the case of time constant, the lower limit is kept as $1/20$

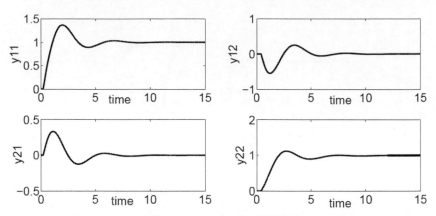

FIGURE 4.5　Response plot of the identified and actual transfer model using optimization Solid line: identified response; dashed line: actual response. Both the responses overlap.

times the initial guess and upper limit as 5 times the initial guess. The lower bound for time delay is taken as 1/4 times the initial guess value and the upper bound as 4 times the initial guess value. The number of iterations taken is 31 and the time taken for convergence is 36.97 s. The sum of the ISE value obtained is 1.2865×10^{-8}. The computational work is done on Intel Core i5/3.10 GHz personal computer. The converged parameters are shown in Table 4.3. Any arbitrary values for the guess values do not give convergence. Fig. 4.5 shows the identified and the actual responses.

Example 2 Consider the transfer function given by Wood and Berry (1973) as in Eq. (4.25).

$$G_p(s) = \begin{bmatrix} \dfrac{12.8\,e^{-s}}{16.7s+1} & \dfrac{-18.9\,e^{-3s}}{21s+1} \\[3ex] \dfrac{6.6\,e^{-7s}}{10.9s+1} & \dfrac{-19.4\,e^{-3s}}{14.4s+1} \end{bmatrix} \tag{4.25}$$

The controller settings are taken from Viswanathan et al. (2001) and given in Eq. (4.26).

$$G_c(s) = \begin{bmatrix} 0.2 + \dfrac{0.05}{s} + 0.6s & 0 \\[3ex] 0 & -0.03 - \dfrac{0.03}{2s} - 0.09s \end{bmatrix} \tag{4.26}$$

The simulation is carried out for 150 s with a sample time of 0.01. The transfer function of the main responses is assumed as a SOPTD model of the form of

TABLE 4.4 Main and interaction responses (Example 2)

Parameters	y_{11}	y_{21}	y_{12}	y_{22}
y_{p1}	1.2493	0.5687	0.2826	1.212
y_{p2}	1.1095	0.0916	0.0297	1.08
y_{m1}	0.8449	−0.3205	−0.1003	0.9335
T	32.75	36.56	36.9050	31.7
τ_e	5.1613	5.6636	5.5541	4.8853
ζ	0.1396	0.3393	0.3253	0.2497
θ	1	7	3	3

Eq. (4.15) and the response model is assumed in the form of Eq. (4.5). The analytical method (YS method) proposed by Yuwana and Seborg (1982) is used to identify the model response. Table 4.4 shows the observed peak values, the minimum valley, calculated period of oscillation, time constant, zeta, and time delay for the main and interaction responses.

The transfer function of the interaction is assumed as 's' times a SOPTD model of the form of Eq. (4.15) and hence the interaction response model of the form Eq. (4.16). The values of k_{p11} and k_{p22} are taken as 1. To obtain the value of k_{p21} and k_{p12}, the Laplace inverse of Eq. (4.16) is taken. Substituting the values of τ_e, ζ, y, t, in each case of interaction response, the values of k_{p21} and k_{p12} are obtained. The obtained responses by using Yuwana and Seborg (1982) method can be written as in Eq. (4.27).

$$y_{11}(s) = \frac{1}{s(26.639s^2 + 1.441s + 1)}e^{-s} \tag{4.27}$$

A good match of the identified model using Yuwana and Seborg (1982) method was not found with the actual response. It has been reported that Yuwana and Seborg method provides poor parameters for processes with large time delays. This is due to the use of a first order Pade approximation for the time delay in the closed loop transfer function denominator (Jutan and Rodriguez, 1983). Hence the effective time constant (τ_e) and zeta (ζ) are calculated by fractional overshoot method:

$$\text{Maximum overshoot /final value} = \exp[-\pi\zeta/(1-\zeta^2)^{0.5}] \tag{4.28}$$

$$\tau_{e,\text{eff}} = T(1-\zeta^2)^{0.5}/(2\pi) \tag{4.29}$$

The obtained values are $\zeta = 0.4044$ and $\tau_e = 4.7671$.

A better match of the response is obtained when the y_{11} is of the form of a SOPTD model with numerator dynamics in Eq. (4.18). Then p is obtained from the Laplace inverse of Eq. (4.18). Substituting the values of k_{p11}, τ_e, ζ, y, t, the value of p is obtained as 1.0616.

We get,

$$y_{11}(s) = \frac{(1 + 1.0616s)}{s(22.7252s^2 + 2.8556s + 1)}e^{-s} \tag{4.30}$$

Similarly for the main response y_{22}, closed loop response obtained by YS method did not match with the actual response. Hence, by using fractional overshoot method, the obtained values are $\zeta = 0.4331$ and $\tau_e = 4.5475$. The response can be written as Eq. (4.31).

$$y_{22}(s) = \frac{1}{s(20.6798s^2 + 3.939s + 1)}e^{-3s} \tag{4.31}$$

$$y_{21}(s) = \frac{4.4832}{(32.0763s^2 + 2.5973s + 1)}e^{-7s} \tag{4.32}$$

$$y_{12}(s) = \frac{3.1235}{(30.848s^2 + 3.6135s + 1)}e^{-3s} \tag{4.33}$$

The fitted response and the actual responses are shown in Fig. 4.6. As mentioned earlier, equations from Eq. (4.30) to Eq. (4.33) are substituted in equations Eq. (4.8) to Eq. (4.11). A graph is plotted between each G_p vs s in each case. Fig. 4.7 shows the G_p vs s graph of the identified responses. Each of the G_p is fitted to a FOPTD model, and the values of k_p, τ, and θ are determined in each case by using *fmincon* in Matlab. The obtained values of k_p, τ, and θ in each transfer function element are shown in Table 4.5.

When these obtained values are taken as initial guess for optimization, a good match of the identified response with the actual response is obtained. The number of iterations taken is 14 and the time taken for convergence is 68.1011 s. The sum of the ISE value is 3.4131×10^{-7}. The converged parameters are shown in Table 4.6. Fig. 4.8 shows the identified and the actual responses. The method can be extended to 3×3 systems or 4×4 systems. In these cases, the response models are identified for step changes in each set point, one at a time. Any arbitrary values for the guess values do not give any convergence.

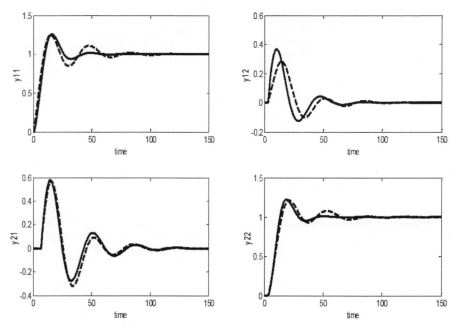

FIGURE 4.6 Comparison plot of fitted response and actual response (Example 2). Solid line: fitted response; dashed line: actual response

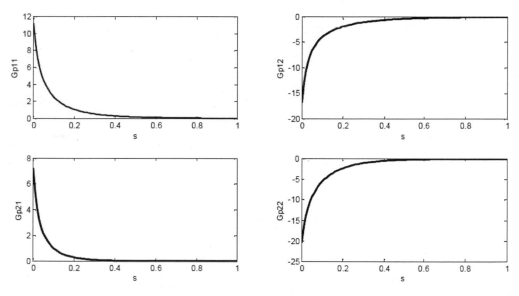

FIGURE 4.7 Plot of Gp vs s curve (Example 2)

Summary

In this chapter, the closed loop reaction curve method for SISO systems proposed by Yuwana and Seborg (1982) were extended to identify multivariable systems. The

TABLE 4.5   Transfer function parameters obtained from Gp vs s curve (Example 2)

	k_p	τ	θ
G_{p11}	11.2633	24.2644	2.8368
G_{p21}	7.2743	24.0863	7.4754
G_{p12}	−16.7834	19.8797	2.9102
G_{p22}	−20.4995	12.7444	4.5955

TABLE 4.6   Converged parameters from optimization using the parameters in Table 4.5 as initial guess (Example 2)

	k_p	τ	θ
G_{p11}	12.7565	16.6364	1.0011
G_{p21}	6.5759	10.8514	7.0034
G_{p12}	−18.8176	20.9070	2.9975
G_{p22}	−19.3505	14.3516	3.0013

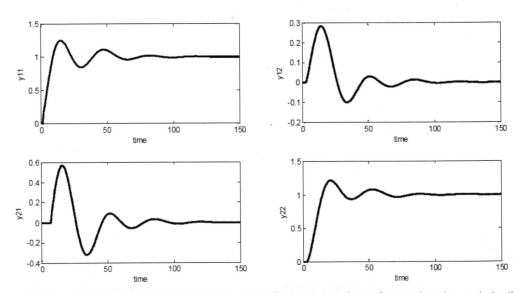

FIGURE 4.8   Closed loop response plot of the identified and actual transfer matrix using optimization identified response (Example 2) Solid line: identified response; dashed line: actual response. Both the responses overlap.

problem associated with this method was brought out. Simulation application was given for the transfer function matrix of 2×2 multivariable systems. Two examples were given. The method involved fitting the main actions and the interactions of the closed loop system for step changes in the set points by a SOPTD transfer functions with (1/s) term and the interaction by SOPTD model. From the analytical expressions, the open loop transfer function matrices were obtained using the closed loop transfer function matrix and the controller transfer function matrix by plotting Gp vs s curve. The obtained model was used with the same PID controllers and the responses/interactions were matched with that of the actual systems. A method to improve the match involved taking the approximate model parameters as the guess values of the least square optimization method. Any arbitrary guess values do not give convergence of the optimization problem.

Problems

1. For the system:

$$G_p = \begin{bmatrix} \dfrac{12.8e^{-4s}}{16.7s+1} & \dfrac{-18.9e^{-3s}}{21s+1} \\ \dfrac{6.6e^{-7s}}{10.9s+1} & \dfrac{19.4e^{-3s}}{14.41s+1} \end{bmatrix}$$

With the PID settings $[k_c = 0.18, \tau_I = 7.0, \tau_D = 1.6]$; $[-0.20, 7.0, 1.6]$ for the diagonal pairing, carry out the closed loop reaction curve method of obtaining the model parameters.

2. For the system:

$$G_p = \begin{bmatrix} \dfrac{22.89e^{-2s}}{4.572s+1} & \dfrac{-11.64e^{-0.4s}}{1.807s+1} \\ \dfrac{4.687e^{-0.2s}}{2.174s+1} & \dfrac{5.80e^{-0.9s}}{1.801s+1} \end{bmatrix}$$

With the PID settings $[k_c = 0.0056, \tau_I = 1.5, \tau_D = 0.37]$; $[0.44, 1.5, 0.37]$ for the diagonal pairing, carry out the closed loop reaction method of obtaining the model parameters.

3. For the system:

$$G_p = \begin{bmatrix} \dfrac{2.5e^{-7s}}{15s+1} & \dfrac{5e^{-3s}}{4s+1} \\ \dfrac{1e^{-4s}}{5s+1} & \dfrac{-4e^{-10s}}{20s+1} \end{bmatrix}$$

With the PID settings $(k_c = 0.35, \tau_I = 10.5, \tau_D = 2.65)$; $(-0.6, 10.5, 2.65)$ for the diagonal pairing, carry out the closed loop reaction curve method of obtaining the model parameters.

5

CRC Method for Identifying SISO Systems by CSOPTD Models

In this chapter, a Closed Loop Reaction Curve (CRC) method is given for identifying analytically the Single Input Single Output (SISO) systems by Critically Damped Second Order Plus Time Delay (CSOPTD) process transfer function models from the closed loop step response. The material presented here will form the basis for the next chapter on identification of multivariable critically damped systems.

5.1 CSOPTD Systems

5.1.1 Introduction

The dynamics of many of the stable processes can be described by that of an over-damped second order system whereas that of the closed loop system is represented by a second order under-damped response. For the purpose of designing PID controllers, many processes are described by a stable second order critically damped transfer function model:

$$y(s)/u(s) = k_p \exp(-Ls)/(\tau s + 1)^2$$

Since the critically damped SOPTD model has only three parameters, the identification may be easier than the four parameters required for the over-damped SOPTD models. The time delay is due to the measurement delay or the actuator delay and/or due to approximation of higher order systems by a simple SOPTD model. In this chapter, the method to identify a SOPTD model with equal time constants is discussed. A simple method of calculating the process steady state gain is applied.

5.1.2   Identification of Critically Damped SOPTD and FOPTD Systems

For the purpose of designing controllers, the process is assumed as given in Eq. (5.1).

$$y(s)/u(s) = k_p \exp(-\theta s)/(\tau s + 1)^2 \tag{5.1}$$

The form of equation in Eq. (5.2) for PID controller is used.

$$u(t) = k_c \left(e + 1/\tau_I \int \mathrm{edt} - \tau_D dy/dt \right) \tag{5.2}$$

Where,

$$e = (y_r - y) \tag{5.3}$$

The closed loop transfer function model is derived as in Eq. (5.4).

$$y/y_r = [k_c k_p (1 + \tau_I s) \exp(-\theta s)] / [(1 + \tau s)^2 \tau_I s + k_c k_p (1 + \tau_I s + \tau_I \tau_D s^2) \exp(-\theta s)] \tag{5.4}$$

For a given step change in the set point, the closed loop response is obtained as shown in Fig. 5.1. Chidambaram and Padmasree (2006) used the u versus time and y versus time profile to get the steady state gain of the system. Since an integral action is present in the controller, the output will reach the desired steady state and hence the input variable of the process will also reach a steady state. The ratio of change in y (output variable deviation) and change in u (manipulated variable deviation) at the steady state gives the steady state gain (k_p). Using the noted values of y_{p1}, y_{m1}, y_{p2}, Δt and y_∞, we can calculate ζ and τ_e of the closed loop

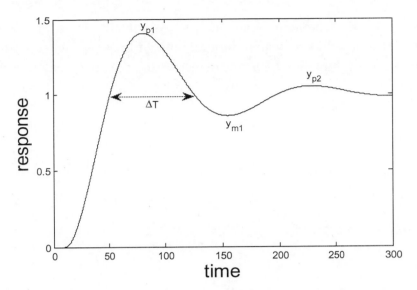

FIGURE 5.1 Typical closed loop response for a step change in set point and associated parameters to be noted

systems described by Eq. (5.5).

$$y(s)/y_r(s) = (1 + \tau_I s)\exp(-\theta_c s)/(\tau_e s^2 + 2\tau \zeta s + 1) \tag{5.5}$$

Where,

$$\tau_e = [\Delta T/(2\pi)](1 - \zeta^2)^{0.5} \tag{5.6}$$

$$\zeta_1 = -\ln(v_1)/\{\pi^2 + [\ln(v_1)]^2\}^{0.5} \tag{5.7}$$

$$\zeta_2 = -\ln(v_2)/\{4\pi^2 + [\ln(v_2)]^2\}^{0.5} \tag{5.8}$$

$$v_1 = (y_\infty - y_{m1})/(y_{p1} - y_\infty) \tag{5.9}$$

$$v_2 = (y_{p2} - y_\infty)/(y_{p1} - y_\infty) \tag{5.10}$$

$$\zeta = 0.5(\zeta_1 + \zeta_2) \tag{5.11}$$

Here θ_c is the noted time delay in the output response of the closed loop system. If the response shows only one peak, then ζ_1 can be used for ζ. From the values of closed loop τ_e and ζ, the real and imaginary parts of the dominant poles of the closed loop system can be obtained as in Eq. (5.12) and Eq. (5.13).

$$\alpha = -\zeta/\tau_e \tag{5.12}$$

$$\beta = (1 - \zeta^2)^{0.5}/\tau_e \tag{5.13}$$

The value of the dominant pole $(\alpha + j\beta)$ is substituted in the characteristic equation of the closed loop system denominator of Eq. (5.4). On equating the real part of the resulting equation to 0 and then equating the imaginary part to 0 we get the two nonlinear algebraic equations in θ and τ given in Eq. (5.14) and Eq. (5.15) for the SOPTD model.

$$\tau^2 \tau_I (\alpha^3 - 3\alpha\beta^2) + [2\tau\tau_I(\alpha^2 - \beta^2) + \tau_I\alpha] + K[1 + \tau_I\alpha + \eta(\alpha^2 - \beta^2)]$$
$$\times \exp(-\theta\alpha)\cos(\theta\beta) + K(\beta\tau_I + 2\eta\alpha\beta)\sin(\beta\theta)\exp(-\theta\alpha) = 0 \qquad (5.14)$$
$$[\tau^2\tau_I(-\beta^3 + 3\alpha^2\beta) + 4\tau\tau_I\alpha\beta + \tau_I\beta] + K(\beta\tau_I + 2\eta\alpha\beta)\cos(\beta\theta)\exp(-\theta\alpha)$$
$$- K[1 + \tau_I\alpha + \eta(\alpha^2 - \beta^2)]\sin(\beta\theta)\exp(-\theta\alpha) = 0 \qquad (5.15)$$

where

$$K = k_c k_p; \quad \eta = \tau_I \tau_D \qquad (5.16)$$

The FOPTD model can be obtained by solving numerically the set of nonlinear algebraic equations given in Eq. (5.17) and Eq. (5.18).

$$\tau\tau_I(\alpha^2 - \beta^2) + [\tau_I\alpha] + K[1 + \tau_I\alpha + \eta(\alpha^2 - \beta^2)]\exp(-\theta\alpha)\cos(\theta\beta)$$
$$+ K(\beta\tau_I + 2\eta\alpha\beta)\sin(\beta\theta)\exp(-\theta\alpha) = 0 \qquad (5.17)$$
$$[\tau\tau_I(2\alpha\beta) + \tau_I\beta] - K[(\alpha^2 - \beta^2)\eta + 1 + \tau_I\alpha]\sin(\beta\theta)\exp(-\theta\alpha)$$
$$+ K[2\eta\alpha\beta + \tau_I\beta)]\cos(\beta\theta)\exp(-\theta\alpha) = 0 \qquad (5.18)$$

The initial guess value for the time delay is considered as the noted delay in the output response (θ_c) and the initial guess value for the time constant is assumed as that of the closed loop system which is calculate as $t_s/8$ (where t_s is the settling time of the output response).

5.2 Simulation Examples

Example 1 The system is assumed as:

$$G(s) = \exp(-10s)/(20s + 1)^2 \qquad (5.19)$$

The closed loop step response similar to Fig. 5.1 is obtained using the PID settings $k_c = 1$, $\tau_I = 20$, $\tau_D = 4$. From the response, the values of y_{p1}, y_{m1}, y_{p2}, Δt, and

y_∞ are noted as 1.405, 0.8571, 1.05, 74.64, and 1. Using Eq. (5.12) and Eq. (5.13), the values of ζ and τ_e are obtained respectively as 0.3147 and 22.5421. Only the value of ζ_1 is used for the calculation of ζ. The values of α and β are calculated as $\alpha = -0.0140$ and $\beta = 0.0421$. The value of deviation in u from the steady state value at the new steady state is noted as 1. Hence k_p is obtained as 1. The initial guess value for θ is considered as the closed loop time delay noted from the response (in this example it is 4.12) and the guess value for the time constant (τ) is considered as $t_s/8$ where t_s is the settling time of the closed loop system. In this example t_s is noted as 300 and hence the guess value for the time constant is 37.5. The numerical solution of Eq. (5.14) and Eq. (5.15) (by *fsolve* in Matlab) gives $\tau = 20.1379$ and $\theta = 9.8493$. These values are close to the actual values of $k_p = 1$, $\theta = 10$, and $\tau = 20$. Different guess values such as half of the above values are also tried. Using both the guess values, and using the *fsolve* routine, give the same converged values for the parameters ($\tau = 20.1379$ and $\theta = 9.8493$). The system is also identified as a FOPTD model. The relevant equations are given by Eq. (5.17) and Eq. (5.18).

The identified FOPTD model parameters are $k_p = 1$, $\tau = 32.6530$, $\theta = 22.7033$. The guess values for the parameters (τ, θ) are assumed as discussed in the previous paragraph. The comparisons of the open loop response of the actual and identified models are shown in Fig. 5.2. The error analysis of the system and the model is also carried out as discussed by Liu and Gao (2012). The sum of the square of the difference between the step response of the actual system and the step response of

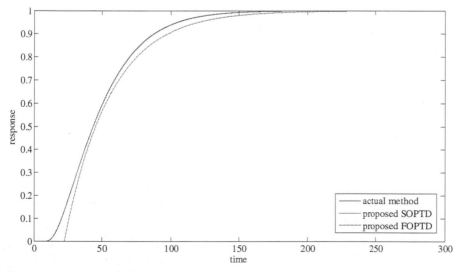

FIGURE 5.2 Open loop step responses of the actual system and the identified SOPTD and FOPTD models (Example 1)

TABLE 5.1 The error analysis (sum of the error squared) of the identified model

Example	SOPTD	FOPTD
1	0.0002133	0.2362
2	0.02734	1.844
3	0.03571	1.738

Sampling time $= 0.01$; Number of points $= 300$

TABLE 5.2 Identified model parameters and PID controller settings

Example	Parameters	Actual	Proposed SOPTD	Proposed FOPTD
1	k_p	1	1	1
	θ	10	9.8493	22.7033
	τ	20	20.1379	32.6530
	k_c	2	2.0446	0.7191
	τ_I	40	40.2758	32.6530
	τ_D	10	10.0685	0
	IAE (servo)	35.37	35.15	47.00
	IAE (reg.)	20.0	19.70	45.33
2	k_p	1	1	1
	θ	2	1.6081	11.2811
	τ_1	1	14.8645	37.1754
	τ_2	10		
	τ_3	20		
	k_c		9.2435	1.6477
	τ_I		29.7290	37.1754
	τ_D		7.4322	0
	IAE (servo)		14.13	22.72
	IAE (reg.)		3.216	22.52
3	k_p	1	1	1
	θ	2	2.1782	11.4844
	τ_N	1.5		
	τ_1	10	14.7104	36.9893
	τ_2	20		
	k_c		6.7535	1.6104
	τ_I		29.4208	36.9893
	τ_D		7.3552	0
	IAE (servo)		15.21	23.33
	IAE (reg.)		4.376	22.94

the model is calculated. The results are given in Table 5.1. PID controllers for the actual SOPTD system, identified SOPTD, and FOPTD models are designed by the SIMC method proposed by Skogestad (2006). The calculated PID settings are given in Table 5.2 along with the identified model parameters.

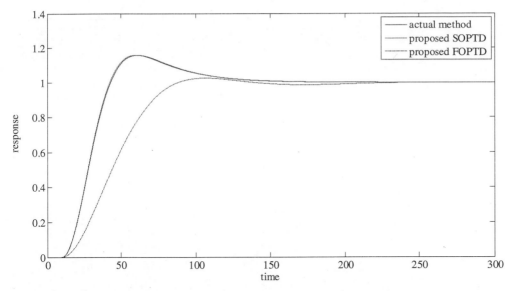

FIGURE 5.3 Closed loop response of the actual system with the controllers designed on the identified models (servo problem) (Example 1)

The closed loop servo responses of the actual system with the PID settings are compared in Fig. 5.3. The control system designed based on the critically damped SOPTD model gives a closed loop performance better than that of the FOPTD model. The IAE values for the closed loop systems are 35.37, 35.15, and 47.0 respectively for the controller designed based on the actual SOPTD system, the identified critically damped SOPTD model, and the identified FOPTD model. Fig. 5.4 shows the comparisons of regulatory responses for unit step change in the load entering, along with u, to the process. The IAE values of the closed loop system are 20.0, 19.70, and 44.77 respectively for the controller designed based on the actual SOPTD system, the identified SOPTD model, and the identified FOPTD model. The control system designed based on the critically damped SOPTD model gives a closed loop performance better than that of the FOPTD model.

The effect of disturbance of the constant value of 0.02 entering the process along with the input (u) is also studied on the estimated model parameters. The change in the steady state value of u is noted as 1 for a step change in y of 1, thus giving k_p as 1. The feedback nature of the closed loop system eliminates the effect of disturbances on the output. The values of y_{p1}, y_{m1}, y_{p2}, Δt, and y_∞ are noted as 1.411, 0.855, 1.052, 74.6, and 1. Using equations Eq. (5.6) to Eq. (5.11), the values of ζ and τ_e are obtained respectively as 0.3148 and 22.5388. The values of α and β are obtained as $\alpha = -0.0140$ and $\beta = 0.0421$. As stated earlier, the initial conditions for the model parameters are obtained ($\tau = t_s/8$ and delay noted in the closed loop response).

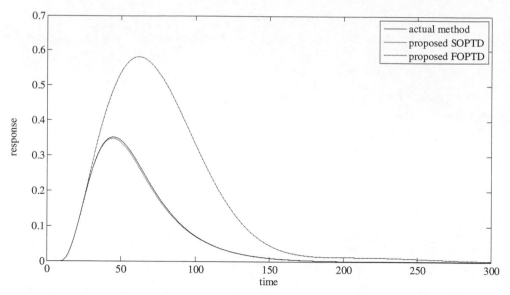

FIGURE 5.4   Comparisons of regulatory responses of controllers on the actual system (Example 1)

The numerical solution of Eq. (5.14) and Eq. (5.15) gives $\tau = 20.1344$ and $\theta = 9.8479$. These values are close to the actual values of the system.

A measurement noise (normal Gaussian noise with 0 mean and standard deviation of 1%) is added to the system output. The noise corrupted signal is used for feedback. The response of the closed loop system for a unit step change in the set point is obtained. A smooth curve can be drawn. In the present work, a first order filter (with time constant = 1) is used in the output. I filtered signal is used only for the model identification (Fig. 5.5b). The noise corrupted signal is used for the feedback control action. From the filtered signal, the values of y_{p1}, y_{m1}, y_{p2}, Δt, and y_∞ are noted as 1.406, 0.8575 1.051, 37.3, and 1. The calculated values of ζ and τ_e are 0.3162 and 22.5308 respectively. The real and imaginary parts of the dominant poles are calculated as $\alpha = -0.014$ and $\beta = 0.0421$. The identified model parameters are $k_p = 1$, $\tau = 20.134$, and $\theta = 9.788$.

Example 2 Consider a third order plus time delay model given by Li et. al. (1991) as:

$$G(s) = \exp(-2s)/[(s+1)(10s+1)(20s+1)] \qquad (5.20)$$

The closed loop response similar to Fig. 5.1 for a unit step change in the set point is obtained using the PID settings $k_c = 4$, $\tau_I = 15$ and $\tau_D = 4$. From the closed loop response, the values of y_{p1}, y_{m1}, y_{p2}, Δt, and y_∞ are noted as 1.476, 0.892,

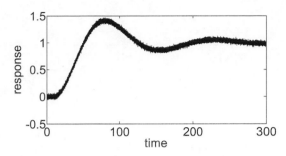

FIGURE 5.5a Noise corrupted output variable versus time behaviour (noise with $\sigma = 0.02$) (Example 1)

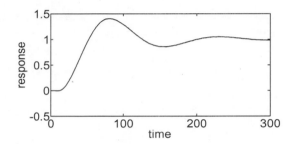

FIGURE 5.5b Filtered output variable for Example 1. Filter time constant $= 1$

1.027, 30.15, and 1. Using Eq. (5.6) to Eq. (5.11), the values of ζ and τ_e are obtained respectively as 0.4205 and 8.7072. The dominant poles are obtained as $\alpha = -0.0483$ and $\beta = 0.1042$. The value of u at the new steady state is noted as 1 for a step change in the set point of 1. Hence k_p is obtained as 1. Numerical solution of Eq. (5.14) and Eq. (5.15) gives $\tau = 14.8645$ and $\theta = 1.6081$. The open loop responses comparisons of the actual and identified model are shown in Fig. 5.6. The system is also identified as a FOPTD model. The identified FOPTD model parameters are obtained from Eq. (5.17) and Eq. (5.18) as $\tau = 37.1754$, $\theta = 11.2811$. The error analysis is given in Table 5.1. The effect of measurement noise is also studied. The estimated model parameters are given in Table 5.3. The values of the model parameters are not changed significantly.

PID controllers for the identified critically damped SOPTD and FOPTD models are designed by the SIMC method proposed by Skogestad (2006). The calculated PID settings are given in Table 5.2 along with the identified model parameters. The closed loop servo responses are shown in Fig. 5.7 and that of the regulatory responses are shown in Fig. 5.8. The IAE values obtained for the closed loop system are given in Table 5.2. The control system designed based on the critically damped

TABLE 5.3   Effect of measurement noise on the identified CSOPTD model parameters

σ (SD)	Example	τ_e	ζ	K_P	τ	τ_d
0	1	22.5421	0.3147	1.0	20.1379	9.8493
0.01		22.5308	0.3162	1.0	20.1341	9.7877
0.02		22.5293	0.3156	1.0	20.1280	9.8187
0	2	8.7072	0.4205	1	14.8645	1.6081
0.01		8.6970	0.4247	1.0	14.8204	1.5607
0.02		8.7021	0.4236	1.0	14.8369	1.5684
0	3	8.4792	0.3997	1	14.7104	2.1782
0.01		8.4905	0.4012	1.0	14.7174	2.1432
0.02		8.4797	0.4024	1.0	14.6968	2.1384

Noise with mean 0 and variance σ^2

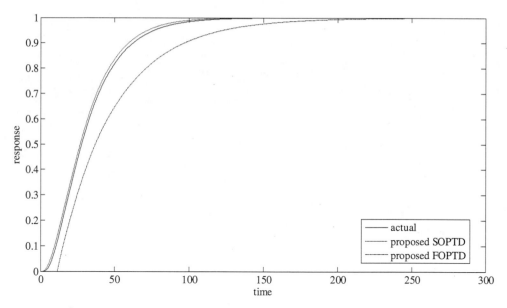

FIGURE 5.6   Comparisons of the open loop step responses of the actual system with those of the identified models (Example 2)

SOPTD (i.e., CSOPTD) model gives a closed loop performance much better than a FOPTD model.

Vivek and Chidambaram (2014) have applied an improved relay auto-tuning method to identify the model parameters of a critically damped SOPTD system. For this example, the relay tuning method proposed by Vivek and Chidambaram (2012) gives a critically damped SOPTD system with $\tau = 14.208$; $\theta = 3.0$, and $k_p = 0.9693$. The PID settings by the IMC method are given by are $k_c = 4.8739$,

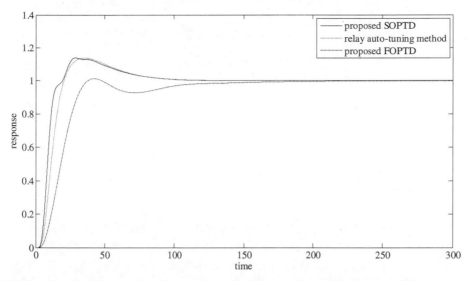

FIGURE 5.7 Comparison of closed loop servo response of the actual system (Example 2) for the controller designed based on proposed SOPTD, relay method, and proposed FOPTD

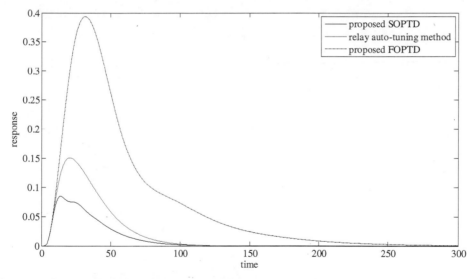

FIGURE 5.8 Comparison of closed loop regulatory response of the actual system (Example 2) for the controller designed based on proposed SOPTD, relay method, and proposed FOPTD

$\tau_I = 28.4166$, and $\tau_D = 7.1041$. The closed loop servo and regulatory performances of the controllers on the original system are compared in Fig. 5.7 and Fig. 5.8. The present method gives the best result. The effect of measurement noise on the estimated model parameters is given in Table 5.3.

Example 3 Consider a system as in Eq. (5.21).

$$G(s) = (1 - 1.5s)\exp(-2s)/[(10s + 1)(20s + 1)] \tag{5.21}$$

The closed loop response similar to Fig. 5.1 for a unit step change in the set point is obtained using the PID setting: $k_c = 4$, $\tau_I = 15$ and $\tau_D = 4$. From the closed loop response, the values of y_{p1}, y_{m1}, y_{p2}, Δt and y_∞ noted as 1.5264, 0.87, 1.036, 29.06 and 1. Using the Eq. (5.6) to Eq. (5.11), the values of ζ and τ_e are obtained respectively as 0.3997 and 8.4792. The dominant poles are obtained as $\alpha = -0.0471$ and $\beta = 0.1081$. The value of u at the new steady state is noted as 1 for a step change in the set point of 1. Hence k_p is obtained as 1. Numerical solution of Eq. (5.14) and Eq. (5.15) gives $\tau = 14.7104$ and $\theta = 2.1782$. The open loop responses comparisons of the actual and identified model are shown in Fig. 5.9. The system is also identified as a FOPTD model using Eq. (5.17) and Eq. (5.18). The identified FOPTD model parameters are $k_p = 1$, $\tau = 36.9893$ and $\theta = 11.4844$. The error analysis is given in Table 5.1. The effect of measurement noise is also studied. The estimated model parameters are given in Table 5.2 and Table 5.3. The values of the model parameters are not changed significantly.

PID controllers for the identified critically damped SOPTD and FOPTD models are designed by the SIMC method proposed by Skogestad (2006). The calculated PID settings are given in Table 5.2 along with the identified model parameters.

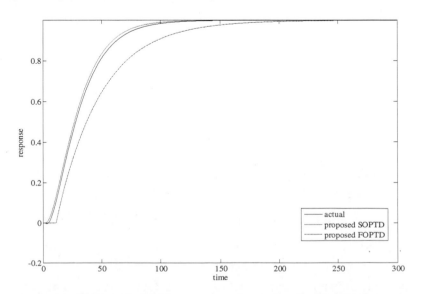

FIGURE 5.9 Comparison of open loop responses (Example 3)

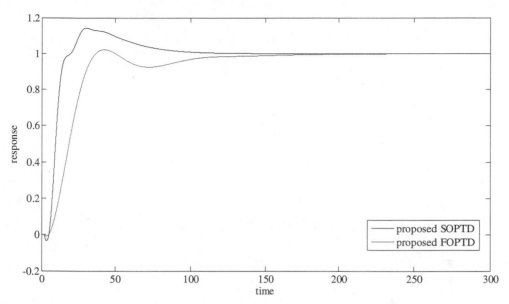

FIGURE 5.10 Comparisons of closed loop servo performances of the controllers on the actual system (Example 3)

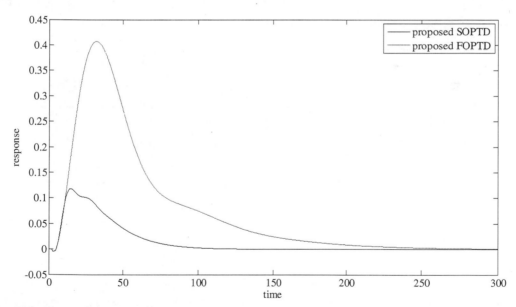

FIGURE 5.11 Comparisons of closed loop regulatory performances of the controllers on the actual system (Example 3)

The closed loop servo responses are shown in Fig. 5.10 and those of the regulatory responses are shown in Fig. 5.11. The IAE values for the closed loop systems are given in Table 5.2. The control system designed based on the critically damped

SOPTD model gives a closed loop performance much better than the FOPTD model. The effect of measurement noise on the estimated model parameters are given in Table 5.3.

5.3 Simulation Study of a Nonlinear Bioreactor

The dimensionless equations for a continuous bioreactor are given by Eq. (5.22) and Eq. (5.33) (Agrawal and Lim, 1984).

$$dX_1/dt = (\mu - D)X_1 \tag{5.22}$$

$$dX_2/dt = (X_{2,f} - X_2)D_1 - (\mu X_1/\gamma) \tag{5.23}$$

where

$$\mu = (\mu_{\max}X_2)/(k_m + X_2) \tag{5.24}$$

Here X_1, X_2 are the dimensionless concentration of biomass cell, substrate; D is the dilution rate; X_{2f} is the feed substrate concentration, and μ is the specific growth rate. The model parameters are given by Agrawal and Lim (1984) as $\gamma = 0.4$ g/g, $\mu_m = 0.4$ h^{-1}, $k_m = 0.4$ h^{-1}, $D_1 = 0.36$ 1/h, and $X_{2,f} = 1.0$. For the values of these parameters, the steady state solution of Eq. (5.22) and Eq. (5.23) gives the steady state condition as $X_{1,s} = 0.22$ and $X_{2,s} = 0.45$. X_1 is considered here as the controlled variable and $X_{2,f}$ is considered as the manipulated variable. With the PI settings, $k_c = 1.2$ and $\tau_I = 6$, the closed loop response for a step change in the set point (from $X_1 = 0.22$ to $X_1 = 0.222$) is considered. From the response, the values of y_{p1}, y_{p2}, y_{m1}, y_{∞}, and ΔT are noted respectively as 1.55, 1.15, 0.70, 1, and 43.835. To get a good accuracy, the scaled response in y is used, calculated as $(X_1 - X_{1,s})/(\Delta X_s)$. The values of ζ and τ_e of the closed loop system are evaluated as 0.1894 and 13.6826 respectively. The values of α and β are calculated as -0.0143 and 0.0717. The gain k_p is obtained (as the ratio of the steady state value for deviation value in y and in u) as 0.4. Hence the open loop model parameters of SOPTD with equal time constant are calculated by the solution of nonlinear algebraic equations as $\tau = 13.0275$ and $\theta = 0.0$.

The two nonlinear algebraic equations for the model parameters τ and θ are calculated by using the *fmincon* optimization routine by defining the objective function as the sum of squared values of the right side of the two equations. As stated earlier, the initial guess values for τ and θ are calculated as the noted delay in the closed loop response and $t_s/8$ respectively. The use of *fmincon* rather than

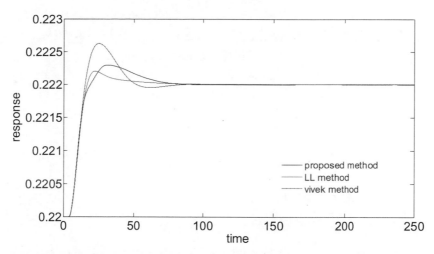

FIGURE 5.12 Comparison of the closed loop servo response of the actual system for the controller designed based on the proposed method, controller designed based on local linearized model and Vivek and Chidambaram method

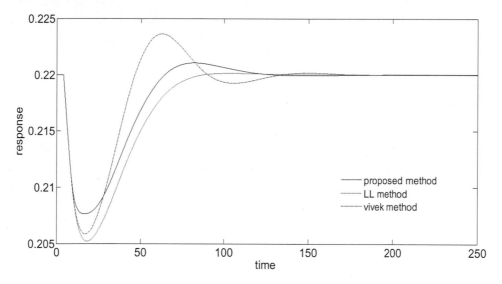

FIGURE 5.13 Comparison of the closed loop regulatory response of the actual system

fsolve is used to solve equations Eq. (5.14) to Eq. (5.16) or Eq. (5.17) and Eq. (5.18) to avoid the negative value for the time delay. For the present problem, we get $\tau = 13.0275$ and $\theta = 0$. The time delay seen in the output is 4. The PID settings by the IMC method are given by $k_c = 5$; $\tau_I = 26.055$; $\tau_D = 6.5132$. The closed loop response is found to be good.

To avoid the value of 0 for the delay for the present problem, the open loop delay is taken as the closed loop delay. Open loop delay is considered as 4. The open loop model parameters (k_p and τ) for SOPTD with equal time constants are calculated by the solution of nonlinear algebraic equations by *fmincon* routine as $\tau = 9.9042$ and $k_p = 0.2899$. The initial guess values for τ and k_p are given as $t_s/8$ and 0 respectively. The PID settings by IMC method are given by $k_c = 8.5410$, $\tau_I = 19.8084$ and $\tau_D = 4.9521$. Vivek and Chidambaram (2012) have reported the SOPTD model parameters by the improved relay auto tune method as $k_p = 0.225$, $\tau = 6.52$, $\theta = 3.843$. The PID settings by IMC method are given by $k_c = 7.54$, $\tau_I = 13.04$, and $\tau_D = 3.26$. The steady state gain for the present method and by the relay method is found to be lesser than the actual gain (calculated by y_∞/u_∞ as 0.4). This is due to the nonlinear behaviour of the bioreactor. The linearized model around the operating point is derived (Vivek and Chidambaram, 2012) as in Eq. (5.25).

$$\Delta X_1(s)/\Delta X_{1,f}(s) = 0.401 \exp(-4s)/(63.2911s^2 + 25.5969s + 1) \qquad (5.25)$$

The PID settings by the IMC method are given by $k_c = 7.97$, $\tau_I = 25.59$, and $\tau_D = 2.4726$. The performances of these three PID controllers are simulated on the nonlinear model equations and the responses are shown in Fig. 5.12 for the servo problem and in Fig. 5.13 for the regulatory problem. The present method gives better performances.

Summary

In this chapter, from the closed loop step response, a method was given for identifying analytically the critically damped SOPTD (CSOPTD) transfer function models of SISO systems.

Problems

For the following systems, carry out the closed loop reaction method for identifying the CSOPTD model. Compare the open loop comparison and also the closed loop comparison with the controller designed by IMC method.

1. $G_p(s) = (1.5s + 1)\exp(-2s)/[(10s + 1)(20s + 1)]$
   with $k_c = 3.0$, $\tau_I = 15$, $\tau_D = 4$

2. $G_p(s) = 1/(s + 1)^8$
   with $k_c = 0.7$, $\tau_I = 5$, $\tau_D = 1.2$

3. $G_p(s) = \exp(-4s)/(63.291s^2 + 25.597s + 1)$
   with $k_c = 7.8$, $\tau_I = 25$, $\tau_D = 6$

6

CRC Method for Identifying TITO Systems by CSOPTD Models

The method suggested in the previous chapter for the identification of Multi Input Multi Output (MIMO) First Order Plus Time Delay (FOPTD) transfer function is extended to identify Critically Damped Second Order Plus Time Delay (CSOPTD) model parameters of higher order MIMO system. The closed loop transfer function is identified as a third order system. The normalized step response curve given by Clark (2005) is used to identify the main responses of the closed loop system as a third order transfer function model. From this model, the open loop model is identified by the Closed Loop Reaction Curve (CRC) method.

6.1 Identification of Multivariable Systems

A step input is given to the y_{r1} and the closed loop main response y_{c11} is obtained. The closed loop step response is assumed to be of third order of the form in Eq. (6.1).

$$Y(s) = \frac{Ke^{-\theta_c s}}{s(s+P)(s^2 + 2\zeta\omega_n s + \omega_n^2)} \tag{6.1}$$

For a step input, the transfer function will be of the form given in Eq. (6.2).

$$G_{pc}(s) = \frac{Y(s)}{U(s)} = \frac{Ke^{-\theta_c s}}{(s+P)(s^2 + 2\zeta\omega_n s + \omega_n^2)} \qquad (6.2)$$

The normalized time τ is given by Eq. (6.3).

$$\tau = \omega_n t \qquad (6.3)$$

$$P = \beta(\zeta\omega_n) \qquad (6.4)$$

Final steady state value is given in Eq. (6.5).

$$y(t) = \frac{AK}{P\omega_n^2} \qquad (6.5)$$

The shapes of the closed loop main responses are compared with the normalized curves given for the third order systems by Clark (2005) where the quadratic term has a damping ratio (ζ) and the most matched curve is selected. A plot is shown in Fig. 6.1. The values of the damping ratio (ζ) and β are noted. The value of ω_n can be obtained from the normalized time τ given by Eq. (6.3). The value of the pole P is obtained from Eq. (6.4). The value of K is obtained from Eq. (6.5). Thus, the closed loop main responses are identified.

The interaction responses y_{c21} and y_{c12} are identified using the Yuwana and Seborg (1982) method since the method given by Clark (2005) does not have normalized plots for responses whose final steady state is 0. The closed loop interaction responses y_{c21} and y_{c12} are assumed of the form given in Eq. (6.6).

$$y_{cij}(s) = \frac{K_{ij}e^{-\theta_{cij}s}}{(\tau_{eij}^2 s^2 + 2\zeta_{ij}\tau_{eij}s + 1)} \qquad (6.6)$$

To identify ζ and τ_e, Yuwana and Seborg (1982) method is used. The value of K is obtained from the Laplace inverse of Eq. (6.6) as shown in Fig. 6.1.

The equation for the transfer function of a closed loop response is given by Melo and Friedly (1992) as in equations Eq. (6.7) to Eq. (6.11).

$$G_p = YY_r^{-1}(I - YY_r^{-1})^{-1}G_c^{-1} \qquad (6.7)$$

$$G_{p11} = \frac{(y_{c12}y_{c21} - y_{c11}y_{c22})G_{c22}s^2 + (y_{c11}G_{c22} - y_{c12}G_{c21})s}{[(y_{c11}y_{c22} - y_{c12}y_{c21})s^2 - (y_{c11} + y_{c22})s + 1][G_{c11}G_{c22} - G_{c21}G_{c12}]} \qquad (6.8)$$

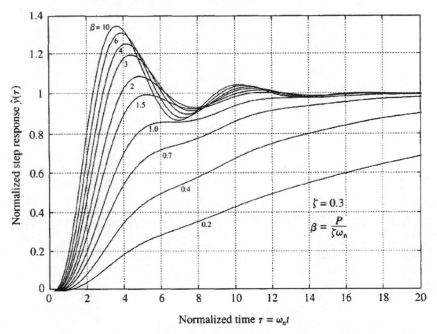

FIGURE 6.1 Normalized step response $\hat{y}(t)$ for the third order system whose quadratic term has a damping ratio $\zeta = 0.3$ (Fig. E.13 in Clark, 2005)

$$G_{p21} = \frac{(y_{c11}y_{c22} - y_{c12}y_{c21})\, G_{c21}s^2 + (y_{c21}G_{c22} - y_{c22}G_{c21})\, s}{[(y_{c11}y_{c22} - y_{c12}y_{c21})\, s^2 - (y_{c11} + y_{c22})\, s + 1]\, [G_{c11}G_{c22} - G_{c21}G_{c12}]} \qquad (6.9)$$

$$G_{p12} = \frac{(y_{c11}y_{c22} - y_{c12}y_{c21})\, G_{c12}s^2 + (y_{c12}G_{c11} - y_{c11}G_{c12})\, s}{[(y_{c11}y_{c22} - y_{c12}y_{c21})\, s^2 - (y_{c11} + y_{c22})\, s + 1]\, [G_{c11}G_{c22} - G_{c21}G_{c12}]} \qquad (6.10)$$

$$G_{p22} \quad \frac{(y_{c12}y_{c21} - y_{c11}y_{c22})\, G_{c11}s^2 + (y_{c22}G_{c11} - y_{c21}G_{c12})\, s}{[(y_{c11}y_{c22} - y_{c12}y_{c21})\, s^2 - (y_{c11} + y_{c22})\, s + 1]\, [G_{c11}G_{c22} - G_{c21}G_{c12}]} \qquad (6.11)$$

The equations for y_{c11}, y_{c12}, y_{c21}, y_{c22}, G_{c11}, G_{c21}, G_{c12}, and G_{c22} are substituted in equations Eq. (6.8) to Eq. (6.11) and the values of G_{p11}, G_{p21}, G_{p12}, and G_{p22} are obtained as a function of s. A graph is plotted between the G_p vs s in each case. G_p is assumed to be the CSOPTD model given in Eq. (6.12).

$$G_p = \frac{k_p e^{-\theta s}}{(\tau s + 1)^2} \qquad (6.12)$$

Suitable curve fitting techniques are applied to solve for k_{p11}, τ_{11} and θ_{11}. The same procedure is repeated to G_{p21}, G_{p12} and G_{p22} in order to obtain k_{p21}, τ_{21} and θ_{21}, k_{p12}, τ_{12}; and θ_{12}, k_{p22}, τ_{22}, and θ_{22}. The transfer function matrix is identified as in

Eq. (6.13).

$$G_p = \begin{bmatrix} G_{p11} & G_{p12} \\ G_{p21} & G_{p22} \end{bmatrix} = \begin{bmatrix} \dfrac{k_{p11}e^{-\theta_{11}s}}{(\tau_{11}s+1)^2} & \dfrac{k_{p12}e^{-\theta_{12}s}}{(\tau_{12}s+1)^2} \\ \dfrac{k_{p21}e^{-\theta_{21}s}}{(\tau_{21}s+1)^2} & \dfrac{k_{p22}e^{-\theta_{22}s}}{(\tau_{22}s+1)^2} \end{bmatrix} \tag{6.13}$$

If further improved parameters are required, these values are used as initial guess values to obtain the transfer function matrix model using any standard optimization method. The CSOPTD model parameters are selected so as to minimize the sum of the squared errors between the model and the actual process responses. The optimization problem is solved by using the Matlab routine *lsqnonlin*.

6.2 Simulation Example

Consider the example (Niederlinski, 1971) given in Eq. (6.14).

$$G_p(s) = \begin{bmatrix} \dfrac{0.5}{(0.1s+1)^2(0.2s+1)^2} & \dfrac{-1.0}{(0.1s+1)(0.2s+1)^2} \\ \dfrac{1.0}{(0.1s+1)(0.2s+1)^2} & \dfrac{2.4}{(0.1s+1)(0.2s+1)^2(0.5s+1)} \end{bmatrix} \tag{6.14}$$

The controller settings G_c are designed by Davison method (1976) (Section 3.13) so as to get under-damped responses.

$$G_{cp}(s) = \begin{bmatrix} 0.2182 + \dfrac{1.4182}{s} & 0.0 \\ 0.0 & 0.0455 + \dfrac{0.2955}{s} \end{bmatrix} \tag{6.15}$$

A unit step input $(A = 1)$ is given in y_{r1} and the responses and the interactions y_{c11} and y_{c21} are obtained. Similarly, a unit step input change is given in y_{r2}, and y_{c22} and y_{c12} are obtained. The simulation is carried out for 30 s with sampling time of 0.01 s. The value of the first peak, second peak and the first minimum are given in Table 6.1 for all the responses.

6.2.1 Identification of Main Response

The closed loop main response y_{11} is compared with the normalized curve. Fig. 6.1 in Clark (2005) whose damping ratio (ζ) is 0.3, and $\beta = 6$ shows a good match.

TABLE 6.1 Noted parameter values of closed loop responses and interactions

Parameters	y_{11}	y_{21}	y_{12}	y_{22}
y_{p1}	1.3015	1.175	−0.2448	1.4528
y_{p2}	1.0728	0.2779	−0.0579	0.779
y_{m1}	0.8507	−0.57	0.1187	1.078
t_{p1}	3.05	1.69	1.69	3.22
t_{p2}	8.69	7.37	7.37	8.88
t_{m1}	5.86	4.54	4.54	6.05
θ_c	0.05	0.02	0.02	0.1

The normalized time for the first peak is 3.8 from the normalized curve. The time for first peak in the main response y_{11} from Fig 6.1 is 3 (= 3.05 − 0.05). From this, we calculate $\omega_n = 1.2667$. Substituting the value of ζ, ω_n, and β, we obtain P as 2.2801. The final steady sate of y_{11} is 1. The value of K from Eq. (6.5) is determined as 3.6585. Hence the identified normalized response may be written as in Eq. (6.16).

$$y_{c11}(s) = \frac{3.6585e^{-0.05s}}{s(s + 2.2801)(s^2 + 0.76s + 1.6046)} \tag{6.16}$$

The actual response and identified response are shown in Fig. 6.1. The same procedure is adopted for identifying y_{22}. Curve E.14 in Clark (2005) shows a good match for which the damping ratio ζ is 0.2 and $\beta = 10$. The normalized time for the first peak is 3.75 and the time for first peak in the actual response plot is 3.12 (= 3.22 − 0.1). From this ω_n is calculated as 1.2019. By substituting the value of ω_n and β, P is found as 2.4038. The final steady state value of y_{22} is 1. The value of K from Eq. (6.5) is 3.4724. The identified normalized response is given by Eq. (6.17).

$$y_{c22}(s) = \frac{3.4724e^{-0.1s}}{s(s + 2.24038)(s^2 + 0.4808s + 1.4446)} \tag{6.17}$$

The actual identified response can be written as in Eq. (6.18) and Eq. (6.19).

$$y_{c11}(s) = \frac{e^{-0.05s}}{s(0.2733s^3 + 0.8310s^2 + 0.9122s + 1.0)} \tag{6.18}$$

$$y_{c22}(s) = \frac{e^{-0.1s}}{s(0.2880s^3 + 0.8307s^2 + 0.7488s + 1.0)} \tag{6.19}$$

The identification of main responses is also carried out with the Yuwana and Seborg (1982) method assuming it to be a SOPTD model of the form in Eq. (6.20).

$$y_{cii}(s) = \frac{K_{ii}e^{-\theta_{cii}s}}{s(\tau_{eii}^2 s^2 + 2\zeta\tau_{eii}s + 1)} \tag{6.20}$$

TABLE 6.2 Closed loop model parameters using Yuwana and Seborg method

Parameters	y_{c11}	y_{c22}
τ_e	0.8758	0.8810
ζ	0.2195	0.2241
K	1.0	1.0

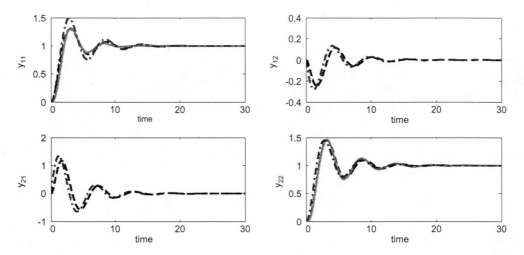

FIGURE 6.2 Comparison of identified and actual closed loop response.

Solid line: identified response; dotted line: actual response; dash–dot line: main response identified using the Yuwana and Seborg method

Where,

$ii = 11$ or 22. The value of τ_e and ζ are given in Table 6.2.

The obtained SOPTD models are given in Eq. (6.21) and Eq. (6.22).

$$y_{c11}(s) = \frac{e^{-0.05s}}{s(0.7670s^2 + 0.3845s + 1)} \tag{6.21}$$

$$y_{c22}(s) = \frac{e^{-0.1s}}{s(0.7762s^2 + 0.3949s + 1)} \tag{6.22}$$

Fig. 6.2 compares the identified main response using Clark (2005) method, actual main response and the main response identified using Yuwana and Seborg (1982) method. It can be seen that main responses identified using Clark (2005) method matches the actual main responses much better than the main responses identified using Yuwana and Seborg (1982) method.

TABLE 6.3 Closed loop model parameters
for interaction response 's'

Parameters	y_{c21}	y_{c12}
τ_e	0.8810	0.8810
ζ	0.2241	0.2241
K	1.6038	−0.3341
θ_c	0.02	0.02

6.2.2 Identification of Interaction Responses

The closed loop interaction responses y_{c21} and y_{c12} are assumed of the form as given in Eq. (6.23).:

$$y_{cij}(s) = \frac{K_{ij}e^{-\theta_{cij}s}}{(\tau_{eij}^2 s^2 + 2\zeta_{ij}\tau_{eij}s + 1)} \tag{6.23}$$

The corresponding transfer function G_{pcij} for unit step input can be written as:

$$G_{pcij}(s) = \frac{sK_{ij}e^{-\theta_{cij}s}}{(\tau_{eij}^2 s^2 + 2\zeta_{ij}\tau_{eij}s + 1)} \tag{6.24}$$

To identify τ_e and ζ, Yuwana and Seborg (1982) method is used. The value of K is obtained from the Laplace inverse of Eq. (6.24) as shown in Table 6.3. The value of closed loop time delay θ_c is noted from the closed loop interaction response. The obtained values of K, ζ, θ_c, and τ_e are shown in Table 6.3.

The identified responses can be written as in Eq. (6.25) and Eq. (6.26).

$$y_{c21}(s) = \frac{1.6038e^{-0.02s}}{(0.7762s^2 + 0.3949s + 1)} \tag{6.25}$$

$$y_{c12}(s) = \frac{-0.3341e^{-0.02s}}{(0.7762s^2 + 0.3949s + 1)} \tag{6.26}$$

But the match of the identified closed loop interaction responses with the actual closed loop interaction responses are not so good as can be seen from Fig. 6.2. Hence, we assume a third order model of the form given in Eq. (6.27) for the closed loop interaction responses.

$$y_{cij}(s) = \frac{K_{ij}e^{-\theta_{cij}s}}{(1 + b_{ij}s)(\tau_{eij}^2 s^2 + 2\zeta_{ij}\tau_{eij}s + 1)} \tag{6.27}$$

The parameters K_{ij}, b_{ij}, τ_{eij}, ζ, and θ_{cij} are obtained by optimization. The initial guess for τ_{eij} and ζ_{ij} parameters are taken from the values obtained using Yuwana

TABLE 6.4 Third order parameters for interaction response obtained from optimization

	Parameters	Initial guess values	Final converged values	Sum of IAE values	Sum of ISE values
21	K_{21}	1.6038	1.5707	0.0794	7.8964×10^{-4}
	τ_{e21}	0.8810	0.8754		
	ζ_{21}	0.2241	0.2172		
	b_{21}	0.4405	0.4854		
	θ_{21}	0.02	0.02		
12	K_{21}	−0.3341	−0.3273	0.0165	3.4281×10^{-5}
	τ_{e21}	0.8810	0.8754		
	ζ_{21}	0.2241	0.2172		
	b_{21}	0.4405	0.4854		
	θ_{21}	0.02	0.02		

Seborg (1982) method and are given in Table 6.4. The initial guess for closed loop time delay is taken as the observed closed loop θ_{cij}. The initial guess value of b is chosen as $\tau_{eij}/2$. The initial guess value of K_{ij} is obtained using the Laplace inverse of Eq. (6.27) as given in Table 6.4. The limits are taken as for K_{ij}; the lower bound is $K_{ij}/20$ and the upper bound is $K_{ij} \times 20$. For τ_{eij}, the limits are from $\tau_{eij}/20$ to $\tau_{eij} \times 5$; for ζ_{ij} the limits are from $\zeta_{ij}/4$ to $\zeta_{ij} \times 4$; for b_{ij} the limits are from $b_{ij}/20$ to $b_{ij} \times 5$; and for θ_{cij} the limits are from $\theta_{cij}/4$ to $\theta_{cij} \times 4$. The initial guesses and final converged values are shown in Table 6.4. The number of iterations taken for convergence in y_{21} is 3 and the elapsed time is 2.9922 s. For the case of y_{12}, the number of iterations taken for convergence is 3 and the time taken is 2.1249 sec.

The actual plot and identified plot for y_{c21} and y_{c12} are given in Fig. 6.3 and Fig. 6.4. The identified equations for y_{c21} and y_{c12} are given in Eq. (6.28) and Eq. (6.29).

$$y_{c21}(s) = \frac{1.5707e^{-0.02s}}{0.3720s^3 + 0.9509s^2 + 0.8657s + 1} \tag{6.28}$$

$$y_{c12}(s) = \frac{-0.3273e^{-0.02s}}{0.3720s^3 + 0.9509s^2 + 0.8657s + 1} \tag{6.29}$$

The equations for y_{c11}, y_{c12}, y_{c21}, y_{c22}, G_{c11}, G_{c21}, G_{c12}, and G_{c22} are substituted in Eq. (6.8) to Eq. (6.11) and the values of G_{p11}, G_{p21}, G_{p12}, and G_{p22} are obtained as a function of s. A graph is plotted between the obtained Gp vs s in each case. The obtained graph is shown in Fig. 6.5.

G_{p11} is assumed to be CSOPTD model, i.e., $G_p = \frac{k_p e^{-\theta s}}{(\tau s+1)^2}$. Using appropriate curve fitting techniques, k_{p11}, τ_{11}, and θ_{11} are solved. The same procedure is

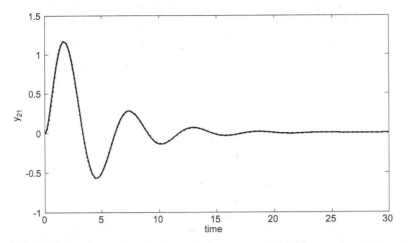

FIGURE 6.3 Comparison plot of actual y_{c21} response and identified third order model response using optimization

Solid line: identified response; dotted line: actual response

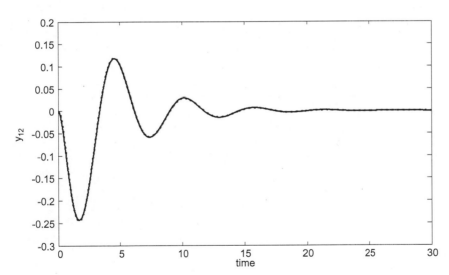

FIGURE 6.4 Comparison plot of actual y_{c12} response and identified third order model response using optimization

Solid line: identified response; dotted line: actual response

repeated for G_{p21}, G_{p12}, and G_{p22} in order to obtain k_{p21}, τ_{21}, and θ_{21}; k_{p12}, τ_{12}, and θ_{12}; and k_{p22}, τ_{22} and θ_{22}. The obtained values of the transfer function matrix are shown in Table 6.5. Using the same controller settings, the main and the interaction responses of the identified transfer function matrix are compared with the actual processes. The obtained graph is shown in Fig. 6.6a. From the figure, it can be seen

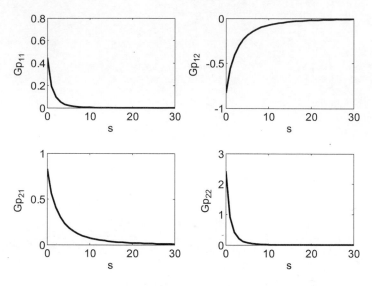

FIGURE 6.5   Plot of Gp vs s

TABLE 6.5   CSOPTD model parameters obtained from Gp vs s curve

	k_p	τ	θ
G_{p11}	0.45	0.3935	0.1656
G_{p21}	0.8317	0.1897	0.0268
G_{p12}	−0.8318	0.1898	0.02768
G_{p22}	2.4495	0.4312	0.2371

that a good match is obtained between the actual and the identified model. A very good result can be obtained if these parameters are used as the initial guess and the transfer function model is identified through an optimization method. The final converged values of the CSOPTD model are shown in Table 6.6. The number of iterations taken is 11 and the time taken for convergence is 35.1938 s. The sum of IAE values is 0.0174 and the sum of ISE values is 2.2620×10^{-5}. Fig. 6.6a and Fig. 6.6b show the comparison of the identified and actual response.

6.2.3   Effect of Measurement Noise

To study the effect of measurement noise, a random signal of standard deviation 0.02 is added to the process responses and interactions. A sample time of 0.01 s is used. The noisy output is used for feedback control action and also for the identification of the model. Fig. 6.7 and Fig. 6.8 show the responses and Fig. 6.9

TABLE 6.6 Converged parameters of CSOPTD model from optimization method

	k_p	τ	θ
G_{p11}	0.5018	0.2276	0.1496
G_{p21}	0.9997	0.2155	0.0726
G_{p12}	−0.9960	0.2104	0.1857
G_{p22}	2.4005	0.4069	0.0811

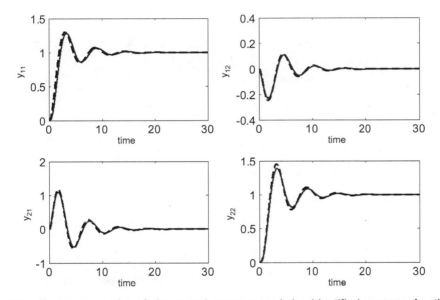

FIGURE 6.6a Comparison plot of the actual process and the identified process for the same controller settings

Solid line: identified response; dotted line: actual response

and Fig. 6.10 show the interactions. For estimation, a smooth curve is drawn. The peak value and the minimum are given in Table 6.8.

The responses are compared with the plot given in Clark (2005). The response y_{11} is assumed of the form Eq. (6.1). Fig. 6.1 shows a good match. For this graph, $\beta = 6$ and $\zeta = 0.3$. At the first peak y_{p1}, the normalized time $\tau = 3.8$. The corresponding time in the actual noise corrupted response curve is 3.12 s. From this ω_n can be obtained, using Eq. (6.3) as 1.2179. The value of P is calculated using Eq. (6.4) as 2.1922. The value of K is obtained using Eq. (6.5) as 3.2516. The obtained response can be written as Eq. (6.30) and rewritten in the standard form

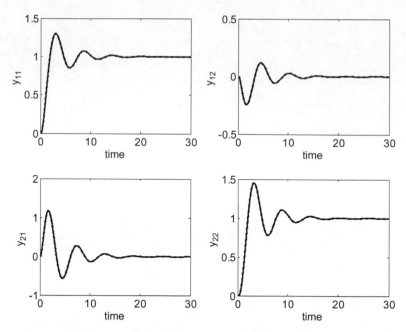

FIGURE 6.6b   Comparison plot of the actual process and the identified process for the same controller settings (optimization based identification)

Solid line: identified response; dotted line: actual response. Both the curves overlap.

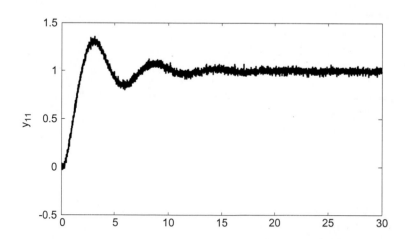

FIGURE 6.7   Response in y_{11} of the noise corrupted signal for a step change in y_{r1}

by Eq. (6.31).

$$y_{c11}(s) = \frac{3.2516}{s(s + 2.1922)(s^2 + 0.7307s + 1.4833)} \tag{6.30}$$

$$y_{c11}(s) = \frac{1}{s(0.3075s^3 + 0.8989s^2 + 0.9488s + 1)} \tag{6.31}$$

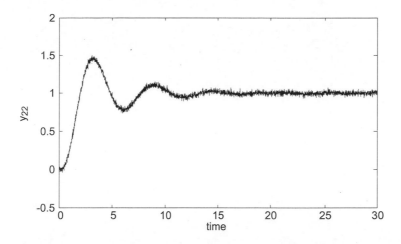

FIGURE 6.8 Response in y_{22} of the noise corrupted signal for a step change in y_{r2}

For y_{22}, the graph E.14 from Clark (2005) with $\zeta = 0.2$ and $\beta = 10$ show a good match. For the peak value 1.4633, the corresponding normalized time τ was 3.75. The response time was 3.265. Hence using Eq. (6.3) $\omega_n = 1.1485$. The value of P obtained using Eq. (6.4) is 2.2970 and the value of K from Eq. (6.5) is 3.0299. The obtained response is given by Eq. (6.32) and rewritten in the standard form by Eq. (6.33).:

$$y_{c22}(s) = \frac{3.0299}{s(s + 2.2970)(s^2 + 0.4594s + 1.3191)} \tag{6.32}$$

$$y_{c22}(s) = \frac{1}{s(0.33s^3 + 0.9097s^2 + 0.7836s + 1)} \tag{6.33}$$

The comparison of the actual response and responses identified using Clark (2005) method are shown in Fig. 6.12. The interaction responses are assumed to be of the form given in Eq. (6.23). The parameters are identified using optimization method. The initial guess values for τ_{eij}, ζ_{ij} are obtained from Yuwana and Seborg (1982) method. Initial guess values of b_{ij} are taken as the corresponding $\tau_{eij}/2$ and initial guess of k_{pij} using Laplace transform of Eq. (6.31) as given in Eq. (6.23). The lower bound and upper bound for optimization are same as mentioned earlier. The initial guess values, converged values, and the sum of IAE and ISE values are shown in Table 6.9.

Fig. 6.13 and Fig. 6.14 show the comparison plot of the actual and identified interaction responses using the optimization method. The identified interaction

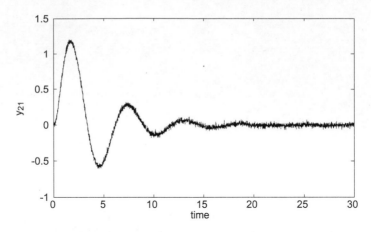

FIGURE 6.9   Interaction in y_{21} of the noise corrupted signal for a step change in y_{r1}

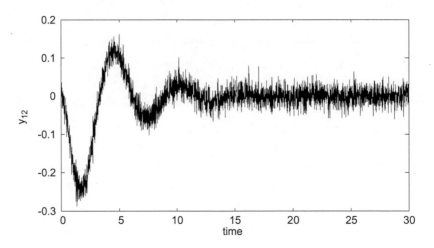

FIGURE 6.10   Interaction in y_{12} of the noise corrupted signal for a step change in y_{r2}

responses from optimization can be written as given in Eq. (6.34) and Eq. (6.35).

$$y_{c21}(s) = \frac{1.5783}{(0.3947s^3 + 0.9580s^2 + 0.8905s + 1)} \qquad (6.34)$$

$$y_{c12}(s) = \frac{1}{(0.3955s^3 + 0.9549s^2 + 0.8988s + 1)} \qquad (6.35)$$

Substituting the values of y_{c11}, y_{c12}, y_{c21}, y_{c22}, G_{c11}, G_{c21}, G_{c12}, and G_{c22} in Eq. (6.8) to Eq. (6.11), the values of G_{p11}, G_{p21}, G_{p12}, and G_{p22} are obtained as a function of s. G_p vs s is plotted on a graph in each case and the model parameters are identified by assuming a CSOPTD model. The obtained graph is shown in Fig. 6.14.

TABLE 6.7 Noted parameters from the closed loop responses (noise corrupted signal)

Parameters	y_{11}	y_{21}	y_{12}	y_{22}
y_{p1}	1.313	1.176	−0.2454	1.4633
y_{p2}	1.086	0.2865	0.06191	1.1230
y_{m1}	0.8365	−0.5785	0.1210	0.7682
t_{p1}	3.12	1.706	1.81	3.265
t_{p2}	9	7.44	7.6	9
t_{m1}	5.99	4.5	4.63	5.99
θ_c	0	0	0	0

TABLE 6.8 Third order parameters for interaction response obtained from optimization for noise corrupted signal

G_{pij}	Parameters	Initial guess values	Final converged values	Sum of IAE values	Sum of ISE values
G_{p21}	K_{21}	1.628	1.5783	0.5168	0.0139
	τ_{e21}	0.8903	0.8754		
	ζ_{21}	0.2198	0.2141		
	b_{21}	0.4451	0.5161		
	θ_{21}	0.0	0.0		
G_{p12}	K_{12}	−0.3609	−0.3301	0.4920	0.0125
	τ_{e12}	0.8996	0.8707		
	ζ_{12}	0.2168	0.2166		
	b_{12}	0.4498	0.5216		
	θ_{12}	0.02	0.0		

As mentioned earlier, by using suitable curve fitting method, k_p, τ, and θ_c are identified in each case. The identified parameters are given in Table 6.9. Fig. 6.15 shows the comparison plot of the actual response and identified response. These parameters can be taken as initial guess for optimization similar to Eq. (4.24) and better parameter estimation can be obtained. Table 6.10 shows the final converged values of the CSOPTD model. The number of iterations is 17 and time taken is 59.228355 s. The sum of IAE values is 1.9731 and ISE value is 0.0501.

TABLE 6.9   CSOPTD model parameters from
Gp vs s curve (noise corrupted output)

	k_p	τ	θ
G_{p11}	0.4378	0.4306	0.0989
G_{p21}	0.8768	0.2138	0.0009
G_{p12}	−0.8798	0.2142	0.0008
G_{p22}	2.5445	0.5105	0.0999

TABLE 6.10 Converged parameters from
optimization method (Noise corrupted output)

	k_p	τ	θ
G_{p11}	0.5036	0.2368	0.1350
G_{p21}	0.9997	0.2169	0.0709
G_{p12}	−0.9933	0.2061	0.0857
G_{p22}	2.4001	0.4061	0.1886

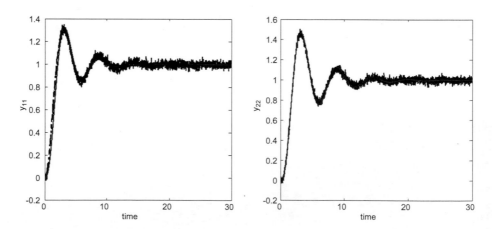

FIGURE 6.11   Comparison of noise corrupted responses and model identified using Clark method
Solid line: identified response; dotted line: noise corrupted response

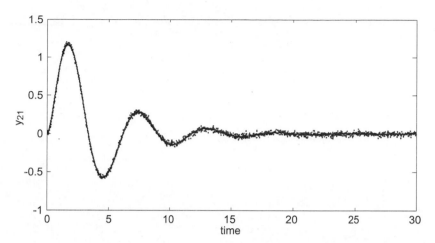

FIGURE 6.12 Comparison of noise corrupted interaction y_{21} and model identified using optimization
Solid line: identified interaction; dotted line: noise corrupted interaction

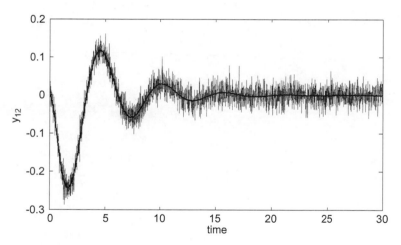

FIGURE 6.13 Comparison of noise corrupted interaction y_{12} and model identified using optimization
Solid line: identified interaction; dotted line: noise corrupted interaction

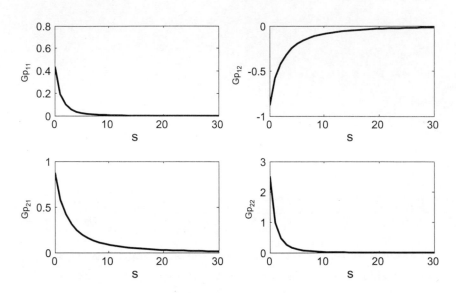

FIGURE 6.14   Plot of Gp vs s for noise corrupted process

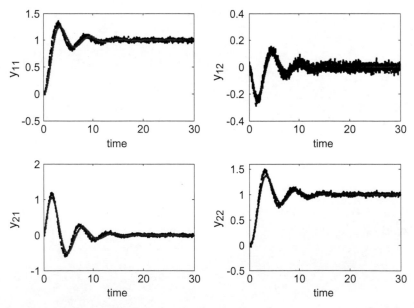

FIGURE 6.15   Comparison plot of the actual noise corrupted process and the identified CSOPTD process for the same controller settings

Solid line: identified process; dotted line: noise corrupted process. Both curves overlap.

Summary

In this chapter, we discussed a method for the identification of closed loop response by a third order transfer function. The closed loop responses were identified by the response curve of the third order system given by Clark (2005). The interactions were identified by using the modified Yuwana and Seborg (1982) method. Using the standard relationship between the closed loop model and the open loop model, the open loop model was identified as a CSOPTD model using curve fitting methods. The proposed method gave a better result for identifying parameters when the responses were identified as third order. The responses of the identified model were compared with the actual closed loop responses using the same controller settings. The effect of measurement noise on model identification was also presented.

Problems

1. Consider the Niederlinski (1971) multivariable transfer function model considered by Viswanathan et al. (2001):

$$G_p(s) = \begin{bmatrix} \dfrac{0.5}{(0.1s+1)^2(0.2s+1)^2} & \dfrac{-1.0}{(0.1s+1)(0.2s+1)^2} \\[3mm] \dfrac{1.0}{(0.1s+1)(0.2s+1)^2} & \dfrac{2.4}{(0.1s+1)(0.2s+1)^2(0.5s+1)} \end{bmatrix}$$

The parameters of the decentralized PI controllers are $K_{c11} = 0.4$, $\tau_{i11} = 0.4$; $K_{c22} = 0.1$, $\tau_{i22} = 0.3$. Carry out the closed loop reaction curve method for identifying the CSOPTD model.

2. Consider the SOPTD distillation column transfer function model matrix (Weischedel and McAvoy, 1980):

$$G(s) = \begin{bmatrix} \dfrac{0.7e^{-s}}{12.35s^2+8.26s+1} & \dfrac{-0.45e^{-s}}{17.04s^2+8.69s+1} \\[3mm] \dfrac{0.35e^{-1.28s}}{3.35s^2+5.67s+1} & \dfrac{-0.5e^{-s}}{1.69s^2+5.06s+1} \end{bmatrix}$$

The decentralized PID controller settings used for the MIMO identification were $k_{c11} = 2$, $\tau_{i11} = 2$, $\tau_{d11} = 0.5$; and $k_{c22} = -3$, $\tau_{i22} = 2$, $\tau_{d22} = 0.5$. Carry out the closed loop reaction curve method for identifying the CSOPTD model.

3. For the system:

$$G(s) = \begin{bmatrix} \dfrac{0.58e^{-s}}{(12.35s^2+8.26s+1)} & \dfrac{-0.45e^{-s}}{(17.04s^2+8.68s+1)} \\[3mm] \dfrac{0.35e^{-1.28s}}{(3.35s^2+5.67s+1)} & \dfrac{-0.48e^{-s}}{(1.69s^2+5.06s+1)} \end{bmatrix}$$

find a suitable input–output pairing. Design approximate decentralized PI controllers. Perform the closed loop reaction curve method to identify the model parameters of the CSOPTD models.

4. For the system:

$$G\left(s\right) = \begin{bmatrix} \dfrac{0.562e^{-s}}{(7.74s+1)\,(7.74s+1)} & \dfrac{-0.516e^{-1.5s}}{(7.1s+1)\,(7.1s+1)} \\[2ex] \dfrac{0.33e^{-1.5s}}{(15.8s+1)\,(0.5s+1)} & \dfrac{-0.394e^{-s}}{(13.8s+1)\,(0.4s+1)} \end{bmatrix}$$

find a suitable input–output pairing. Design approximate decentralized PI controllers. Perform the closed loop reaction curve method to identify the model parameters of the CSOPTD models.

5. For the system:

$$G\left(s\right) = \begin{bmatrix} \dfrac{0.471e^{-s}}{(30.7s+1)\,(30.7s+1)} & \dfrac{0.495e^{-2s}}{(28.5s+1)\,(28.5s+1)} \\[2ex] \dfrac{0.749e^{-1.7s}}{(57s+1)\,(57s+1)} & \dfrac{-0.832e^{-s}}{(50.5s+1)\,(50.5s+1)} \end{bmatrix}$$

find a suitable input–output pairing. Design approximate decentralized PI controllers. Perform the closed loop reaction curve method to identify the model parameters of the CSOPTD models.

7

Identification of Stable MIMO System by Optimization Method

The majority of existing techniques for identification are based on the frequency domain approach. For any optimization method, the selection of initial guess values plays an important role in computational time and convergence. In this chapter, a simple and generalized method for obtaining reasonable initial guess values for the First Order Plus Time delay (FOPTD) transfer function model parameters are discussed. A method to obtain the upper and lower bounds for the parameters to be used in the optimization routine is also presented. The method gives a quick and guaranteed convergence. The standard *lsqnonlin* routine is used for solving the optimization problem in Matlab. This method is applied to FOPTD and higher order transfer function models of multivariable systems.

7.1 Identification of Decentralized Controlled Systems

7.1.1 Identification Method

Consider an n-input and n-output multivariable system. $G(s)$ and $G_C(s)$ are process transfer function matrix and decentralized controller matrix with compatible dimensions, expressed in Eq. (7.1) and Eq. (7.2).

$$G_p(s) = \begin{bmatrix} g_{p11} & g_{p12} \cdots & g_{p1n} \\ g_{p21} & g_{p22} \cdots & g_{p2n} \\ \vdots & \ddots & \vdots \\ g_{pn1} & g_{pn1} \cdots & g_{pnn} \end{bmatrix} \qquad (7.1)$$

$$G_c(s) = \begin{bmatrix} g_{c11} & 0 \cdots & 0 \\ 0 & g_{c22} \cdots & 0 \\ \vdots & \ddots & \vdots \\ 0 & 0 \cdots & g_{cnn} \end{bmatrix} \qquad (7.2)$$

The controller parameters can be chosen arbitrarily for the multivariable systems such that the closed loop system is stable with reasonable responses.

Consider a decentralized TITO multivariable system as shown in Fig. 3.2. The process transfer function models are identified by FOPTD models. A FOPTD model is given in Eq. (7.3).

$$g_{mij}(s) = \frac{K_{mpij}e^{-\theta_{mij}s}}{(\tau_{mij}s + 1)} \qquad i = 1, 2; \ j = 1, 2 \qquad (7.3)$$

In this case, a known magnitude of step change is introduced in the set point y_{r1} with all the remaining set points unchanged and all other loops kept under closed loop operation. From the prescribed step change in the set point y_{r1}, we obtain the main response y_{11} and interaction response y_{21}. Similarly, the same magnitude of step change is introduced in set point y_{r2}, and we obtain the main response y_{22} and interaction response y_{12}. The response matrix of the TITO system can be expressed as Eq. (7.4).

$$Y(t) = \begin{bmatrix} y_{11} & y_{12} \\ y_{21} & y_{22} \end{bmatrix} \qquad (7.4)$$

The first column in the response matrix in Eq. (7.4) contains the responses (main and interaction) obtained by the step change in the set point y_{r1} in the first loop. The second column contains the responses (main and interaction) obtained by the step change in set point y_{r2} in the second loop. From these step responses, the initial guess values of the model parameters are obtained.

In any optimization method, the selection of initial guess values plays a vital role. A simple method is proposed here. The initial guess values of time delay for each transfer function model can be taken to be the same as the corresponding closed loop time delay values. The initial guess values for time constant are obtained from the settling time (t_s) of the main and interaction responses. From the settling time (to reach 98% of final steady state value and remains within the limit), the closed loop time constant (τ_c) can be assumed as $(t_s/4)$. The time constant of the open loop system is assumed to be lesser than that of the closed loop time constant. In principle, for SISO systems, the closed loop response is faster than the open loop response. In MIMO systems, due to the interaction among the loops, usually the closed loop response is assumed to be slower than the open loop response. This assumption is made only to obtain the initial guess values of the time constant of the open loop system for the optimization problem. Hence, the initial guess value of the time constant for the model (Pramod and Chidambaram, 2000; 2001) is to be assumed as $(t_s/8)$.

The initial guess values of the main loop process gains (K_{p11} and K_{p22}) are obtained by the following method.

For the lower order model systems, the value (Chidambaram, 1998) of $k_p k_{c,\max}$ is taken as 4 and the design value for $k_p k_{c,des} = 2$. Hence, the guess value for k_p is taken as $2/k_c$. Similarly, for the higher order model systems, the value (Chidambaram, 1998) of $k_p k_{c,\max}$ is 1 and the design value for $k_p k_{c,des} = 0.5$. Hence, the guess value for k_p is taken as $0.5/k_c$. The higher order model and lower order model are distinguished by checking the initial dynamics of the main responses of the actual process. If the initial slope of the main response is sluggish (0 slope), then we consider the system as a higher order model. Similarly, if the initial slope of the main response has a steep response (non-zero slope), then we consider the system as a lower order model.

The initial guess value of the cross loop process gains (K_{p21} and K_{p12}) are obtained by the following method.

The Laplace transforms of interaction response $y_{21}(s^*)$ and $y_{12}(s^*)$ can be expressed as in Eq. (7.5) and Eq. (7.6).

$$y_{21}(s^*) = \int_0^\infty y_{21}(t)e^{-ts^*}\,dt \tag{7.5}$$

$$y_{12}(s^*) = \int_0^\infty y_{12}(t)e^{-ts^*}\,dt \tag{7.6}$$

To evaluate Eq. (7.5) and Eq. (7.6), we need to assume the value of s^*. In the present case, the value of s^* is considered as $8/t_s$. Here t_s is the settling time of the closed loop response. For, $t \geq t_s$, the value of e^{-8} is close to 0. Then the

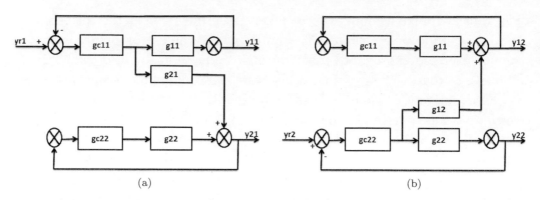

FIGURE 7.1   Simplified diagram for TITO system (a) for Eq. (7.8) and (b) for Eq. (7.10)

integral value does not change further. So, in order to reduce the computation time of the integral, the value of s^* is taken as $8/t_s$. The upper bound for t is t_s. The numerical values for $y_{21}(s^*)$ and $y_{12}(s^*)$ is obtained by substituting the value of s^* in Eq. (7.5) and Eq. (7.6). The closed loop transfer function for the cross-loops is assumed by considering the interaction from the main loop taken as disturbance to the interacted loop. Some signals are ignored for simplicity, as shown in Fig. 7.1, in order to avoid g_{p12} (Fig. 7.1a) in deriving Eq. (7.7) and to avoid g_{p21} (Fig. 7.1b) in deriving Eq. (7.9). Thus, we get Eq. (7.7).

$$\frac{y_{21}(s^*)}{y_{r1}(s^*)} = \frac{g_{c11}(s^*)g_{p21}(s^*)}{(1 + g_{c11}(s^*)g_{p11}(s^*))(1 + g_{c22}(s^*)g_{p22}(s^*))} \tag{7.7}$$

Substitute the value for $y_{r1}(s)$ as $1/s^*$ and also substitute the controller settings and the guess values of model parameters (time delays, time constants and main loop process gains) in Eq. (7.7). So we get Eq. (7.8).

$$y_{21}(s^*) = \frac{\left(1/s^*\right)K_{c11}\left(1 + \left(1/\tau_{i11}s^*\right) + \tau_{d11}s^*\right)\frac{K_{p21}e^{-\theta_{21}s^*}}{(\tau_{21}s^*+1)}}{\left[\left(1 + K_{c11}\left(1 + \left(1/\tau_{i11}s^*\right) + \tau_{d11}s^*\right)\frac{K_{p11}e^{-\theta_{11}s^*}}{(\tau_{11}s^*+1)}\right) \\ \left(1 + K_{c22}\left(1 + \left(1/\tau_{i22}s^*\right) + \tau_{d22}s^*\right)\frac{K_{p22}e^{-\theta_{22}s^*}}{(\tau_{22}s^*+1)}\right)\right]} \tag{7.8}$$

Similarly, the initial guess values for K_{p12} is obtained by Eq. (7.9).

$$\frac{y_{12}(s^*)}{y_{r2}(s^*)} = \frac{g_{c22}(s^*)g_{p12}(s^*)}{(1 + g_{c22}(s^*)g_{p22}(s^*))(1 + g_{c11}(s^*)g_{p11}(s^*))} \tag{7.9}$$

Substituting the value for $y_{r2}(s)$ as $1/s^*$ and also substituting the controller settings and the guess values of model parameters (time delays, time constants, and main loop process gains) in Eq. (7.9), we get Eq. (7.10).

$$y_{12}(s^*) = \frac{(1/_{s^*})K_{c22}(1 + (1/_{\tau_{i22}s^*}) + \tau_{d22}s^*)\frac{K_{p12}e^{-\theta_{12}s^*}}{(\tau_{12}s^*+1)}}{\left[\begin{array}{l}(1 + K_{c22}(1 + (1/_{\tau_{i22}s^*}) + \tau_{d22}s^*)\frac{K_{p22}e^{-\theta_{22}s^*}}{(\tau_{22}s^*+1)}) \\ (1 + K_{c11}(1 + (1/_{\tau_{i11}s^*}) + \tau_{d11}s^*)\frac{K_{p11}e^{-\theta_{11}s^*}}{(\tau_{11}s^*+1)})\end{array}\right]} \tag{7.10}$$

The guess values for K_{p21} and K_{p12} are to be obtained from Eq. (7.8) and Eq. (7.10). After obtaining all the guess values of the model, the same magnitude of step change is introduced to the model set point y_{mr1} with all other remaining set points kept unchanged and all other loops under closed loop condition. From the prescribed step change in set point y_{mr1} the main response y_{m11} and interaction response y_{m21} are obtained. Similarly, the same magnitude of step change is introduced in set point y_{mr2} the main response y_{m22} and interaction response y_{m12} are obtained. The closed loop responses are to be compared with the main and interaction responses of the actual process. To obtain the optimized model parameters $(K_{mpij}, \tau_{mij}, \theta_{mij})$, the optimization problem is formulated so as to select the model parameters in order to minimize the sum of squared errors between the model and the actual process responses:

$$SSQ = \text{minimize } f = \sum_{i=1}^{2}\sum_{j=1}^{2}\sum_{k=1}^{n} [y_{mij}(t_k) - y_{ij}(t_k)]^2 \Delta t \quad i = 1, 2; \; j = 1, 2$$

$$\tag{7.11}$$

TITO systems have two main and two interaction responses. In the objective function, n is the number of data points in the response of the process and the model. First get the error between responses of each transfer function model with the associated process response for various times (to form an error vector for each model). Further take the square of each of the errors and multiply by a fixed sampling time Δt. Sum of the elements of each vector gives an ISE value. There are four ISE values. Sum of these four ISE values is considered as a scalar objective function as given in Eq. (7.11). The model parameters are selected by minimizing this scalar function. In Eq. (7.11), the sampling time Δt need not be included since it is a constant value. In this present case, the optimization problems are solved by using the routine *lsqnonlin* in Matlab and the routine implements the Trust–Region–Reflective algorithm. Although the optimization problem is formulated and solved using the standard routine *lsqnonlin* in Matlab, the convergence and computational time depends completely on the initial guesses. The main focus of the present work is to propose the above simple method to get the initial guess values of the model parameters. The computational time is found drastically reduced by using the proposed guess values.

In *lsqnonlin* routine, the values used for the convergence parameters are $\text{TolX} = 1.0 \times 10^{-3}$, $\text{TolFun} = 1.0 \times 10^{-6}$ and the number of iterations $= 400$. The closed loop multivariable system is simulated using Simulink. Fourth order Runge–Kutta method (Gupta, 1995) with fixed step size is used for solving integration of governing differential equations. All the computational work for the identification of multivariable systems is performed in Core 2 Quad (Intel, 3.00 GHz) personal computer. To evaluate the proposed method, a known transfer function model matrix is considered with suitable decentralized PI/PID controllers. The closed loop main and interaction responses of the actual process are obtained by the earlier discussed method. From the responses, the model parameters are obtained by the proposed method and the closed loop main and interaction responses are compared with the actual system with the same decentralized PI/PID controller settings.

7.2   Simulation Examples

In this section, three simulation examples (2×2 FOPTD, 2×2 higher order system and 3×3 FOPTD systems) are considered to show the effectiveness of the proposed method.

Example 1 Consider the Wood and Berry (1973) distillation process. The transfer function matrix of WB column is given by Eq. (7.12).

$$G_p(s) = \begin{bmatrix} \dfrac{12.8e^{-s}}{16.7s + 1} & \dfrac{-18.9e^{-3s}}{21s + 1} \\[2ex] \dfrac{6.6e^{-7s}}{10.9s + 1} & \dfrac{-19.4e^{-3s}}{14.4s + 1} \end{bmatrix} \tag{7.12}$$

The decentralized PID controller settings used for the identification are (Ham and Kim, 1998) $K_{c11} = 0.396$, $\tau_{i11} = 5.926$, $\tau_{d11} = 1.070$; $K_{c22} = -0.124$, $\tau_{i22} = 7.092$, $\tau_{d22} = 1.849$.

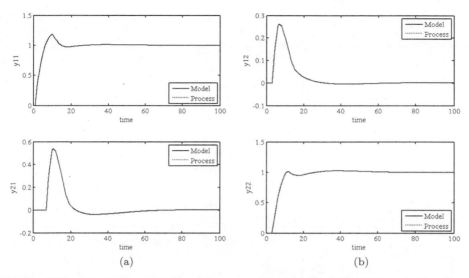

FIGURE 7.2 Closed loop response of the process and the identified FOPTD transfer function model (Example 1) using PID controller settings (Ham and Kim, 1998)
The model and process response curves overlap. The responses (a) are obtained by step change in y_{r1} and the responses (b) are obtained by step changes in y_{r2}.

Here the FOPTD multivariable process is taken to show the effectiveness of this method by the direct comparison of the converged parameters of the model with the actual process. If we take a higher order system, we cannot compare the model parameters as such. The comparison of the process and the model can be made only by matching the response. The closed loop main and interaction responses are obtained by the previously mentioned method using Simulink and it is shown in Fig. 7.2. The FOPTD model is assumed and the initial guess values for the model parameters are obtained by the earlier mentioned methods (Table 7.1). Since the responses show a finite non-zero initial slope, $k_p k_{c,des} = 2$ is used for calculating the guess values of process gain k_{p11} and k_{p22}. The optimization routine *lsqnonlin* is used in Matlab and the sample time of 0.01 s is considered.

For all the simulation examples in the optimization routine, we fix the lower bound value for time constant as $1/20^{\text{th}}$ of the guess value and the upper bound as 5 times of the initial guess value. For time delays, the lower bound is specified as $1/4^{\text{th}}$ of the guess value and the upper bound is specified as 4 times of the initial guess value. In case the initial guess value of process gain is calculated as negative, we fix the upper bound as $1/20^{\text{th}}$ of the guess value and lower bound as 20 times of the guess value. Similarly, for positive guess values of the process gain, we fix the lower bound as $1/20^{\text{th}}$ of the guess value and upper bound as 20 times of the

TABLE 7.1   Model identification results under with and without noise using PID controller (Mei et al., 2005)

Model (ij)		K_{pij}	τ_{ij}	θ_{ij}	IAE values	ISE values	Time (N.I.)
		• **Model identification results without noise**					
11	Guess	5.0505	6.2500	1.0000	6.41×10^{-4}	1.17×10^{-8}	152 (25)
	Converged	12.8009	16.7031	1.0002			
21	Guess	5.2391	8.7500	7.0000	1.60×10^{-3}	7.24×10^{-8}	
	Converged	6.6004	10.9020	7.0001			
12	Guess	−7.2450	6.2500	3.0000	9.10×10^{-4}	1.16×10^{-8}	
	Converged	−18.9062	21.0104	3.0002			
22	Guess	−16.1290	7.5000	3.0000	1.60×10^{-3}	7.05×10^{-8}	
	Converged	−19.4026	14.4027	3.0001			
		• **Model identification results with noise**					
11	Guess	5.0505	5.0000	1.0000	0.1017	6.24×10^{-4}	68 (21)
	Converged	12.5038	16.0582	1.0447			
21	Guess	5.6799	7.5000	7.0000	0.1810	1.90×10^{-3}	
	Converged	6.4410	10.3176	6.9856			
12	Guess	−8.6169	6.2500	3.0000	0.0969	3.62×10^{-4}	
	Converged	−18.2553	20.1191	2.9997			
22	Guess	−16.1290	6.2500	3.0000	0.2110	2.40×10^{-3}	
	Converged	−19.1683	14.0428	2.9041			

N.I.: number of iterations for convergence (in brackets); Time: computational time in seconds

guess value. These upper and lower bound values used in the optimization routine are found to give quick convergence and also reduce the computational time.

The initial guess values, final converged parameters, number of iterations, computational time, and IAE and ISE values are listed in Table 7.1. The converged model parameters match well with the actual FOPTD system. The closed loop main responses and interaction responses using the identified model parameters with the same controller settings are also shown in Fig. 7.2. Fig. 7.2 shows a good match between the model and the actual process. The computational time for optimization is considerably reduced from 89 minutes (Viswanathan et al., 2001) to 4 minutes and 17 seconds (maximum computational time) because of the proposed method for obtaining the initial guess values of the model parameters. The identification method is also carried out by using the closed loop main and interaction responses obtained by different PI/PID settings. The converged model parameters show a good agreement with that of the actual process parameters. The number of iterations, computational time, ISE* (Integral Squared Error) and IAE*

TABLE 7.2 Effect of changing the controller settings on the number of iteration and computational time (Example 1)

Parameters $(K_{c11}, \tau_{i11}, \tau_{d11};$ $K_{c22}, \tau_{i22}, \tau_{d22})$	Vu et al. (2009) (0.7, 17.27; −0.13, 16.08)	Loh et al. (1993) (0.868, 3.246; −0.0868, 10.4)	Viswanathan et al. (2001) (0.2, 4.0, 3.0; −0.03, 2.0, 3.0)
N.I.	16	21	28
Computational time (s)	94	132	257
IAE* values	0.0229	0.0046	0.0066
ISE* values	6.51×10^{-6}	3.35×10^{-7}	4.40×10^{-7}

*: sum of main and interaction response values; N.I.: number of iterations

(Integral Absolute Error) values are given in Table 7.2. (* refers to the sum of ISE values for the main action and the interactions.)

The effect of initial guess values on the number of iterations, computational time, and IAE and ISE values is also studied by perturbing the guess values by +10% and separately by −10%. The present method shows a good agreement with the actual process parameters. The number of iterations, computational time, ISE* and IAE* values are listed in Table 7.3 for the perturbed guess values. The effect of measurement noise on the guess values and model parameters is also considered by adding a random noise (SD = 0.05) to the actual main and interaction responses and a sample time of 0.1 s is considered. The noisy output is used for the feedback control action and also for the model identification. To obtain the initial guess values, a smooth curve is first drawn. Because of the integral action in the controller, the main response reaches to 1 and the interaction reaches to 0 in steady state. Hence, a horizontal is line drawn for the main response at the value of 1 and a horizontal line is drawn at the value of 0 in the interaction response in the steady state portion only. In the initial dynamics part, a smooth curve is passed through a midpoint of the response. Alternatively, one can use the least squares method proposed by Savitzky and Golay (1964) which filters the noise and gives a smooth curve. From this smooth curve, the initial guess values are calculated by the proposed method.

The identified model parameters are closer to those obtained without measurement noise. The closed loop main responses and interaction responses using the identified model with the same controller settings are shown in the Fig. 7.3, with the corrupted process responses. The initial guess values, final converged values, number of iterations, computational time, and IAE and ISE values are also listed in Table 7.1. The proposed method is robust for the controller settings and for the measurement noise.

TABLE 7.3 Effect of initial guess values of the model parameters on the number of iterations and computational time (Example 1) using PID controller settings (Ham and Kim, 1998)

Variation in guess values	N.I.	IAE* values	ISE* values	Computational time (s)
+10%	25	0.004	1.88×10^{-7}	152
−10%	25	0.0047	1.62×10^{-7}	154

*: sum of main and interaction response value; N.I.: number of iterations

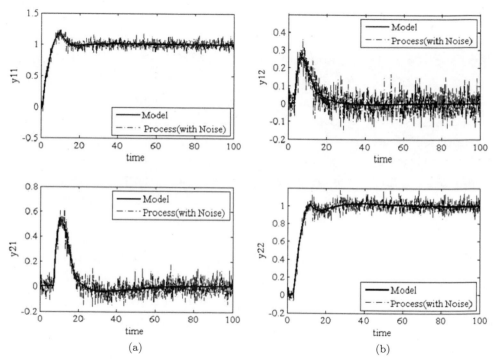

(a)   (b)

FIGURE 7.3 Closed loop response of the process and the identified FOPTD transfer function model for Example 1 (with noise) using PID controller settings (Ham and Kim, 1998)
The responses (a) are obtained by step change in y_{r1} and the responses (b) are obtained by step changes in y_{r2}.

Example 2 Consider the Niederlinski (1971) multivariable transfer function model presented by Viswanathan et al. (2001) given in Eq. (7.31).

$$
G_p(s) = \begin{bmatrix} \dfrac{0.5}{(0.1s+1)^2(0.2s+1)^2} & \dfrac{-1.0}{(0.1s+1)(0.2s+1)^2} \\ \dfrac{1.0}{(0.1s+1)(0.2s+1)^2} & \dfrac{2.4}{(0.1s+1)(0.2s+1)^2(0.5s+1)} \end{bmatrix} \tag{7.13}
$$

TABLE 7.4 FOPTD model identification results with and without noise (Example 2)

Model (ij)		K_{pij}	τ_{ij}	θ_{ij}	IAE values	ISE values	Time (N.I.)
FOPTD Model identification results for without noise:							
11	Guess	1.2500	1.2500	0.1000	0.0177	1.18×10^{-4}	69 (22)
	Converged	0.5114	0.3554	0.2801			
21	Guess	2.2314	1.2500	0.1000	0.0515	7.46×10^{-4}	
	Converged	0.9906	0.2998	0.2247			
12	Guess	−2.2314	1.2500	0.1000	0.0228	9.39×10^{-4}	
	Converged	−0.9843	0.3283	0.1754			
22	Guess	5.0000	1.2500	0.1000	0.0415	3.92×10^{-4}	
	Converged	2.4518	0.6369	0.4000			
FOPTD Model identification results for with noise:							
11	Guess	1.2500	1.0000	0.1000	0.0492	3.64×10^{-4}	30 (23)
	Converged	0.5074	0.3354	0.3000			
21	Guess	2.2793	1.0000	0.1000	0.1222	2.00×10^{-3}	
	Converged	0.9813	0.3316	0.1917			
12	Guess	−2.4484	1.2500	0.1000	0.0348	2.22×10^{-4}	
	Converged	−0.9793	0.2371	0.2611			
22	Guess	5.0000	1.0000	0.1000	0.0771	6.84×10^{-4}	
	Converged	2.5160	0.6608	0.4000			

N.I.: number of iterations; time in seconds

Here the higher order TITO process transfer function model is to be approximated by FOPTD models. Decentralized PI controllers are used for this identification process. The decentralized PI controllers parameters are $K_{c11} = 0.4$, $\tau_{i11} = 0.4$; $K_{c22} = 0.1$, $\tau_{i22} = 0.3$.

The closed loop responses of the actual process are obtained by previously mentioned methods using Simulink. The initial guess values are obtained by the proposed method (Table 7.4 gives the guess values). Since the response shows a higher order model behaviour (initial 0 slope), $k_p k_{c,des} = 0.5$ is considered for calculating the guess value for the process gains. The initial guess values, final converged parameters, number of iterations, computational time, and IAE and ISE values are listed in Table 7.4. The closed loop main and interaction responses using the identified FOPTD models with the same controller settings are shown in the Fig. 7.4 with the actual process responses. This figure shows a good match between the FOPTD model and the actual process. The computational time is significantly reduced to 1 minute and 26 seconds (maximum computational time) when compared to 80 minute as reported by Viswanathan et al. (2001).

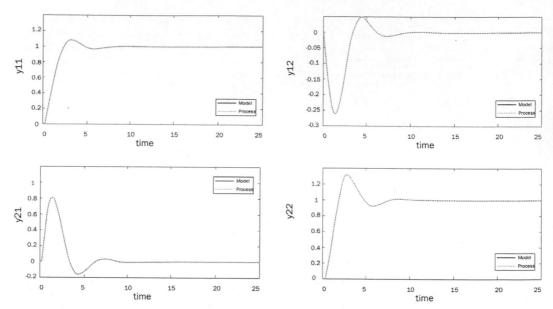

FIGURE 7.4   Closed loop responses of the process and the identified FOPTD model (Example 2) The model and process response curves overlap. The responses (a) are obtained by step change in y_{r1} and the responses (b) are obtained by step changes in y_{r2}.

Since the dynamics of the higher order system of Example 2 is fast (settling time is about 10 minutes), the computational time for the optimization problem is lesser than that required for Example 1 (settling time is about 70 minutes for Example 1). The effect of initial guess values on the number of iterations, computational time, and IAE and ISE values is also studied by changing the guess values by +10% and separately by −10%. The present method shows a good agreement with the actual process parameters for these perturbed guess values. The number of iterations, computational time, and IAE and ISE values are listed in Table 7.5 for these perturbed guess values.

The effect of measurement noise is also evaluated on the identified FOPTD model parameters by adding a random noise (SD = 0.05) to the actual main and interaction responses and the sample time of 0.1 s is considered. The noisy output is used for the feedback control action and for the model identification. To obtain the initial guess values for FOPTD models, a smooth curve is first drawn. As discussed earlier, the initial guess values for the model parameters are obtained from the smooth curve. The identified model parameters are closer to those obtained without measurement noise. The closed loop main responses and interaction responses using the identified FOPTD model with the same

TABLE 7.5 Effect of initial guess values of the model parameters on the number of iterations and computational time (Example 2)

Guess Variation	N.I.	IAE values	ISE values	Computational Time
+10%	13	0.1226	0.0014	48
−10%	24	0.2089	0.0028	86

N.I.: number of iterations; time in seconds

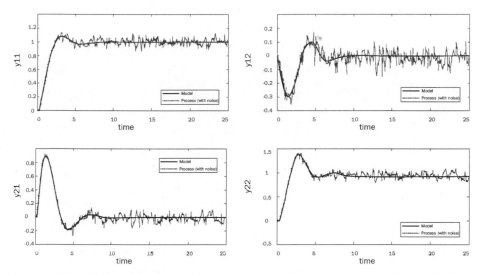

FIGURE 7.5 Closed loop responses of the corrupted process and the identified FOPTD transfer function model with noise (Example 2)
The responses (a) are obtained by step change in y_{r1} and the responses (b) are obtained by step changes in y_{r2}.

controller settings are shown in Fig. 7.5, with the actual corrupted process main and interaction responses. The initial guess values, final converged values, number of iterations, computational time, and IAE and ISE values for the FOPTD model identification with noise are also listed in Table 7.4.

Example 3 Consider the distillation column process transfer function matrix reported by Ogunnaike et al. (1983) and Ham and Kim (1998) as in Eq. (7.14).

$$
G = \begin{bmatrix}
\dfrac{0.66e^{-2.6s}}{6.7s+1} & \dfrac{-0.61e^{-3.5s}}{8.64s+1} & \dfrac{-0.0049e^{-1.0s}}{9.06s+1} \\[2mm]
\dfrac{1.11e^{-6.5s}}{3.25s+1} & \dfrac{-2.3e^{-3s}}{5.0s+1} & \dfrac{-0.01e^{-1.2s}}{7.09s+1} \\[2mm]
\dfrac{-34.68e^{-9.2s}}{8.15s+1} & \dfrac{46.2e^{-9.4s}}{10.9s+1} & \dfrac{0.85e^{-0.83s}}{6.60s+1}
\end{bmatrix}
\tag{7.14}
$$

The system is interactive as indicated by its steady state relative gain array matrix given in Eq. (7.15).

$$\Lambda = \begin{bmatrix} 2.0451 & -0.7554 & -0.2896 \\ -0.6730 & 1.8667 & -0.1938 \\ -0.3721 & -0.1113 & 1.4834 \end{bmatrix} \tag{7.15}$$

Decentralized PI/PID controllers are used for the MIMO identification. The PI controller parameters (Vu et al., 2010a) are $K_{c11} = 1.57$, $\tau_{i11} = 5.96$; $K_{c22} = -0.31$, $\tau_{i22} = 4.81$; $K_{c33} = 6.1$, $\tau_{i33} = 9.60$. The step responses of the closed loop system are shown in Fig. 7.6a for a step change in the set point of y_{1r} and in Fig. 7.6b for the step change in y_{2r} and in Fig. 7.6c for step change in y_{3r}. The interactions are significant.

The FOPTD model is assumed and the initial guess values for the model parameters are obtained by using step responses (Table 7.6 contains the guess values). The guess values for the main-loop (diagonal) process gains k_{p11}, k_{p22} and k_{p33} are calculated based on the corresponding controller gains as discussed in Example 1. Since the responses of the actual process show a lower order model behaviour (initial 0 slope), the $k_p k_{c,des}$ value is considered as 2 for calculating the guess value of the process gains (g_{11}, g_{22}, and g_{33}). The non-diagonal process gains (interactions) are calculated using the equations similar to Eq. (7.8) and Eq. (7.10). To simplify the calculations, at a time only two loops are considered and the interaction coming from the remaining loop is considered as negligible. For example, the guess value of the process gain (non-diagonal) k_{p21} is estimated by considering the interaction comes from main loop only (similar to Fig. 7.7a). The interaction to and also from the third loop are assumed to be neglected. This procedure gives explicitly the guess value of required process gain. Similarly, for getting the guess values of k_{p12}, figure similar to Fig. 7.7b is considered. To get the guess values for the gain k_{p31} and k_{p32}, Fig. 7.7a and Fig. 7.7b, respectively, are considered. Similarly, the guess values for the other gains k_{p13} and k_{p23} are calculated. The sign of the guess values of the steady state gains are important. With these guess values, the convergence of the parameters of the model is achieved as shown in Table 7.6.

The responses obtained from g_{13} and g_{23} are of values which are of very low order of magnitude compared with other responses. Hence, the responses obtained from all the models are normalized between 0 to 1. This normalized data is used for the optimization routine. For simulating the model, the sample time is considered as 0.01 sec. The initial guess values, final converged parameters, number of

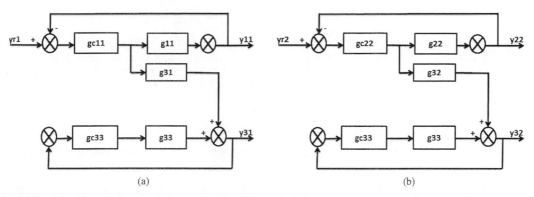

FIGURE 7.6 The action and interaction for Example 3 for step changes in the set points In each figure, at steady state, the sum of the three responses will be equal to 1.

iterations, and IAE and ISE values are presented in Table 7.6. The converged model parameters match well with the actual FOPTD model parameters. Computational time for the optimization is 23 minutes and 20 seconds and the number of iterations is 29. The closed loop main and interaction responses using the identified model parameters with the same controller settings are also shown in Fig. 7.8. Fig. 7.8 shows a good match between the model and the actual process. For this example (Example 3), since the number of parameters to be identified is larger (27 parameters) and the dynamics of the closed loop system takes larger settling time (Fig. 7.8), the computational time for the optimization problem is higher (about 23 minutes) when compared to Example 1 and Example 2.

The identification method is also carried out by using the closed loop main and interaction responses obtained by different PI/PID settings. The converged model parameter shows a good agreement with the actual process parameters. The number of iterations, computational time, ISE* and IAE* values are listed in Table 7.7. The effect of initial guess values on the number of iterations, computational time, and IAE and ISE values is also studied by perturbing the guess values by +10% and separately by −10%. The model parameters obtained by the proposed method shows a good agreement with the actual process parameters for these perturbed guess values. Number of iterations, computational time, and IAE, and ISE values are listed in Table 7.8.

The initial guess values for the model parameters can also be obtained by assuming the discrete time ARX and OE models (Lyzell, 2009) and by solving the linear regression equations to get the parameter estimations. The guess values for time delays are identified from the impulse response plot (by using the Matlab command 'cra'). After getting the estimate for the time delays, a standard routine

TABLE 7.6   FOPTD model identification results for Example 3 using the PI controller (Vu and Lee, 2010a)

Model (ij)		K_{pij}	τ_{ij}	θ_{ij}	IAE values	ISE values
11	Guess	1.2739	10.0000	2.6000	1.75×10^{-3}	6.36×10^{-8}
	Converged	0.6598	6.6958	2.6007		
21	Guess	4.3027	12.5000	6.5000	6.85×10^{-3}	1.44×10^{-6}
	Converged	1.1093	3.2468	6.5012		
31	Guess	−22.6872	10.0000	9.2000	5.01×10^{-2}	6.58×10^{-5}
	Converged	−34.6606	8.1438	9.1991		
12	Guess	−1.7594	12.5000	3.5000	1.06×10^{-3}	2.12×10^{-8}
	Converged	−0.6096	8.6365	3.5021		
22	Guess	−6.4516	10.0000	3.0000	3.55×10^{-3}	1.76×10^{-7}
	Converged	−2.2988	4.9947	3.0009		
32	Guess	28.3333	12.5000	9.4000	8.62×10^{-3}	1.80×10^{-6}
	Converged	46.1734	10.8923	9.4012		
13	Guess	−0.0030	10.0000	1.0000	6.09×10^{-5}	1.55×10^{-10}
	Converged	−0.0049	9.0634	0.9996		
23	Guess	−0.0064	8.7500	1.2000	1.50×10^{-4}	5.95×10^{-10}
	Converged	−0.0100	7.0942	1.2007		
33	Guess	0.3279	6.2500	0.8000	1.11×10^{-3}	2.81×10^{-8}
	Converged	0.8500	6.5995	0.8300		

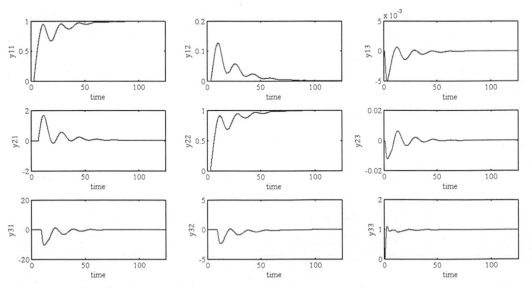

FIGURE 7.7   Simplified diagram for 3×3 multivariable system for calculating guess values for k_{p31} and k_{p32}

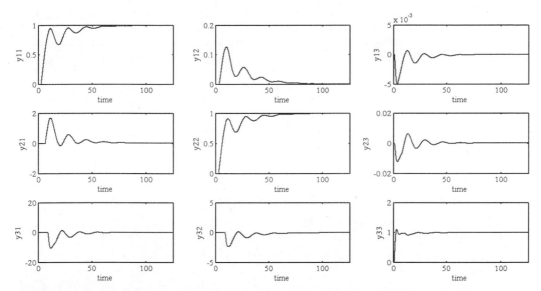

FIGURE 7.8 Closed loop response of the process and the identified FOPTD transfer function model for Example 3 using PI controller (Vu and Lee, 2010a).
Solid line: model, Dashed line: process. Model and process response curves overlap.

TABLE 7.7 Effect of changing controller settings on the number of iterations and computational time (Example 3)

Parameters $(K_{c11}, \tau_{i11}, \tau_{d11}; K_{c22}, \tau_{i22}, \tau_{d22}; K_{c33}, \tau_{i33}, \tau_{d33})$	N.I.	IAE* values	ISE* values	Computational Time (minutes)
Ham and Kim (1998) (2.794, 4.230, 1.102; –0.524, 5.635, 1.110; 3.835, 6.006, 1.069)	50	0.6222	0.0096	30.53
Chien et al. (1999) (1.08, 4.25; –0.2333, 3.32; 2.78, 5.24)	42	0.056	4.08E–05	33.23

*: sum of main and interaction response values; N.I.: number of iterations

TABLE 7.8 Effect of the initial guess values on IAE, ISE, computational time, and number of iterations

Variation in guess values	N. I	IAE* values	ISE* values	Computational Time (minutes)
–10%	25	0.0939	4.32E–04	20
+10%	27	0.0476	2.37E–05	21.31

*: sum of main and interaction response values; N.I.: number of iterations

TABLE 7.9   Guess values obtained by ARX and OE models and the final converged model parameters by *lsqnonlin* (Example 3)

Model		K_{pij}	τ_{ij}	θ_{ij}	IAE values	ISE values	Time (N.I.)
ARX Model							
11	Guess	8.9424	9.0750	0.9900	0.4470	0.0394	37 (11)
	Converged	17.0097	19.6764	0.4689			
21	Guess	12.7846	36.8000	6.9900	0.4070	0.0617	
	Converged	7.8409	10.8091	7.8595			
12	Guess	−7.3675	12.0110	3.0000	0.1606	0.0042	
	Converged	−29.3309	29.2621	2.5819			
22	Guess	−28.1234	33.7490	3.0000	0.2570	0.0138	
	Converged	−23.8196	19.1251	2.5529			
OE Model							
11	Guess	6.3269	7.0472	0.9900	0.4844	0.0334	31 (10)
	Converged	8.6570	8.2931	0.9751			
21	Guess	6.3795	163.1587	6.9900	0.9123	0.3452	
	Converged	6.1981	163.0393	10.2617			
12	Guess	−17.0640	18.3385	3.0000	0.1517	0.0041	
	Converged	−16.7263	18.0798	3.0744			
22	Guess	−9.5539	5.8685	3.0000	0.2595	0.0136	
	Converged	−10.9623	5.9882	3.4727			

N.I.: number of iterations; time in seconds

'*arx*' in Matlab is used to get the higher order ARX model. For solving the linear regression equations, the MIMO model is first converted into a MISO (multi-input and single-output) model. The identified higher order ARX model is reduced to FOPTD model. The guess values for the OE (output error) model are also obtained by fixing the variable *maxiter* as 0. The OE model solves the linear regression equation (Lyzell, 2009) to get the initial values of the transfer function model parameters. The guess values obtained by these methods are used in the nonlinear optimization (*lsqnonlin*) to get the optimal model parameters. For Example 1, the initial guess values, final model parameters, number of iterations, computational time, and IAE and ISE values are listed in Table 7.9. The guess values obtained by OE model are not in the order of magnitude of the actual process parameters and the final converged parameters of the model are not closer to the true parameters. The guess values obtained by the ARX model are in the same order of magnitude of the true parameters. With these guess values also, satisfactory results could not be obtained. The method of getting the initial guess values discussed in Section 7.1.1 is simpler to use and it gives the satisfactory results also.

The proposed work considers only the identification of FOPTD model for 2×2 systems and 3×3 systems. The identification of transfer function models with a zero and under-damped SOPTD transfer function models of the multivariable systems need further study.

Summary

In this chapter, a simple and generalized method for obtaining reasonable initial guess values for the FOPTD transfer function model parameters was discussed. The method to obtain the upper and lower bound for the parameters to be used in the optimization routine was also presented. The method gave a quick and guaranteed convergence. The standard *lsqnonlin* routine was used for solving the optimization problem in Matlab. This method was applied to FOPTD and higher order transfer function models of multivariable systems.

Problems

1. For the system,

$$
G_p = \begin{bmatrix} \dfrac{12.8e^{-4s}}{16.7s+1} & \dfrac{-18.9e^{-3s}}{21s+1} \\[2ex] \dfrac{6.6e^{-7s}}{10.9s+1} & \dfrac{-19.4e^{-3s}}{14.4s+1} \end{bmatrix}
$$

With the PID settings $(k_c = 0.18, \tau_I = 7.0, \tau_D = 1.6)$; $(-0.20, 7.0, 1.6)$ for the diagonal pairing, carry the closed loop optimization method of obtaining the FOPTD model parameters.

2. For the system,

$$
G_p = \begin{bmatrix} \dfrac{22.89e^{-2s}}{4.572s+1} & \dfrac{-11.64e^{-0.4s}}{1.807s+1} \\[2ex] \dfrac{4.687e^{-0.2s}}{2.174s+1} & \dfrac{5.80e^{-0.9s}}{1.801s+1} \end{bmatrix}
$$

With the PID settings $(k_c = 0.0056, \tau_I = 1.5, \tau_D = 0.37)$; $(0.44, 1.5, 0.37)$ for the diagonal pairing, carry the closed loop optimization method of obtaining the FOPTD model parameters.

3. For the system,

$$
G_p = \begin{bmatrix} \dfrac{2.5e^{-7s}}{15s+1} & \dfrac{5e^{-3s}}{4s+1} \\[2ex] \dfrac{1e^{-4s}}{5s+1} & \dfrac{-4e^{-10s}}{20s+1} \end{bmatrix}
$$

With the PID settings $(k_c = 0.35, \tau_I = 10.5, \tau_D = 2.65)$; $(-0.6, 10.5, 2.65)$ for the diagonal pairing, carry the closed loop optimization method of obtaining the FOPTD model parameters.

8

Identification of Centralized Controlled Multivariable Systems

For systems with significant interactions, a centralized control system is preferred over a decentralized control system. The closed loop identification of model parameters of FOPTD models of a multivariable system under the control of centralized PI control system is carried out by an optimization method. The method of getting the initial guess values of the model parameters is given. The closed loop performances of the original system are compared with the closed loop of the identified model with the same centralized controllers.

8.1 Identification Method

The method used here is basically the one proposed in the previous chapter wherein the control system was a decentralized control system whereas in the present work a centralized PI control system is considered. Let $G(s)$ and $G_c(s)$ be the transfer function matrix (of size $n \times n$) for the process and the centralized control system as given in Eq. (8.1) and Eq. (8.2).

$$G(s) = [g_{pij}] \tag{8.1}$$

$$G_c(s) = [g_{c,ij}] \tag{8.2}$$

Here $i = 1, 2, \ldots, N$ and $j = 1, 2, \ldots, N$. The block diagram in Fig. 3.3 (Chapter 3) shows a centralized TITO multivariable system. The process transfer function models are given by FOPTD model as in Eq. (8.3).

$$g_{mij}(s) = \frac{K_{mpij}e^{-\theta_{mij}s}}{(\tau_{mij}s + 1)} \quad i = 1, 2; \ j = 1, 2 \tag{8.3}$$

With a suitable set of the controller parameters, the closed loop system gives a stable response. A unit step change is introduced in the set point y_{r1}. The main response y_{11} and the interaction response y_{21} are obtained. Similarly, a unit step change is given separately to y_{r2} and the main response y_{22} and the interaction response y_{12} are obtained. Thus, we have Eq. (8.4).

$$Y(t) = [y_{ij}]; \quad i = 1, 2; \ j = 1, 2 \tag{8.4}$$

The guess values for the model parameters can be obtained from the response matrix. Let us consider the proposed method for the calculation of the guess values for the centralized control system of a TITO system. The guess value for k_{pii} is considered as $1/k_{cii}$. The guess values of the interaction gains (k_{p21} and k_{p12}) are to be calculated. The method given in the previous chapter is suitably modified. The Laplace transforms of interaction response $y_{21}(s*)$ and $y_{12}(s*)$ are given by the expression by Rajapandiyan and Chidambaram (2012a) and Wang et al. (2008) as in Eq. (8.5) and Eq. (8.6).

$$y_{21}(s^*) = \int_0^\infty y_{21}(t)e^{-ts^*}dt \tag{8.5}$$

$$y_{12}(s^*) = \int_0^\infty y_{12}(t)e^{-ts^*}dt \tag{8.6}$$

As discussed in the previous chapter, $s^* = 8/t_s$. The upper limit for t is considered as t_s. The numerical values for $y_{21}(s^*)$ and $y_{12}(s^*)$ are obtained from the above equations. The closed loop transfer function of the interaction loops can be found out by ignoring the interaction from the other loop, that is, for the calculation of k_{p21}, the block g_{p12} is ignored (Fig. 8.1) for deriving Eq. (8.5) and for the calculation of k_{p12}, the block g_{p21} is ignored (Fig. 8.2) for deriving Eq. (8.6).

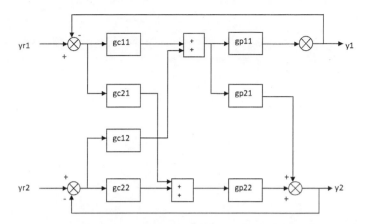

FIGURE 8.1   Modified TITO system with the centralized controllers for Eq. (8.7)

The initial guess for k_{P21} is obtained from the equation as in Eq. (8.7).

$$\frac{y_{21}(s^*)}{y_{r1}(s^*)} = \frac{k_{p21}g_{p21}g_{c11} + k_{p22}g_{p22}g_{c21}}{1 + k_{p11}g_{p11}g_{c11} + k_{p21}g_{p21}g_{c12} + k_{p22}g_{p22}g_{c22}} \\ \qquad\qquad +k_{p11}g_{p11}k_{p22}g_{p22}[g_{c11}g_{c22} - g_{c12}g_{c21}]} \tag{8.7}$$

where g_{cij} (controller settings) is given by Eq. (8.8).

$$g_{cij} = k_{cij}\left(1 + \frac{1}{\tau_{Iij}s^*} + \tau_{Dij}s^*\right) \tag{8.8}$$

Substituting the value for y_{r1} as $1/s^*$, and the guess values of model parameters and the controller settings in Eq. (8.21), we will get the guess value for k_{p21}. Thus k_{p12} is obtained from Eq. (8.9).

$$\frac{y_{12}(s^*)}{y_{r2}(s^*)} = \frac{k_{p11}g_{p11}g_{c12} + k_{p12}g_{p12}g_{c22}}{1 + k_{p11}g_{p11}g_{c11} + k_{p12}g_{p12}g_{c21} + k_{p22}g_{p22}g_{c22}} \\ \qquad\qquad +k_{p11}g_{p11}k_{p22}g_{p22}[g_{c11}g_{c22} - g_{c12}g_{c21}]} \tag{8.9}$$

By the method proposed in the previous chapter, the minimization problem is formulated to get the model parameters in order to minimize the sum of squared errors between the closed loop model responses and the closed loop process responses for a step change in the corresponding set point (one at a time) as in

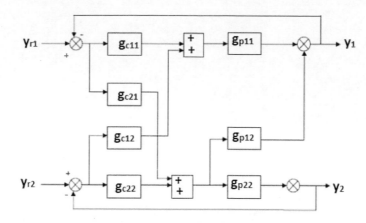

FIGURE 8.2   Modified TITO system with the centralized controllers for Eq. (8.9)

Eq. (8.10).

$$SSQ = \underset{(K_{p_{mij}},\tau_{mij},\theta_{mij})}{\text{minimize}} f = \sum_{i=1}^{2}\sum_{j=1}^{2}\sum_{k=1}^{n}[y_{mij}(t_k) - y_{ij}(t_k)]^2 \Delta t \qquad (8.10)$$

where k is the sampling instant. Here, n is the number of data points in the response of the process and the model. All the computational work is performed in Intel Core i5/3.10 GHz personal computer.

8.2   Simulation Examples

We will consider four simulation examples studied by Rajapandiyan and Chidambaram (2012a, 2012b).

Example 1 Consider the transfer function matrix (Chien et al., 1999; Rajapandiyan and Chidambaram, 2012a, 2012b) as given in Eq. (8.11).

$$G(s) = \begin{bmatrix} \dfrac{22.89e^{-0.2s}}{4.572s+1} & \dfrac{-11.64e^{-0.4s}}{1.807s+1} \\[2ex] \dfrac{4.689e^{-0.2s}}{2.174s+1} & \dfrac{5.8e^{-0.4s}}{1.801s+1} \end{bmatrix} \qquad (8.11)$$

The relative gain λ_{11} is calculated as 0.7087 and hence $\lambda_{12} = 0.2913$; $\lambda_{21} = \lambda_{12}$; $\lambda_{22} = \lambda_{11}$. The centralized PID controller settings used for the identification are

TABLE 8.1   FOPTD model estimation (Example 1)

Model ij		k_{pij}	τ_{ij}	θ_{ij}	SSA	SSQ
11	Guess	6.0827	2	0.2	5.4459×10^{-5}	4.0247×10^{-10}
	Converged	22.8916	4.5727	0.2001		
21	Guess	11.3819	1.875	0.2	3.1951×10^{-5}	3.6365×10^{-10}
	Converged	4.6894	2.174	0.2		
12	Guess	−5.6777	3.125	0.4	2.4183×10^{-4}	5.0492×10^{-9}
	Converged	−11.6378	1.8067	0.4		
22	Guess	11.8624	1.875	0.4	5.3035×10^{-5}	3.2403×10^{-10}
	Converged	5.7998	1.8011	0.4		

Computation time: 67.52 s; number of iterations: 26

(Vijaykumar et al., 2012) as in Eq. (8.12).

$$G_c(s) = \begin{bmatrix} 0.1644 + \dfrac{0.0424}{s} & 0.1403 + \dfrac{0.0383}{s} \\ -0.1922 - \dfrac{0.0538}{s} & 0.0843 + \dfrac{0.0764}{s} \end{bmatrix} \tag{8.12}$$

The process considered is a FOPTD multivariable system. Here, k_p, $k_{cdes} = 1$ is used (Chidambaram, 1998) for calculating the guess values of process gain k_{p11} and k_{p22}. The Matlab optimization routine *lsqnonlin* is used and the sample time of 0.01 s is considered.

The lower and upper bounds for the parameters are selected by the method discussed in the previous chapter. For the gains of the off-diagonal elements, the initial guess of k_{Pij} is fixed from negative infinity to positive infinity. The guess values, final converged parameters, SSQ (between the closed loop model responses and the closed loop process responses for a step change in the corresponding set point) values, SSA (between the closed loop model responses and the closed loop process responses for a step change in the corresponding set point) values, number of iterations, and computational time are listed in Table 8.1. Fig. 8.3 shows a good match between the model and the actual process.

The effect of measurement noise on initial guess values and model parameters is studied by adding random noise (standard deviation 0.01) to the main and interaction responses of the actual process. Sample time of 0.01 s is used. The closed loop response is shown in Fig. 8.4. The initial guess values and the final converged values are as in Eq. (8.13).

$$G(s)(init) = \begin{bmatrix} \dfrac{6.0827e^{-0.2s}}{2s+1} & \dfrac{-5.6619e^{-0.4s}}{3.125s+1} \\ \dfrac{11.3916e^{-0.2s}}{1.875s+1} & \dfrac{11.8624e^{-0.4s}}{1.875s+1} \end{bmatrix} \qquad (8.13)$$

$$G(s)(con) = \begin{bmatrix} \dfrac{22.798e^{-0.204s}}{4.5518s+1} & \dfrac{-11.596e^{-0.4s}}{1.8051s+1} \\ \dfrac{4.691e^{-0.204s}}{2.1622s+1} & \dfrac{5.801e^{-0.403s}}{1.8024s+1} \end{bmatrix} \qquad (8.14)$$

The number of iterations taken is 23 and the computational time is 59.37 s. If we use 0.1 as the initial guess value for k_p, time constant and delay, then the minimization method does not converge to any value. This shows the importance of the present method.

The method is applied to different controller settings as in Eq. (8.15).

$$G_c(s) = \begin{bmatrix} 0.2402 + \dfrac{0.0688}{s} & 0.1073 + \dfrac{0.0327}{s} \\ -0.1792 - \dfrac{0.0556}{s} & 0.0838 + \dfrac{0.0643}{s} \end{bmatrix} \qquad (8.15)$$

The guess values and the converged values of the parameters are given by Eq. (8.16).

$$G(s)(guess) = \begin{bmatrix} \dfrac{4.1632e^{-0.2s}}{2.125s+1} & \dfrac{-4.1976e^{-0.4s}}{4.375s+1} \\ \dfrac{10.3569e^{-0.2s}}{2s+1} & \dfrac{11.933e^{-0.4s}}{1.25s+1} \end{bmatrix} \qquad (8.16)$$

$$G(s)(conv) = \begin{bmatrix} \dfrac{22.89e^{-0.2s}}{4.572s+1} & \dfrac{-11.64e^{-0.4s}}{1.807s+1} \\ \dfrac{4.689e^{-0.2s}}{2.174s+1} & \dfrac{5.8e^{-0.4s}}{1.801s+1} \end{bmatrix} \qquad (8.17)$$

The number of iterations taken for convergence is 11 and the computational time is 30.86 s. Fig. 8.5 shows a good match. The present identification method is not affected by the values of controller settings used. When the guess value for each of the model parameters is considered as 0.1, then the optimization method converges to values which are entirely different from the actual values. The obtained model parameters are $k_{p11} = 7.7907$, $\tau_{11} = 1.8768$, $\theta_{11} = 0.4392$,

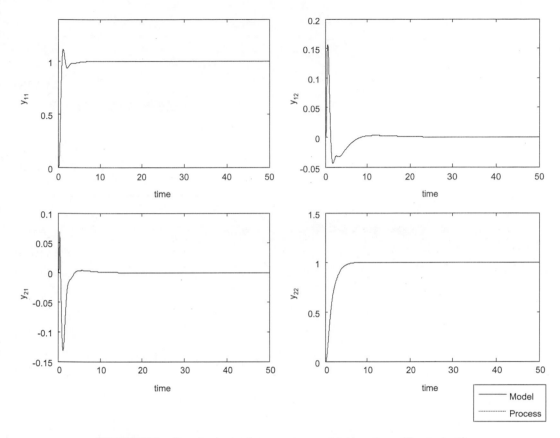

FIGURE 8.3 Comparison of responses and interactions (Example 1)

$k_{p21} = 3.4093$, $\tau_{21} = 1.4266$, $\theta_{21} = 0.0471$, $k_{p12} = -4.3959$, $\tau_{12} = 1.3861$, $\theta_{12} = 0.0666$, $k_{p22} = 5.2673$, $\tau_{22} = 1.6468$, $\theta_{22} = 0.3737$. These parameters do not give any satisfactory response.

Example 2 Consider the system studied by Wood and Berry (1973) and by Rajapandiyan and Chidambaram (2012a, 2012b) as given in Eq. (8.18).

$$G_p(s) = \begin{bmatrix} \dfrac{12.8e^{-s}}{16.7s+1} & \dfrac{-18.9e^{-3s}}{21s+1} \\ \dfrac{6.6e^{-7s}}{10.9s+1} & \dfrac{-19.4e^{-3s}}{14.4s+1} \end{bmatrix} \qquad (8.18)$$

The relative gain λ_{11} is calculated as 2.0094 and hence $\lambda_{12} = -1.0094$; $\lambda_{21} = \lambda_{12}$; $\lambda_{22} = \lambda_{11}$. This system is a highly interactive one. The centralized controller settings used for the identification are (Rajapandiyan and Chidambaram, 2012a)

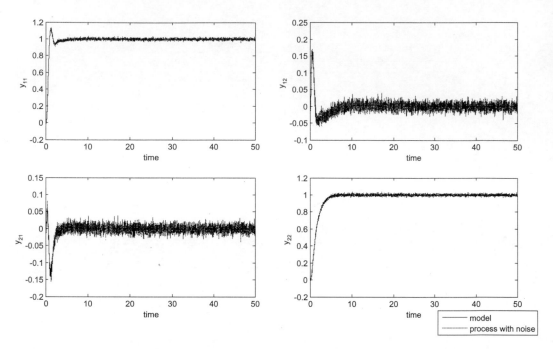

FIGURE 8.4   Comparison of the responses and interactions with noise (Example 1)

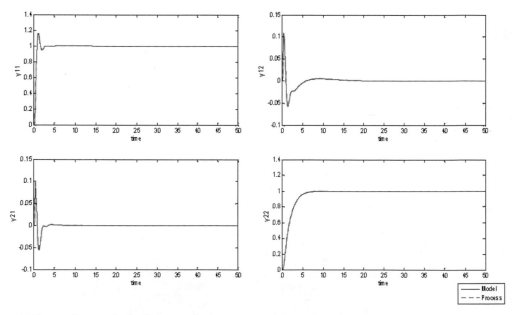

FIGURE 8.5   Comparison of the responses and the interactions (Example 1)
Different PI settings as in Eq. (8.14).

TABLE 8.2 FOPTD model estimation (Example 2)

Model ij		k_{pij}	τ_{ij}	θ_{ij}	SSA	SSQ
11	guess	5.8928	15	1	5.1668×10^{-4}	5.9858×10^{-9}
	converged	12.8	16.7	1		
21	guess	6.06	16.25	7	3.2×10^{-3}	4.3799×10^{-7}
	converged	6.6	10.90	7		
12	guess	−9.6524	22.5	3	5.4170×10^{-5}	6.6976×10^{-11}
	converged	−18.9	21.00	3		
22	guess	−13.8313	11.25	3	3.2212×10^{-4}	4.5529×10^{-9}
	converged	−19.4	14.40	3		

Computation time: 198.43 s; number of iterations: 26

are derived from Eq. (8.19).

$$G_c(s) = \begin{bmatrix} 0.1697 + \dfrac{0.0173}{s} & -0.0172 - \dfrac{0.0140}{s} \\ 0.0161 + \dfrac{0.0048}{s} & -0.0723 - \dfrac{0.0096}{s} \end{bmatrix} \qquad (8.19)$$

Table 8.2 gives the guess values of the parameters. Assume $k_p, k_{cdes} = 1$ for the guess values of process gain k_{p11} and k_{p22}. A sample time of 0.01 s is considered. Fig. 8.6 shows a good match.

The effect of measurement noise on guess values and model parameters is studied. The identified model parameters are found to be closer to that obtained without noise. The same controller settings are used. The closed loop response is shown in Fig. 8.7. The initial guess values and the, final converged values are as in Eq. (8.20).

$$G(s)(init) = \begin{bmatrix} \dfrac{5.8928e^{-1s}}{15s+1} & \dfrac{-7.9297e^{-3s}}{15s+1} \\ \dfrac{5.6611e^{-7s}}{15s+1} & \dfrac{-13.8313e^{-3s}}{12.5s+1} \end{bmatrix}$$

$$G(s)(conv) = \begin{bmatrix} \dfrac{12.7768e^{-s}}{16.685s+1} & \dfrac{-18.8475e^{-3.015s}}{20.963s+1} \\ \dfrac{6.5814e^{-7.012s}}{10.871s+1} & \dfrac{-19.3615e^{-3.007s}}{14.378s+1} \end{bmatrix} \qquad (8.20)$$

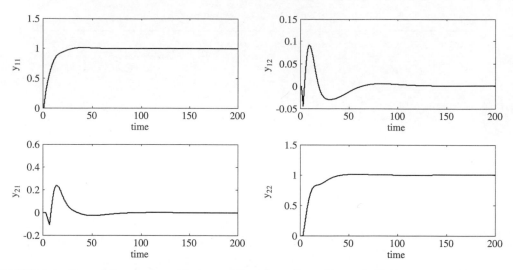

FIGURE 8.6 Comparisons of responses and the interactions (Example 2) Original system and the identified model coincide.

Different controllers settings are considered to get the closed loop responses and interactions.

$$G_c(s) = \begin{bmatrix} 0.0885 + \dfrac{0.0131}{s} & -0.1016 - \dfrac{0.0146}{s} \\ 0.0284 + \dfrac{0.0045}{s} & -0.1039 - \dfrac{0.0099}{s} \end{bmatrix} \tag{8.21}$$

The guess values and the final converged parameters are given by Eq. (8.22).

$$G(s)(init) = \begin{bmatrix} \dfrac{11.299e^{-1s}}{18.75s + 1} & \dfrac{-17.403e^{-3s}}{25s + 1} \\ \dfrac{3.6325e^{-7s}}{19.375s + 1} & \dfrac{-9.6246e^{-3s}}{16.875s + 1} \end{bmatrix}$$

$$G(s)(conv) = \begin{bmatrix} \dfrac{12.8e^{-1s}}{16.7s + 1} & \dfrac{-18.9e^{-3s}}{21s + 1} \\ \dfrac{6.6e^{-7s}}{10.9s + 1} & \dfrac{-19.4e^{-3s}}{14.4s + 1} \end{bmatrix} \tag{8.22}$$

The number of iterations taken for convergence is 21 and the computational time is 203.85 s. Fig. 8.8 shows a good match. The present identification method is not affected by the values of controller settings used. When the guess values for each of the parameters are considered as 0.1, then the optimization method does not converge to any value.

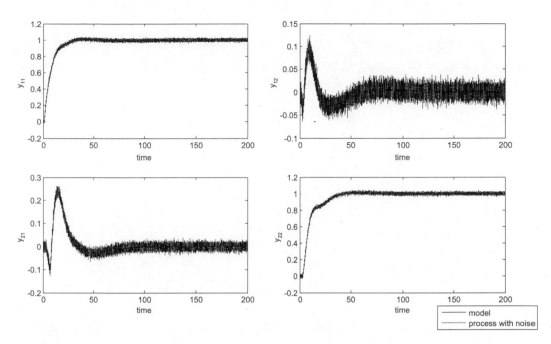

FIGURE 8.7 Comparison of the responses and the interactions for effect of noise (Example 2)

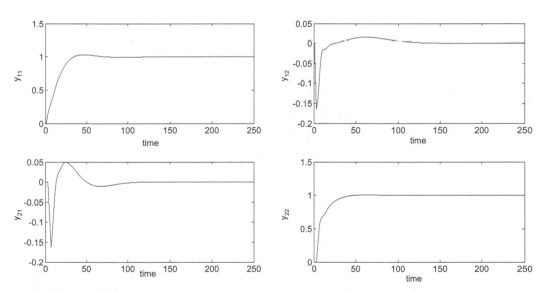

FIGURE 8.8 Comparison of the responses and the interactions for different settings of the controllers Eq. (8.19)
Both the curves coincide.

Example 3 Consider the multivariable model considered by Rajapandiyan and Chidambaram (2012a, 2012b) and by Niederlinski (1971) given in Eq. (8.23).

$$G(s) = \begin{bmatrix} \dfrac{0.5}{(0.1s+1)^2\,(0.2s+1)^2} & \dfrac{-1.0}{(0.1s+1)\,(0.2s+1)^2} \\ \dfrac{1.0}{(0.1s+1)\,(0.2s+1)^2} & \dfrac{2.4}{(0.1s+1)\,(0.2s+1)^2\,(0.5s+1)} \end{bmatrix} \qquad (8.23)$$

The centralized PID controller settings used for the identification are as in Eq. (8.24).

$$G_c(s) = \begin{bmatrix} 0.2182 + \dfrac{1.6364}{s} & 0.0909 + \dfrac{0.6818}{s} \\ -0.0909 - \dfrac{0.6818}{s} & 0.0455 + \dfrac{0.3409}{s} \end{bmatrix} \qquad (8.24)$$

The process will be identified as a FOPTD multivariable system. Fig. 8.9 is considered for getting the guess values for the model parameters. For calculating the guess values of process gain k_{p11} and $k_{p22}, k_p k_{cdes} = 0.5$ is used (since the response shows a higher order model behaviour, i.e., initial slope 0). A sample time of 0.01 s is considered.

The guess values, final converged parameters, number of iterations, computational time, SSQ and SSA values are given in Table 8.3. Fig. 8.9 shows a good match. The computational time for optimization is 19.2301 s (maximum computational time). The results obtained are given in Table 8.3. When the guess value for gains, time constants and delays are given as 0.1, the optimization method does not converge to any value.

Example 4 Consider the multivariable SOPTD transfer function model proposed by Garcia et al. (2005) and Rajapandiyan and Chidambaram (2012a, 2012b) given in Eq. (8.25).

$$G_p(s) = \begin{bmatrix} \dfrac{e^{-0.1s}}{s^2+2s+1} & \dfrac{0.3e^{-0.01s}}{2s^2+3s+1} & \dfrac{0.3e^{-0.01s}}{2s^2+3s+1} \\ \dfrac{0.3e^{-0.01s}}{2s^2+3s+1} & \dfrac{e^{-0.1s}}{s^2+2s+1} & \dfrac{0.3e^{-0.01s}}{2s^2+3s+1} \\ \dfrac{0.3e^{-0.01s}}{2s^2+3s+1} & \dfrac{0.3e^{-0.01s}}{2s^2+3s+1} & \dfrac{e^{-0.1s}}{s^2+2s+1} \end{bmatrix} \qquad (8.25)$$

TABLE 8.3 FOPTD model estimation (Example 3)

Model ij		k_{pij}	τ_{ij}	θ_{ij}	SSA	SSQ
11	guess	2.2917	0.5	0.1	0.0267	3.1559×10^{-4}
	converged	0.4985	0.3415	0.2821		
21	guess	7.1068	0.875	0.1	0.0446	5.7822×10^{-4}
	converged	0.9813	0.2813	0.2394		
12	guess	−6.4370	0.875	0.1	0.0057	8.5295×10^{-6}
	converged	−1.0077	0.3206	0.2063		
22	guess	10.9987	0.75	0.1	0.0361	6.0381×10^{-4}
	converged	2.4018	0.6424	0.4		

Computation time: 19.23 s; number of iterations: 16

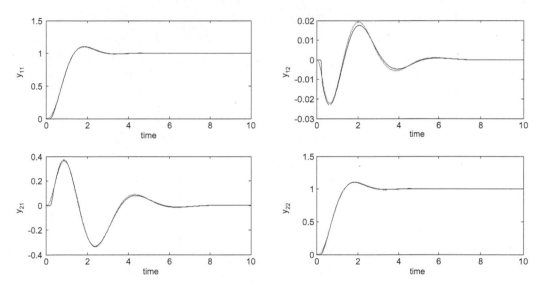

FIGURE 8.9 Comparison of responses and interactions (Example 3)
Solid line: original system; dashed line: identified model. Both the responses are very close to each other.

The centralized PID controller settings used for the identification are given in Eq. (8.26).

$$G(s) = \begin{bmatrix} 0.2321 + \dfrac{1.1607}{s} & -0.0536 - \dfrac{0.2679}{s} & -0.0536 - \dfrac{0.2679}{s} \\[3mm] -0.0536 - \dfrac{0.2679}{s} & 0.2321 + \dfrac{1.1607}{s} & -0.0536 - \dfrac{0.2679}{s} \\[3mm] -0.0536 - \dfrac{0.2679}{s} & -0.0536 - \dfrac{0.2679}{s} & 0.2321 + \dfrac{1.1607}{s} \end{bmatrix} \qquad (8.26)$$

The process is to be identified as a FOPTD multivariable system. Similar to Fig. 8.2, (2×2 sub-system) is considered for getting the guess values for the model parameters as in Eq. (8.27).

$$\frac{y_{ij}(s^*)}{y_{rj}(s^*)} = \frac{k_{p_{ii}}g_{p_{ii}}g_{c_{ij}} + k_{p_{ij}}g_{p_{ij}}g_{c_{jj}}}{1 + k_{p_{ii}}g_{p_{ii}}g_{c_{ii}} + k_{p_{ij}}g_{p_{ij}}g_{c_{ji}} + k_{p_{jj}}g_{p_{jj}}g_{c_{jj}} + k_{p_{ii}}g_{p_{ii}}k_{p_{jj}}g_{p_{jj}}[g_{c_{ii}}g_{c_{jj}} - g_{c_{ij}}g_{c_{ji}}]}$$

(8.27)

In Eq. (8.25), we have to consider the set of indices ($i = 1, j = 3; i = 2, j = 3; i = 3, j = 1; i = 3, j = 2$) for the initial values of $k_{p13}, k_{p23}, k_{p,31}$, and $k_{p,32}$. For calculating the guess values of process gains k_{p11}, k_{p22}, and $k_{p33}, k_p k_{cdes} = 0.5$ is used (since the response shows a higher order model behaviour, i.e., initial slope 0). A sample time of 0.01 s is considered.

The guess values, final converged parameters, SSQ, and SSA values are listed in Table 8.4. The converged model parameters match satisfactorily well with the actual FOPTD system. The closed loop main responses and interaction responses using the identified model parameters with the same controller settings are also shown in Fig. 8.10. We observe that Fig. 8.10 shows a good match. The computational time for optimization is 19.2301 s (maximum computational time). The results obtained are given in Table 8.4. When the guess value for gains, time constants and delays are given as 0.1, the optimization method converged to values which do not show any acceptable match with the actual parameters.

8.3   Another Method of Getting Guess Values

In the method proposed in Section 8.2 the initial guess values of k_{p12} and k_{p21} was obtained by using the Laplace transforms of interaction response $y_{21}(s^*)$ and $y_{12}(s^*)$ given by the expression in Eq. (8.28a) and Eq. (8.28b).

$$y_{21}(s^*) = \int_0^\infty y_{21}(t)e^{-ts^*} dt \qquad (8.28a)$$

$$y_{12}(s^*) = \int_0^\infty y_{12}(t)e^{-ts^*} dt \qquad (8.28b)$$

For initial guess of k_{p21} is obtained from Eq. (8.29a).

$$y_{21}/y_{r1} = (k_{p21}g_{p21}g_{c11} + k_{p22}g_{p22}g_{c21})/D_1 \qquad (8.29a)$$

The value of k_{p12} is obtained from Eq. (8.29b).

$$y_{12}/y_{r2} = (k_{p11}g_{p11}g_{c12} + k_{p12}g_{p12}g_{c22})/D_1 \qquad (8.29b)$$

Where,

$$D_1 = [1 + k_{p11}g_{p11}g_{c11} + k_{p12}g_{p12}g_{c21} + k_{p22}g_{p22}g_{c22} + k_{p11}g_{p11}k_{p22}g_{p22}(g_{c11}g_{c22} - g_{c12}g_{c21})]$$
$$(8.30)$$

Here s^* is equal to $8/t_s$. This process of obtaining the initial guess values for K_p is tedious as it requires the calculation of the area under the closed loop response curve and the Laplace transform. Davison (1976) suggested the method of tuning multivariable PI controllers for systems under stability.

$$\text{The } K_C \text{ matrix is given by } K_c = \delta[G_p(s = 0)]^{-1} \qquad (8.31)$$

Where,

$G_p(s = 0)$ is the process gain K_P.

The initial guess for process gain can be obtained from the matrix $K_p = \delta K_c^{-1}$

For stable systems, Davison (1976) reported that the values of δ is in the range of 0.1 to 0.4.

This method of obtaining the initial guess value of process gain from the inverse of the proportional controller gain matrix is used in the present chapter. The guess value for process time constant (τ) is taken as $t_s/8$ and for time delay as

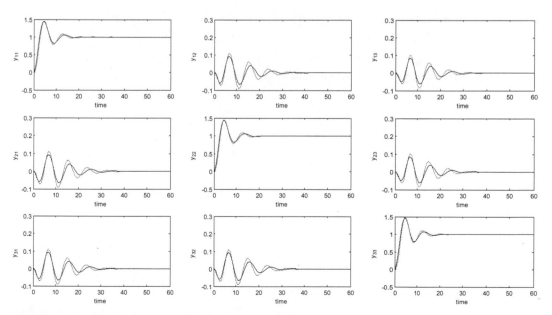

FIGURE 8.10 Comparison of responses and interactions (Example 4)
Solid line: original system; dashed line: identified model

TABLE 8.4   FOPTD model estimation (Example 4)

Model ij		k_{pij}	τ_{ij}	θ_{ij}	SSA	SSQ
11	Guess	2.1539	2.5	0.1	0.9094	0.0733
	Converged	1.6914	3.4098	0.1		
21	Guess	0.3705	4.375	0.01	0.5239	0.0118
	Converged	0.4225	4.375	0.01		
31	Guess	0.3777	4.375	0.01	0.4491	0.0082
	Converged	0.7109	7.6243	0.01		
12	Guess	0.3705	4.375	0.01	0.4573	0.0082
	Converged	0.6013	6.5750	0.01		
22	Guess	2.1539	2.5	0.1	0.9138	0.0738
	Converged	1.6501	3.3041	0.1		
32	Guess	0.3777	4.375	0.01	0.4495	0.0082
	Converged	0.6861	7.3359	0.01		
13	Guess	0.4436	4.375	0.01	0.4245	0.0074
	Converged	0.6313	6.8016	0.01		
23	Guess	0.4436	4.375	0.01	0.4345	0.0078
	Converged	0.5299	5.6273	0.01		
33	Guess	2.1539	2.5	0.1	1.0234	0.0836
	Converged	1.7803	3.5885	0.1		

Computation time: 147.45 s; number of iterations: 8

the time delay obtained for closed loop responses as given by Rajapandiyan and Chidambaram (2012a). The upper bound for K_p is 50 times of the guess value and the lower bound is 1/50 times of the guess value. For time constant, upper bound is 5 times of the guess value and lower bound is 1/20 times of the guess value. Time delay upper bound is 4 times of the guess value and the lower bound is 1/4 times of the guess value. As discussed in the previous chapter, the optimization problem is formulated to obtain the model parameter. Using least square method in Matlab (*lsqnonlin*) with trust region reflective algorithm and $TolX = 1xe^{-3}$.

Example 5 Let us consider the system matrix by Wood and Berry (1973) as given in Eq. (8.32).

$$G_p = \begin{bmatrix} \dfrac{12.8e^{-s}}{16.7s + 1} & \dfrac{-18.9e^{-3s}}{21s + 1} \\ \dfrac{6.6e^{-7s}}{10.9s + 1} & \dfrac{-19.4e^{-3s}}{14.4s + 1} \end{bmatrix} \qquad (8.32)$$

TABLE 8.5 Estimation values (Example 5)

Model ij		K_{pij}	τ_{ij}	θ_{ij}	IAE	ISE
11	Guess	1.8086	15	1	0.0011	$3.6421e^{-8}$
	Converged	12.8	16.7	1		
21	Guess	0.4028	16.5	7	0.0036	$2.3355e^{-7}$
	Converged	6.6	10.9	7		
12	Guess	−0.4303	22.5	3	0.0038	$4.1263e^{-7}$
	Converged	−18.9	21	3		
22	Guess	−4.2452	11.25	3	0.0015	$3.4986e^{-8}$
	Converged	-19.4	14.4	3		

Computation time: 1 235 s; number of iterations: 26

Using PI controller settings given by Vijaykumar et al. (2012), we derive Eq. (8.33).

$$G_c = \begin{bmatrix} 0.1697 + \dfrac{0.0173}{s} & -0.0172 - \dfrac{0.014}{s} \\ 0.0161 + \dfrac{0.0048}{s} & -0.0723 - \dfrac{0.0096}{s} \end{bmatrix} \tag{8.33}$$

Using Davison method (Section 3.13) with $\delta = 0.3$; $K_p = 0.3 * K_c^{-1}$

$$\text{Thus} \quad K_p = \begin{bmatrix} 1.8086 & -0.4303 \\ 0.4028 & -4.2452 \end{bmatrix} \tag{8.34}$$

Table 8.5 shows the guess values, final converged parameters, number of iterations, computational time, and IAE and ISE values. Fig. 8.11 shows an excellent match between the process and the model response. The results obtained are given in Table 8.5.

When $\delta = 0.7$ to 0.8 is converging, $K_p = K_c^{-1} * \delta$.
Here

$$K_c = \begin{bmatrix} 0.1697 & -0.0172 \\ 0.0161 & -0.0723 \end{bmatrix}, \quad K_p = \begin{bmatrix} Kp11 & Kp12 \\ Kp21 & Kp22 \end{bmatrix} \tag{8.35}$$

$$K_p = 0.8 * K_c^{-1}$$

$$K_p = \begin{bmatrix} 4.8224 & -1.14736 \\ 1.074 & -11.3205 \end{bmatrix} \tag{8.36}$$

The Davison method is simple to calculate as compared to other methods. Also, we got a good response which completely matches with actual process response. Table

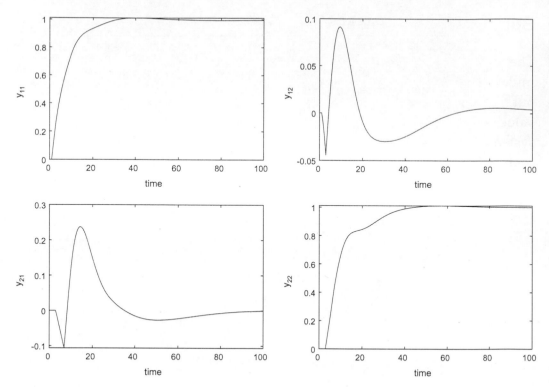

FIGURE 8.11   Comparison between process and model (Example 5)
Solid line: model; dotted line: process. Both the curves are the same.

TABLE 8.6   Estimation values for Davison method (Example 5)

Model ij		K_{pij}	τ_{ij}	θ_{ij}	IAE	ISE
11	Guess	4.8224	15	1	$5.0126e^{-04}$	$5.9901e^{-09}$
	Converged	12.8	16.7	1		
21	Guess	1.074	16.5	7	0.0031	$4.3769e^{-07}$
	Converged	6.6	10.9	7		
12	Guess	−1.14736	22.5	3	$5.281e^{-05}$	$6.7007e^{-11}$
	Converged	−18.9	21	3		
22	Guess	−11.3205	11.25	3	$3.2050e^{-04}$	$4.5492e^{-09}$
	Converged	−19.4	14.4	3		

Computation time: 1,622.43 s; number of iterations: 30

8.6 has all the guess values and converged values of K_p, τ, and θ, IAE, ISE, number of iterations, and time related to Example 5. The method takes a longer time for identification than the previous method.

Summary

The closed loop identification of model parameters of FOPTD models of a multivariable system under the control of centralized PI control system was carried out by an optimization method. A method was given for getting the initial guess values of the model parameters. The closed loop performances of the original system were compared with the closed loop of the identified model with the same centralized controllers.

Problems

1. Consider the Niederlinski (1971) multivariable transfer function model discussed by Viswanathan et al. (2001):

$$G_p(s) = \begin{bmatrix} \dfrac{0.5}{(0.1s+1)^2(0.2s+1)^2} & \dfrac{-1.0}{(0.1s+1)(0.2s+1)^2} \\ \dfrac{1.0}{(0.1s+1)(0.2s+1)^2} & \dfrac{2.4}{(0.1s+1)(0.2s+1)^2(0.5s+1)} \end{bmatrix}$$

Design suitable centralized PI controllers. Carry out the closed loop optimization method for identifying FOPTD models.

2. Consider the 2×2 (SOPTD) distillation column transfer function model matrix (Weischedel and McAvoy, 1980):

$$G(s) = \begin{bmatrix} \dfrac{0.7e^{-s}}{12.35s^2+8.26s+1} & \dfrac{-0.45e^{-s}}{17.04s^2+8.69s+1} \\ \dfrac{0.35e^{-1.28s}}{3.35s^2+5.67s+1} & \dfrac{-0.5e^{-s}}{1.69s^2+5.06s+1} \end{bmatrix}$$

Design suitable centralized PID controllers settings. Carry out the closed loop closed loop optimization method for identifying FOPTD models.

3. For the system:

$$G(s) = \begin{bmatrix} \dfrac{0.58e^{-s}}{(12.35s^2+8.26s+1)} & \dfrac{-0.45e^{-s}}{(17.04s^2+8.68s+1)} \\ \dfrac{0.35e^{-1.28s}}{(3.35s^2+5.67s+1)} & \dfrac{-0.48e^{-s}}{(1.69s^2+5.06s+1)} \end{bmatrix}$$

Find a suitable input–output pairing. Design approximate centralized PI controllers. Perform the closed loop optimization method to identify the model parameters of FOPTD models.

4. For the system:

$$G(s) = \begin{bmatrix} \dfrac{0.562e^{-s}}{(7.74s+1)\,(7.74s+1)} & \dfrac{-0.516e^{-1.5s}}{(7.1s+1)\,(7.1s+1)} \\[3mm] \dfrac{0.33e^{-1.5s}}{(15.8s+1)\,(0.5s+1)} & \dfrac{-0.394e^{-s}}{(13.8s+1)\,(0.4s+1)} \end{bmatrix}$$

Find suitable input–output pairing. Design approximate centralized PI controllers. Perform the closed loop optimization method to identify the model parameters of FOPTD models.

5. For the system:

$$G(s) = \begin{bmatrix} \dfrac{0.471e^{-s}}{(30.7s+1)\,(30.7s+1)} & \dfrac{0.495e^{-2s}}{(28.5s+1)\,(28.5s+1)} \\[3mm] \dfrac{0.749e^{-1.7s}}{(57s+1)\,(57s+1)} & \dfrac{-0.832e^{-s}}{(50.5s+1)\,(50.5s+1)} \end{bmatrix}$$

Find suitable input–output pairing. Design approximate centralized PI controllers. Perform the closed loop optimization method to identify the model parameters of a FOPTD model.

9

Identification of Multivariable SOPTD Models by Optimization Method

In this chapter, the identification method for a First Order Plus Time Delay (FOPTD) model, discussed in Chapter 7, is extended to identify a Second Order Plus Time Delay (SOPTD) model of multivariable systems. This chapter discusses the need for step-up and step-down excitation in the set points to obtain the parameter convergence with significantly reduced computational time.

9.1 Introduction

It should be observed that the time domain identification method is easier to formulate and it is conceptually simple. Viswanathan et al. (2001) presented a method for the closed loop identification of FOPTD and SOPTD models by using genetic algorithm (GA) for TITO systems. The computational time for the identification of a SOPTD TITO system using the GA (Viswanathan et. al, 2001) ranges from 1 hour 55 minutes to 5 hours 28 minutes (for different controller settings and test timings). Further, the converged values from GA are not reliable. The converged values are refined by using a local optimizer (Gupta, 1995) (Broyden–Fletcher–Goldfarb–Shannon method). For the SOPTD model identification, Viswanathan et al. (2001) reported the success rate for 10 trials ranging from 25% to 100% to get the global minimum. There is no guarantee to get the exact transfer function model (global minimum) in a single trial. If we obtain the least IAE values also, some deviations are found among the time constant and the damping coefficient (similar to what we get in any model reduction technique). They provide success rate, in which how many times, they obtained the global minima (exact model parameters) or local accepted minima are not clearly specified. It is to be inferred that the attainment of global minimum (in SOPTD model identification test) is difficult from the step response. The computational time given above is for one trial. The computational time is higher (about 20 to 50 hours) considering together all the 10 trials. Viswanathan et al. (2001) have presented the upper and lower bounds for the parameters in the GA arbitrarily for each of the simulation examples.

For the limit values of process gains, Viswanathan et al. (2001) used a method reported by Papastathopoulo and Luyben (1990). It requires the process input data also for obtaining the guess values of steady state gains. Hence, there is a need to propose a method for obtaining reasonable initial guess values for the model parameters. In Chapter 7, a method has been proposed for getting the initial guess values of FOPTD model parameters using only the closed loop step responses of multivariable systems. The computational time for the time domain identification is shown to be reduced significantly (from 89 minutes (for GA) to 4 minutes 17 seconds for TITO systems). The method is extended to identify the SOPTD models and brings the need for step-up and step-down excitation in the set points to obtain the parameter convergence with significantly reduced computational time. The number of parameters to be estimated increases (from 12 to 16 parameters for TITO systems) as the model order is changed from FOPTD to SOPTD. This poses problems in the convergences of the optimization method.

The present method differs from the conventional step test. The identification method not only considers the dynamics of the step (up) change but also considers the dynamics by removing the step change after the system reaches steady state conditions (combined step-up and step-down). For any optimization method, the selection of the initial guess values plays an important role in computational time and convergence. In the present method, a simple and generalized method is proposed for obtaining the reasonable initial guess values for the parameters of SOPTD transfer function models. A method to obtain the upper and lower bounds to be used in the optimization routine is also presented. This method is applied to 2×2 SOPTD, higher order and 3×3 SOPTD transfer function matrix models of multivariable systems.

9.2 Identification Method

Consider a TITO system. $G(s)$ and $G_C(s)$ are process transfer function matrix and decentralized controller matrix with compatible dimensions, respectively, expressed as in Eq. (9.1) and Eq. (9.2).

$$G(s) = \begin{bmatrix} g_{11} & g_{12} \\ g_{21} & g_{22} \end{bmatrix} \tag{9.1}$$

$$G_c(s) = \begin{bmatrix} g_{c11} & 0 \\ 0 & g_{c22} \end{bmatrix} \tag{9.2}$$

The controller parameters can be chosen arbitrarily for multivariable systems such that the closed loop response of the system is stable, with reasonable responses. Consider a decentralized TITO multivariable system as shown in Fig. 3.2 (Chapter 3). The process transfer function models are identified as SOPTD models given in Eq. (9.3).

$$g_{m,ij}(s) = \frac{k_{pm,ij}e^{-\theta_{m,ij}s}}{(\tau_{m,ij}^2 s^2 + 2\varepsilon_{m,ij}\tau_{m,ij}s + 1)} \quad i = 1,2; \; j = 1,2 \tag{9.3}$$

Initially, a unit magnitude of step change is introduced in the set point y_{r1}. After the step response reaches its new steady state, the step excitation is removed from the set point y_{r1} while all the remaining set points are unchanged and all loops are kept under closed loop operation. From the prescribed manner of excitation in set point y_{r1} we obtain the main response y_{11} and the interaction response y_{21}. Similarly, the same magnitude of combined step-up and step-down excitation is introduced in the set point y_{r2} and we obtain the main response y_{22} and interaction response y_{12}. The problem associated with using only the step change excitation is discussed later in this chapter in the simulation example section.

The response matrix of the TITO system can be expressed as in Eq. (9.4).

$$Y(t) = \begin{bmatrix} y_{11} & y_{12} \\ y_{21} & y_{22} \end{bmatrix} \tag{9.4}$$

The first column in the response matrix in Eq. (9.4) contains the responses (main and interaction) obtained for the combined step-up and step-down excitations in set point y_{r1} in the first loop. The second column contains the responses obtained for the step-up and step-down excitations in set point y_{r2} in the second loop. From these step responses, the initial guess values of the model parameters are obtained by the method given in the next section.

As stated earlier, the selection of initial guess values plays a vital role in any optimization method. The combined dynamics of step-up and step-down responses are used in the optimization routine for the identification of the transfer function model matrix. The initial guess values for the time delay can be taken as the corresponding value of the closed loop time delay. The initial guess values for the time constant are obtained from the settling time (t_s) of main and interaction responses of the step-up response only. From the settling time (to reach 98% of the final steady state value and remain within the limit), the closed loop time constant (τ_c) can be assumed as $(t_s/4)$. The time constant of the open loop system is assumed to be lesser than that of the closed loop time constant. In principle, for SISO systems, the closed loop response is faster than the open loop response. In MIMO systems, due to the interaction among the loops, usually the closed loop response is assumed to be slower than the open loop response. This assumption is made only to obtain the initial guess values of the time constant of the open loop system for the optimization problem. Hence, the initial guess value of the time constant for the model is to be assumed as $(t_s/8)$. The initial guess value for the damping coefficient is taken as 1.

The initial guess values of the main loop process gains $(k_{p11}$ and $k_{p22})$ are obtained by the following method. For lower order model systems, the value (Chidambaram, 1998) of $k_p k_{c,max}$ is taken as 4 and the design value for $k_p k_{c,des}$ as 2. Hence, the guess value for k_p is taken as $2/k_c$. Similarly, for the higher order model systems, the value of $k_p k_{c,max}$ is taken as 1 and the design value for $k_p k_{c,des}$ is taken as 0.5. Hence, the guess value for k_{pi} is taken as $0.5/k_{ci}$. The higher order model and the lower order model are distinguished by checking the initial dynamics of the main responses of the actual process. If the initial slope of the main response is sluggish (0 slope), then we consider the system as a higher order model. Similarly, if the initial slope of the main response has a steep response (non-zero slope), then we consider the system as a lower order model.

The initial guess value of the cross loop process gains $(k_{p21}$ and $k_{p12})$ are obtained by the following method. The Laplace transforms of interaction responses $y_{21}(s^*)$ and $y_{12}(s^*)$ can be expressed as in Eq. (9.5) and Eq. (9.6).

$$y_{21}(s^*) = \int_0^\infty y_{21}(t)e^{-ts^*}\, dt \qquad (9.5)$$

$$y_{12}(s^*) = \int_0^\infty y_{12}(t)e^{-ts^*}\, dt \qquad (9.6)$$

To evaluate Eq. (9.5) and Eq. (9.6), we need to assume the value of s^*. In the present case, the value of s^* is considered as $8/t_s$. Here t_s is the settling time of the closed loop response. For $t \geq t_s$, the value of e^{-8} is close to 0. Then the integral value does not change further. So, to reduce the computation time of the integral, the value of s^* is taken as $8/t_s$. The upper bound for t is t_s. The numerical values for $y_{21}(s^*)$ and $y_{12}(s^*)$ is obtained by substituting the value of s^* in Eq. (9.5) and Eq. (9.6).

The closed loop transfer functions for the cross-loops are assumed by considering the interaction from the main loop taken as a disturbance to the interacted loop. Some signals are ignored for simplicity, as shown in Fig. 7.1, in order to avoid g_{12} (Fig. 7.1a) in deriving Eq. (9.7) and to avoid g_{21} (Fig. 7.1b) in deriving Eq. (9.9). Thus, we obtain Eq. (9.7).

$$\frac{y_{21}(s^*)}{y_{r1}(s^*)} = \frac{g_{c11}(s^*)g_{m21}(s^*)}{([1+g_{c11}(s^*)g_{m11}(s^*)])([1+g_{c22}(s^*)g_{m22}(s^*)])} \tag{9.7}$$

Substitute the value for $y_{r1}(s)$ as $1/s^*$ and also substitute the controller settings and the guess values of model parameters into Eq. (9.7) to obtain Eq. (9.8).

$$y_{21}(s^*)$$
$$= \frac{\left(\frac{1}{s^*}\right)k_{c11}\left(1+\left(\frac{1}{\tau_{I11}s^*}\right)+\tau_{d11}s^*\right)\frac{k_{pm21}e^{-\theta_{m21}s^*}}{(\tau_{m21}^2 s^{*2}+2\varepsilon_{m21}\tau_{m21}s^*+1)}}{\left[\left(1+k_{c11}\left(1+\left(\frac{1}{\tau_{I11}s^*}\right)+\tau_{d11}s^*\right)\frac{k_{mp11}e^{-\theta_{m11}s^*}}{(\tau_{m11}^2 s^{*2}+2\varepsilon_{m11}\tau_{m11}s^*+1)}\right)\left\{\left(1+k_{c22}\left(1+\left(\frac{1}{\tau_{I22}s^*}\right)+\tau_{d22}s^*\right)\frac{k_{mp22}e^{-\theta_{m22}s^*}}{(\tau_{m22}^2 s^{*2}+2\varepsilon_{m22}\tau_{m22}s^*+1)}\right)\right\}\right]} \tag{9.8}$$

Similarly, the initial guess values for k_{p12} is obtained from Eq. (9.9).

$$\frac{y_{12}(s^*)}{y_{r2}(s^*)} = \frac{g_{c22}(s^*)g_{m12}(s^*)}{(1+g_{c22}(s^*)g_{m22}(s^*))(1+g_{c11}(s^*)g_{m11}(s^*))} \tag{9.9}$$

Substitute the value for $y_{r2}(s)$ as $1/s^*$ and also substitute the controller settings and the guess values of model parameters (time delays, time constants, damping constants and main-loop process gains) in Eq. (9.9). Thus we obtain Eq. (9.10).

$$y_{12}(s^*)$$
$$= \frac{\left(\frac{1}{s^*}\right)k_{c22}\left(1+\left(\frac{1}{\tau_{I22}s^*}\right)+\tau_{d22}s^*\right)\frac{k_{mp12}e^{-\theta_{m12}s^*}}{(\tau_{m12}^2 s^{*2}+2\varepsilon_{m12}\tau_{m12}s^*+1)}}{\left[\left(1+k_{c22}\left(1+\left(\frac{1}{\tau_{I22}s^*}\right)+\tau_{d22}s^*\right)\frac{k_{mp22}e^{-\theta_{m22}s^*}}{(\tau_{m22}^2 s^{*2}+2\varepsilon_{m22}\tau_{m22}s^*+1)}\right)\left\{\left(1+k_{c11}\left(1+\left(\frac{1}{\tau_{I11}s^*}\right)+\tau_{d11}s^*\right)\frac{k_{mp11}e^{-\theta_{m11}s^*}}{(\tau_{m11}^2 s^{*2}+2\varepsilon_{m11}\tau_{m11}s^*+1)}\right)\right\}\right]} \tag{9.10}$$

The guess value for k_{mp21} and k_{mp12} are to be obtained by substituting the other known guess values and controller parameters in Eq. (9.8) and Eq. (9.10). After obtaining all the guess values of the model, the same manner of step-up and step-down excitation is introduced to the set point y_{mr1} of the model. All remaining set points are unchanged and all other loops are kept under closed loop condition. From the prescribed excitation in set point y_{mr1} the main response y_{m11} and interaction response y_{m21} of the model are obtained. Similarly, the combined step-up and step-down excitation is introduced in the set point y_{mr2}, and the main response y_{m22} and interaction response y_{m12} of the model are obtained. These four (main and interaction) responses are compared with the actual process responses. To obtain the model parameters, the optimization problem is formulated so as to select the model parameters in order to minimize the sum of squared errors between the model and the actual process responses (obtained from combined step-up and step-down test).

$$SSQ = \text{minimize} f = \sum_{i=1}^{2} \sum_{j=1}^{2} \sum_{k=1}^{n} \left[y_{mij}\left(t_k\right) - y_{ij}\left(t_k\right) \right]^2 \Delta t \quad i = 1, 2; \ j = 1, 2$$

(9.11)

TITO systems have two main and two interaction responses. In the objective function, n is the number of data points in the response of the process and the model. First get the error between responses of each transfer function model with the associated process response for various times (to form an error vector for each model). Next, take the square of each of the errors and multiply by a fixed sampling time (Δt). The sum of the elements of each vector gives an ISE value. There are four ISE values. The sum of these four ISE values is considered as a scalar objective function as in Eq. (9.11). The model parameters are selected by minimizing this scalar function. In Eq. (9.11), the sampling time Δt need not be included since it is a constant value. In this example, the optimization problems are solved by using the routine *lsqnonlin* in Matlab and the routine implements the Trust–Region–Reflective algorithm. Although the optimization problem is formulated and solved using the standard routine in Matlab, the convergence and computational time depends on initial guesses. The main focus of this example is to propose the above simple method to get the initial guess values of the model parameters.

In the optimization routine, the values of parameters are considered as TolX $= 1e^{-3}$, TolFun $= 1e^{-6}$ and maximum number of iterations is 400. The closed loop multivariable system is simulated using Simulink. Fourth-order Runge–Kutta method (Gupta, 1995) with a fixed step size is used for solving the governing initial value ordinary differential equations. All of the computational work for the identification of the multivariable systems was performed using Core 2 Quad (Intel/3.00 GHz) personal computer. To evaluate the proposed method, a known transfer function model matrix is considered with suitable decentralized PI/PID controllers to get a reasonable and stable closed loop response. The closed loop main and interaction responses of the actual process are obtained by the earlier discussed method. From the closed loop responses, the model parameters are obtained by the proposed method. The closed loop main and interaction responses of the model with the decentralized PI/PID controllers are compared with the actual system with the same decentralized PI/PID controller settings.

9.3 Simulation Examples

In this section, three simulation examples are considered to show the effectiveness of the proposed method.

Example 1 Consider the 2×2 (SOPTD) distillation column transfer function model matrix (Weischedel and McAvoy, 1980) as given in Eq. (9.12).

$$
G(s) = \begin{bmatrix} \dfrac{0.58e^{-s}}{12.35s^2 + 8.26s + 1} & \dfrac{-0.45e^{-s}}{17.04s^2 + 8.69s + 1} \\ \dfrac{0.35e^{-1.28s}}{3.35s^2 + 5.67s + 1} & \dfrac{-0.48e^{-s}}{1.69s^2 + 5.06s + 1} \end{bmatrix} \tag{9.12}
$$

The decentralized PID controller settings used for the MIMO identification (Viswanathan et al., 2001) were $k_{c11} = 2$, $\tau_{i11} = 2$, $\tau_{d11} = 0.5$; $k_{c22} = -3$, $\tau_{i22} = 2$, $\tau_{d22} = 0.5$.

For all the simulation examples, in the optimization routine, the lower bound value for time constant is fixed as $1/20$ times of the guess value and upper bound value as five times the initial guess value. Similarly, for time delays, the lower bound is specified as $1/4$ times of the guess value and the upper bound is specified as 4 times the initial guess value. For the damping coefficient, the lower bound is taken as 0 and the upper bound is taken as 5. In case the initial guess value of process gain is calculated as negative, the upper bound is fixed as $1/20$ times of the guess value and the lower bound as 20 times the guess value. Similarly, for positive guess values of the process gain, the lower bound is fixed as $1/20$ times of the guess value and the upper bound as 20 times the guess value. These parameter settings in the optimization routine are found to ensure the convergence and also to reduce the computational time.

Since the response showed a near 0 initial slope, $k_p k_c = 0.5$ was used for calculating the guess values of process gains k_{p11} and k_{p22}. SOPTD model is assumed and the initial guess values for the model parameters are obtained by using the step responses. Initially, the step-up (only) signal is used as the set point excitation for the identification of the SOPTD model. The guess values are obtained by the earlier mentioned method (Table 9.1). The final converged parameters are obtained from the optimization routine. These identified model parameter values deviate from the actual process parameter values. In the estimated SOPTD models, the process gain values nearly matched with actual process gain values. But the time constant, time delay and damping co-efficient have significant deviations (like we get in any model reduction technique). The converged SOPTD transfer function model parameters obtained from the step response data are reported in Table 9.4. By using the step response, we obtain the satisfactory closed loop responses of the model, but the parameters deviate with the true values.

Cordons (2005) reported that for the same controller, different models show the same closed loop behaviour even though they have different open loop behaviour. Two examples are considered to show the validity of this statement. The effect of proportional gain value on the model's mismatch are also analysed. Cordons (2005) reported that models can only have their quality evaluated for a particular set of experimental conditions. In the present example, a combined step-up and step-down excitation is performed to get the exact model parameters and these compared with the model parameters obtained from the conventional step test. The proposed method (combined step-up and step-down with the guess values) gives the lowest sum of ISE value.

The final converged parameters, IAE and ISE values, are also listed in Table 9.1 for the combined step-up and step-down responses. The converged model parameters match well with the actual SOPTD system parameters. The Nyquist plots for the subsystems (g_{11}, g_{12}, g_{21}, and g_{22}) are shown in Fig. 9.1 for the original system and the identified model for the step (up) changes and for the original system and the identified model by the combined step-up and step-down changes in the set points. Fig. 9.1 shows a better fit of the open loop model (identified by combined step-up and step-down method) with that of the actual system. Fig. 9.2

TABLE 9.1 Model identification results using PID controller (Example 1)

Model (ij)		k_{pij}	τ_{ij}	θ_{ij}	ε_{ij}	IAE values	ISE values
11	Guess	0.2500	5.0000	1.0000	1.0000		
	Converged*	0.5141	2.2175	1.9342	1.5865	0.4499	1.78×10^{-02}
	Converged**	0.5798	3.5232	0.9939	1.1709	0.0100	2.66×10^{-06}
	Actual	**0.5800**	**3.5143**	**1.0000**	**1.1752**		
21	Guess	0.2017	6.2500	1.2800	1.0000		
	Converged*	0.3038	1.0894	1.7442	2.1350	0.2370	2.80×10^{-03}
	Converged**	0.3498	1.8870	1.2439	1.5061	0.0155	1.50×10^{-05}
	Actual	**0.3500**	**1.8300**	**1.2800**	**1.5492**		
12	Guess	–	6.2500	1.0000	1.0000		
	Converged*	–	3.6363	1.1944	1.1002	0.242	2.10×10^{-03}
	Converged**	–	4.1289	0.9982	1.0491	0.010	2.25×10^{-06}
	Actual	**–**	**4.1279**	**1.0000**	**1.0514**		
22	Guess	–	6.2500	1.0000			
	Converged*	–	1.5547	0.8511	1.4886	0.339	3.60×10^{-03}
	Converged**	–	1.3518	0.9748	1.8758	0.025	3.04×10^{-05}
	Actual	**–**	**1.3000**	**1.0000**	**1.9462**		
				Total value*		1.2689	0.0263
				Total value**		0.0618	$5.031 \ 10^{-5}$

*: step (only) response; **: combined step-up and step-down response

shows the closed loop time response matching of the actual system with that of the identified SOPTD system (by combined step-up and step-down method). The computational time for optimization is 11 minutes and 11 seconds (maximum computational time for 2×2 systems) because of the proposed method for obtaining the initial guess values for the model parameters.

The computational time reported by Viswanathan et al. (2001) varies from 1 hour 55 minutes to 5 hours 28 minutes for one trial and they have used a minimum of 10 trials, as discussed earlier. Viswanathan et al. (2001) did not report the values of the converged model parameters for their method and have given only the ISE value as 8.87E–08. The present method (combined step-up and step-down excitation along with the proposed guess values of the parameters) gives the best fit that corresponds to the global optimal. The proposed initial guess values are important to achieve faster local convergence to local optima whereas only the combined step-up and step-down excitation along with the proposed initial guess values lead to a faster global convergence of the optimization problem.

The identification method is also carried out by using the closed loop main and interaction responses obtained by PI controllers ($k_{c11} = 1$, $\tau_{i11} = 3$; $k_{c22} = -1$, $\tau_{i22} = 2$). Here also the step-up (only) response poses the same problem in

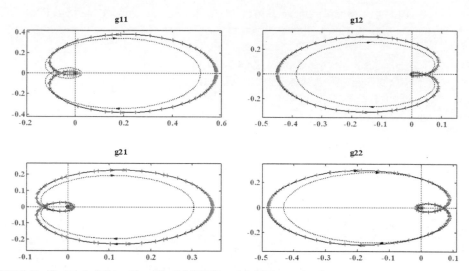

FIGURE 9.1   Nyquist plot for identified SOPTD model (Example 1)
Solid line: actual system;
++: Identified SOPTD model for combined step-up and step-down change) combined step-up and step-down excitation; dotted line: Identified SOPTD model for step-up (only) excitation.

the global convergence. But by using the step-up and step-down responses, the identified model parameters match well with the actual process parameters. The initial guess values, final converged parameters, and IAE and ISE values are listed in Table 9.2 for both the tests. The effect of initial guess values (for the two controller settings) on the number of iterations, computational time, and IAE and ISE values is also studied by changing the guess values +10% and separately by −10%. The present method shows a good agreement with the actual process parameters for these perturbed guess values also. The number of iterations, computational time ISE* and IAE* values are listed in Table 9.3 for this perturbed guess values.

The effect of measurement noise on the guess values and model parameters is also considered by adding a random noise (SD = 0.05) to the actual main and interaction responses, and the sample time of 0.1 s is considered. The noisy output is used for the feedback control action and also for the model identification. To obtain the initial guess values, a smooth curve is drawn first. Because of the integral action in the controller, the main response reaches to 1 and the interaction reaches to 0 in steady state. Hence, a horizontal line is drawn for the main response at 1 and a horizontal line is drawn at 0 in the interaction response in the steady state portion only. In the initial dynamics part, a smooth curve is passed through the midpoint of the response. Alternatively, one can use the least squares method proposed by Savitzky and Golay (1964) which filters the noise and gives a smooth curve. From that smooth curve, the initial guess values can be calculated by the present method.

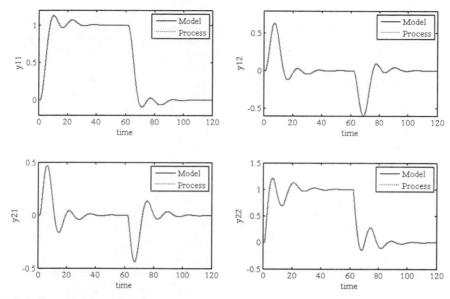

FIGURE 9.2 Closed loop response of the process and the identified SOPTD transfer function model (Example 1)
Obtained from combined step-up and step-down responses using PID controller. Model and process curves overlap.

The proposed method is robust for the controller settings and for the measurement noise. The identified model parameters were closer to those obtained without measurement noise. The initial guess values, final converged values, number of iterations, computational times, and IAE and ISE values are also listed in Table 9.4. The closed loop main responses and interaction responses using the identified model with the same controller settings are shown in the Fig. 9.3, along with the corrupted process responses.

As discussed in Chapter 5, if we identify the above systems by critically damped SOPTD (i.e., CSOPTD) models, then identification will be simple (using only the step input as n^{th} set point) since we need to identify only 3 parameters. Hence, there will no need for using step-up and step-down input in the set point.

Example 2 Consider the Niederlinski multivariable transfer function model presented by Viswanathan et al. (2001) given in Eq. (9.13).

$$G(s) = \begin{bmatrix} \dfrac{0.5}{(0.1s+1)^2(0.2s+1)^2} & \dfrac{-1.0}{(0.1s+1)(0.2s+1)^2} \\ \dfrac{1.0}{(0.1s+1)(0.2s+1)^2} & \dfrac{2.4}{(0.1s+1)(0.2s+1)^2(0.5s+1)} \end{bmatrix} \qquad (9.13)$$

TABLE 9.2   Model identification results under using PI controller settings (Example 1)

Model (ij)		k_{pij}	τ_{ij}	θ_{ij}	ε_{ij}	IAE values	ISE values
11	Guess	0.5000	7.5000	1.0000	1.0000		
	Converged*	0.7802	4.2211	0.3029	1.4287	0.2093	1.90E–03
	Converged**	0.5801	3.5138	1.0004	1.1746	0.0081	7.25E–07
	Actual	0.5800	3.5143	1.0000	1.1752		
21	Guess	0.5151	8.7500	1.2800	1.0000		
	Converged*	0.5384	2.1056	0.3203	2.3266	0.1165	4.21E–04
	Converged**	0.3500	1.9701	1.1846	1.4582	0.0204	1.94E–05
	Actual	0.3500	1.8300	1.2800	1.5492		
12	Guess	–	10.0000	1.0000	1.0000		
	Converged*	–	4.6489	0.4664	1.3975	0.2403	1.80E–03
	Converged**	–	4.1354	0.9924	1.0492	0.0134	2.07E–06
	Actual	–	4.1279	1.0000	1.0514		
22	Guess	–	7.5000	1.0000	1.0000		
	Converged*	–	1.6360	8.4989	0.2849	0.1513	1.40E–03
	Converged**	–	1.4619	0.9145	1.7549	0.0224	2.14E–05
	Actual	–	1.3000	1.0000	1.9462		
					Total value*	0.7174	0.0055
					Total value**	0.0643	4.3895E–05

$k_{c,11} = 1$, $\tau_{I,11} = 3$; $k_{c,22} = -1$, $\tau_{I,22} = 2$
*: Only step response; **: combined step-up and step-down response

TABLE 9.3   Effect of initial guess values of the model parameters on the number of iterations and computational time for combined step-up and step-down responses (Example 1)

Controller	PID Controller			PI controller		
Perturbation in Guess values	−10%	0%	+10%	−10%	0%	+10%
IAE* values	0.0477	0.0618	0.0618	0.0519	0.0643	0.0643
ISE* values	3.49E–05	5.03E–05	5.03E–05	2.64E–05	4.35E–05	4.35E–05
Number of iterations	33	41	40	33	37	44
Computational time (s)	323	390	379	509	567	671

*: sum of main and interaction responses

The higher order TITO process transfer function models are to be approximated as SOPTD transfer function models. The decentralized PI controllers are used for this identification process. The controller parameters for the decentralized PI controllers were (Viswanathan et al., 2001) PI^a; $k_{c11} = 0.4$, $\tau_{i11} = 0.9$; $k_{c22} = 0.15$, $\tau_{i22} = 1.5$.

TABLE 9.4 Model identification results using PID controller with measurement noise for combined step-up and step-down change (Example 1)

Model (ij)		K_{pij}	τ_{ij}	θ_{ij}	ε_{ij}	IAE values	ISE values
11	Guess	0.2500	5.0000	1.0000	1.0000		
	Converged	0.5948	3.8334	0.7693	1.0948	0.3386	2.90E–03
21	Guess	0.2152	5.0000	1.3000	1.0000		
	Converged	0.3673	1.2488	1.5696	2.2692	0.2057	8.86E–04
12	Guess	−0.1824	5.0000	1.0000	1.0000		
	Converged	−0.4653	4.3886	0.7786	0.9779	0.4083	3.40E–03
22	Guess	−0.1667	5.0000	1.0000	1.0000		
	Converged	−0.4988	1.7106	0.7267	1.5988	0.3889	5.70E–03

: sum of main and interaction values (IAE) = 1.3415, (ISE*) = 1.29E–02

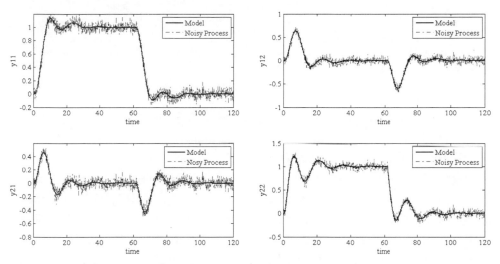

FIGURE 9.3 Closed loop response of the noise corrupted process output with the identified SOPTD model obtained from combined step-up and step-down response (Example 1)

The closed loop step responses of the actual process (with PI[a] controller) are obtained by using Simulink. Since the responses of the actual process show a higher order model behaviour (initial 0 slope), the $k_p k_{c,des}$ value is considered as 0.5 for calculating the guess value of the process gains (main loop). Then the other initial guess values are obtained by the earlier mentioned methods. The initial guess values, final converged parameters, and IAE and ISE values are listed in Table 9.5. In the identification of the higher order model into the SOPTD model also, the step (only) response identification poses a problem to attain the global minimum (least ISE value). In higher order transfer function matrix model, we can verify the better fit of the models based on IAE and ISE values. For the Nyquist plots for the

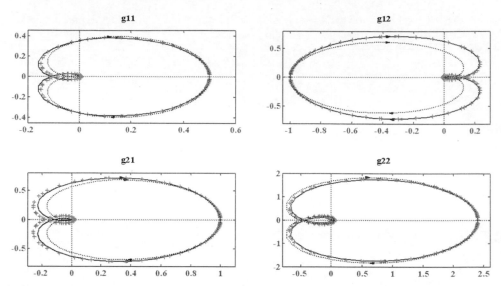

FIGURE 9.4   The Nyquist plot for the actual system (Example 2)
Solid line: identified model; ++ line: combined step-up and step-down method; dotted line: step (up)
only method

subsystems (g_{11}, g_{12}, g_{21}, and g_{22}). Fig. 9.4 shows that the identified model by step (up) only method is not adequate.

To obtain the global minima (least ISE values), the combined step-up and step-down responses are used. The initial guess values, final converged model parameters, and IAE and ISE values are listed in Table 9.5. Nyquist plot (Fig. 9.4) shows a good match between the estimated SOPTD model by the step-up and step-down method with that of the actual process. Fig. 9.5 shows the closed loop time response of the actual system and the identified model. The computational time is significantly reduced to 2 minutes and 20 seconds (maximum computational time) when compared to 51 minutes as reported by Viswanathan et al. (2001). The SOPTD transfer function model obtained from the combined step-up and step-down excitation has lower IAE and ISE values compared to the model obtained from step response only.

Further, the SOPTD model identification is carried with another set of PI[b]; $k_{c11} = 0.4$, $\tau_{i11} = 0.4$; $k_{c22} = 0.1$, $\tau_{i22} = 0.3$ (under-damped response). The closed loop step response adequately matches the actual process response using the PI[b] controller. But the final converged model parameters deviate from the parameters obtained by the PI[a] controller settings. By using the step response, we obtain a satisfactory closed loop response of the model. In the step (only) response method, the estimated model parameters are found; it depends on the controller settings.

TABLE 9.5 Model identification results under using PI[a] controller (Example 2)

Model (ij)		k_{pij}	τ_{ij}	θ_{ij}	ε_{ij}	IAE values	ISE values
11	Guess	1.2500	3.1250	0.1000	1.0000		
	Converged*	0.4999	0.3557	0.0498	0.7682	0.0132	5.67E–05
	Converged**	0.5005	0.2818	0.1049	0.8734	0.0065	9.50E–06
21	Guess	1.9825	3.1250	0.1000	1.0000		
	Converged*	1.0008	0.2494	0.0250	0.9629	0.0145	5.85E–05
	Converged**	1.0002	0.2061	0.0859	1.0254	0.0118	4.84E–05
12	Guess	−1.9825	3.1250	0.1000	1.0000		
	Converged*	−1.0015	0.1826	0.0250	1.2708	0.0112	8.79E–05
	Converged**	−0.9992	0.2267	0.0667	0.9727	0.0038	2.24E–06
22	Guess	3.3333	3.1250	0.1000	1.0000		
	Converged*	2.3959	0.4415	0.2020	0.9183	0.0165	8.46E–05
	Converged**	2.4015	0.4110	0.1905	0.9901	0.0107	2.93E–05
			Total values*			0.0554	2.87E–04
			Total value**			0.0328	1.09E–04

*: step response; **: step-up and step-down response

By using the PI[b] controller also, the combined step-up & step-down excitation provides the lower ISE value compared to that of the values obtained by the model from the step (only) excitation. The initial guess values, final converged parameters, and IAE and ISE values are listed in Table 9.6 for the combined step-up & step-down responses and for the step (only) response.

The effect of initial guess values on the number of iterations, computational time, IAE* and ISE* values is also studied by changing the guess values by +10% and by −10% separately using both the PI controllers. The present method shows a good agreement with the actual process parameters for these perturbed guess values also. The results obtained for this perturbed guess values are given in Table 9.7.

Example 3 Consider the 3×3 (SOPTD) transfer function matrix model (Garcia et al., 2005) given in Eq. (9.14).

$$G_p = \begin{bmatrix} \dfrac{e^{-0.1s}}{s^2+2s+1} & \dfrac{0.3e^{-0.01s}}{2s^2+3s+1} & \dfrac{0.3e^{-0.01s}}{2s^2+3s+1} \\[3mm] \dfrac{0.3e^{-0.01s}}{2s^2+3s+1} & \dfrac{e^{-0.1s}}{s^2+2s+1} & \dfrac{0.3e^{-0.01s}}{2s^2+3s+1} \\[3mm] \dfrac{0.3e^{-0.01s}}{2s^2+3s+1} & \dfrac{0.3e^{-0.01s}}{2s^2+3s+1} & \dfrac{e^{-0.1s}}{s^2+2s+1} \end{bmatrix} \tag{9.14}$$

TABLE 9.6   Model identification results using PI[b] controller (Example 2)

Model (ij)		k_{pij}	τ_{ij}	θ_{ij}	ε_{ij}	IAE values	ISE values
11	Guess	1.2500	1.2500	0.1000	1.0000		
	Converged*	0.4974	0.2861	0.1155	0.8401	0.0186	5.96E–05
	Converged**	0.5002	0.2792	0.1050	0.8772	0.0008	1.04E–05
21	Guess	2.4658	1.2500	0.1000	1.0000		
	Converged*	1.0009	0.1981	0.0999	1.0298	0.0203	9.26E–05
	Converged**	0.9996	0.2293	0.0650	0.9537	0.0066	9.52E–06
12	Guess	−2.4658	1.2500	0.1000	1.0000		
	Converged*	−1.0157	0.1313	0.1003	1.5929	0.0147	4.52E–05
	Converged**	−1.0001	0.2373	0.0539	0.9459	0.0035	1.74E–06
22	Guess	5.0000	1.2500	0.1000	1.0000		
	Converged*	2.3896	0.4785	0.1200	0.8915	0.0209	5.80E–05
	Converged**	2.3979	0.4030	0.1910	1.0035	0.0093	2.35E–05
				Total value*		0.0745	2.55E–04
				Total value**		0.0202	1.308E–04

*: step response; **: combined step-up and step-down response

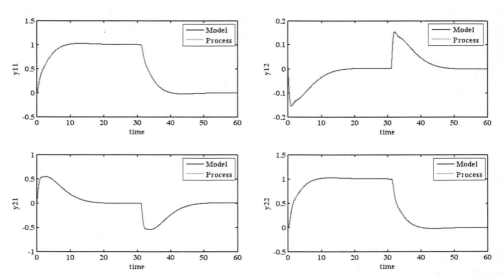

FIGURE 9.5   Closed loop response of the process and the identified SOPTD transfer function model (Example 2)
Obtained from combined step-up and step-down responses using PI[a] controller. Model and process curves overlap.

The PI controller parameters are obtained by using BLT method (Luyben, 1986) with $F = 2$ and PI parameters were $K_{cii} = 4.6979$, $\tau_{Iii} = 2.3612$. The guess values for the main-loop (diagonal) process gains K_{p11}, K_{p22} and K_{p33} were calculated based on the corresponding controller gains as discussed in Example 1.

TABLE 9.7 Effect of initial guess values of the model parameters on the number of iterations and computational time for combined step-up and step-down change (Example 2)

Mode of Controller	PI[a] Controller			PI[b] controller		
Perturbation in Guess values	−10%	0%	+10%	−10%	0%	+10%
IAE* values	0.0463	0.0275	0.0347	0.0297	0.0294	0.0257
ISE* values	1.97E−04	4.52E−05	7.67E−05	4.30E−05	6.01E−05	3.26E−05
Number of iterations	37	46	42	27	26	25
Computational time (s)	105	140	121	130	128	125

*: sum of main and interaction response values

Since, the responses of the actual process showed a higher-order model behaviour (initial 0 slope), the $K_p K_{c,des}$ value was considered as 0.5 for calculating the guess value of the process gains $(K_{p11}, K_{p22}$ and $K_{p33})$ The non-diagonal process gains (interactions) were calculated using equations similar to Eq. (9.8) and Eq. (9.10). To simplify the calculations, only two loops were considered at a time, and the interaction coming from the remaining loop was considered as negligible. For example, the guess value of the process gain (non-diagonal) K_{p21} was estimated by considering the interaction coming from the main loop only (similar to Fig. 7.1a in Chapter 7). The interaction to and from the third loop were assumed to be negligible. This procedure gave explicitly the guess value of the required process gain. Similarly, to obtain the guess values of K_{p12}, an approach similar to Fig. 7.1b was considered. To obtain the guess values for the gains K_{p31} and K_{p32} the corresponding Fig. 7.7 was considered. Similarly, the guess values for the other gains K_{p13} and K_{p23} were calculated. For simulating the model, the sample time was considered as 0.01 s.

With these guess values here, also initially, the step (only) change is used as the set point excitation for the identification of the SOPTD model. The identified model parameters from step (only) response deviate from the actual process parameter values and are listed in Table 9.8. The Nyquist plots for the subsystems (Fig. 9.6) show that the identified model by this method is not adequate. To converge to the true values, a combined step-up and step-down change in the set point is considered and the output responses are used for the identification test. The proposed method (combined step-up and step-down with the guess values) gives the lowest sum of ISE value. The final converged parameters and IAE and ISE values are also listed in Table 9.8 for the combined step-up and step-down responses. The converged model parameters match well with the actual SOPTD system parameters. The Nyquist plot (Fig. 9.6) also shows a better match

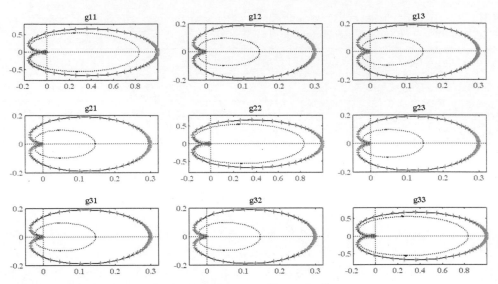

FIGURE 9.6   The Nyquist plot for the actual system (Example 3)
Solid line: identified model; ++ line: combined step-up step-down method; dotted line: step (up) only method

of the identified model by the proposed method with the actual system. The computational time for the optimization was 3 minutes 17 seconds, and the number of iterations was 13. The closed loop main and interaction responses using the identified model parameters with the same controller settings are shown in Fig. 9.7. Fig. 9.7 shows a good match between the model and the actual process.

The identification method was also carried out using the closed loop main and interaction responses obtained with different PI/PID settings. The converged model parameters showed good agreement with the actual process parameters. The number of iterations, computational times, and IAE and ISE values are listed in Table 9.9. The effects of initial guess values on the number of iterations, computational time, and IAE and ISE values were also studied by perturbing the guess values by +10% and separately by −10%. The model parameters obtained by the proposed method showed good agreement with the actual process parameters for these perturbed guess values. The number of iterations, computational times, and IAE and ISE values are listed in Table 9.9.

9.4   Method of Identifying CSOPTD Models

Consider a decentralized TITO system as shown in Fig. 3.2 (Chapter 3). The process transfer functions matrix Gp(s) is controlled by a decentralized PI/PID controller

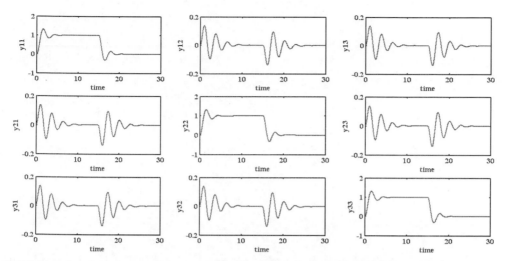

FIGURE 9.7   Closed loop response of the process and the identified SOPTD transfer function model (Example 3)

Obtained from combined step-up and step-down responses using PI (BLT, $F = 2$) controller. Solid line: model; dashed line: process. Model and process response curves overlap.

matrix $G_c(s)$. $G_p(s)$ and $G_c(s)$ matrix are given as in Eq. (9.15).

$$G_p(s) = \begin{bmatrix} g_{p11} & g_{p12} \\ g_{p21} & g_{p22} \end{bmatrix} \quad G_c(s) = \begin{bmatrix} g_{c11} & 0 \\ 0 & g_{c22} \end{bmatrix} \tag{9.15}$$

The second order transfer functions and decentralized PI controller settings are considered as in Eq. (9.16) and Eq. (9.17).

$$g_{p_{ij}}(s) = \frac{K_{p,ij}e^{-\theta_{ij}s}}{\tau_{p,ij}^2 s^2 + 2\varepsilon_{p,ij}\tau_{p,ij}s + 1} \quad i = 1,2; \ j = 1,2 \tag{9.16}$$

$$g_{c_{ij}}(s) = K_{C_{ij}} + \frac{K_{I_{ij}}}{s} \quad i = 1,2; \ j = 1,2 \tag{9.17}$$

The process transfer function matrix elements are identified as CSOPTD multivariable systems. Critically damped means ($\varepsilon_{p,ij} = 1$) and remaining parameters guess values can be obtained as discussed now.

The CSOPTD model has only three parameters as in the FOPTD model. A unit step input excitation is introduced in set point y_{s1} and the remaining kept unchanged in the system. Then we obtained the main response as y_{11} and interaction response as y_{21}. Similarly, a unit step input excitation is introduced into set point y_{s2} and all the remaining kept unchanged, and we obtain the main response as y_{22} and interaction response as y_{12}. The closed loop responses are used

TABLE 9.8 SOPTD model identification results using the PI (BLT, F = 2) controller (Example 3)

Model (ij)		K_{pij}	τ_{ij}	ε_{ij}	θ_{ij}	IAE values	ISE values
11	Guess	0.1064	1.0000	1.0000	0.1000		
	Converged*	0.8337	0.8593	1.0810	0.1749	0.1018	3.60×10^{-03}
	Converged**	0.9992	1.0018	0.9952	0.0984	0.0073	8.65×10^{-06}
	Guess	0.0352	1.4142	1.0607	0.01		
21	Converged*	0.1452	0.9993	0.9509	0.0392	0.0534	4.03×10^{-04}
	Converged**	0.2982	1.4096	1.0535	0.0101	0.0030	7.20×10^{-07}
	Guess	0.0352	1.4142	1.0607	0.01		
31	Converged*	0.1452	0.9993	0.9509	0.0392	0.0534	4.03×10^{-04}
	Converged**	0.2982	1.4096	1.0535	0.0101	0.0030	7.20×10^{-07}
	Guess	0.0352	1.4142	1.0607	0.01		
12	Converged*	0.1452	0.9993	0.9509	0.0392	0.0534	4.03×10^{-04}
	Converged**	0.2982	1.4096	1.0535	0.0101	0.0030	7.20×10^{-07}
	Guess	0.1064	1.0000	1.0000	0.1000		
22	Converged*	0.8337	0.8593	1.0810	0.1749	0.1018	3.60×10^{-03}
	Converged**	0.9992	1.0018	0.9952	0.0984	0.0073	8.65×10^{-06}
	Guess	0.0352	1.4142	1.0607	0.01		
32	Converged*	0.1452	0.9993	0.9509	0.0392	0.0534	4.03×10^{-04}
	Converged**	0.2982	1.4096	1.0535	0.0101	0.0030	7.20×10^{-07}
	Guess	0.0352	1.4142	1.0607	0.01		
13	Converged*	0.1452	0.9993	0.9509	0.0392	0.0534	4.03×10^{-04}
	Converged**	0.2982	1.4096	1.0535	0.0101	0.0030	7.20×10^{-07}
	Guess	0.0352	1.4142	1.0607	0.01		
23	Converged*	0.1452	0.9993	0.9509	0.0392	0.0534	4.03×10^{-04}
	Converged**	0.2982	1.4096	1.0535	0.0101	0.0030	7.20×10^{-07}
	Guess	0.1064	1.0000	1.0000	0.1000		
33	Converged*	0.8337	0.8593	1.0810	0.1749	0.1018	3.60×10^{-03}
	Converged**	0.9992	1.0018	0.9952	0.0984	0.0073	8.65×10^{-06}

*: step response; **: combined step-up and step-down response

to obtain the guess values for time constants, time delays, and process gains of all the transfer functions for optimization (Haritha et al. 2021).

In the previous chapter, we have reported that the initial guess values of the main response process gains $(K_{p,ii})$ of higher order and lower order are different. The higher order and lower order models are differentiated based on the initial slope of the main responses of the actual process. If the initial slope of the response is sluggish (0 slope), then we consider the system as a higher order system. Similarly, if the initial slope of response is steep (non-zero slope), then we consider the system as a lower order system. For higher order systems, the value

TABLE 9.9 Effect of changing controller settings and effect of guess values on the number of iterations and computational time obtained from combined step-up and step-down responses (Example 3)

	PI (BLT, $F = 3$) $(K_{cii} = 3.132, \tau_{Iii} = 3.5417)$			(Chen & Seborg, 2002) $(K_{cii} = 5.25, \tau_{Iii} = 1.08, \tau_{Dii} = 0.271)$		
Perturbation in Guess values	-10%	0%	$+10\%$	$+10\%$	0%	$+10\%$
IAE* values	0.078	0.078	7.84×10^{-03}	0.0273	8.88×10^{-06}	0.0150
ISE* values	1.14×10^{-04}	1.14×10^{-04}	9.13×10^{-07}	1.79×10^{-05}	2.13×10^{-12}	4.99×10^{-06}
Number of iterations	10	10	27	18	13	32
Computational time (s)	203	195	554	228	171	393

*: sum of main and interaction response values

of $K_p K_{c,max} = 1$ and the design value is $K_p K_{c,des} = 0.5$. Hence, the initial guess value $K_{p,ii} = 0.5/K_{c,ii}$. For lower order systems, the value of $K_p K_{c,max} = 4$ and the design value is $K_p K_{c,des} = 2$. Hence, the initial guess value $K_{p,ii} = 2/K_{c,ii}$.

The initial guess values of interactive response process gains K_{p12} and K_{p21} are obtained as follows:

The Laplace transforms of interaction response $y_{21}(s^*)$ and $y_{12}(s^*)$ are given by Eq. (9.18) and Eq. (9.19).

$$y_{21}(s^*) = \int_0^\infty y_{21}(t)e^{-ts^*} dt \tag{9.18}$$

$$y_{12}(s^*) = \int_0^\infty y_{12}(t)e^{-ts^*} dt \tag{9.19}$$

To minimize the computation time, the value of s^* is considered as $s^* = 8/t_s$ and the upper bound of t is taken as settling time (t_s). For $t \geq t_s$, e^{-8} leading to 0 depicts no further change in integral value. The initial guess value of interaction response was calculated by considering the interaction loop from the main loop only and neglecting the other interaction loop shown in Fig. 7.1a (Chapter 7). In Fig. 7.1a, g_{p12} is neglected for deriving Eq. (9.20). Similarly, in Fig. 7.1b, g_{p21} is neglected for deriving Eq. (9.21).

The initial guess value of K_{p21} was calculated from Eq. (9.20) by substituting $y_{s1(s^*)}$ as $1/s^*$. $y_{21}(s^*)$ value obtained from Eq. (9.18). and using the guess values of the other model parameters and controllers settings:

$$\frac{y_{21}(s^*)}{y_{s1}(s^*)} = \frac{g_{p21}g_{c11}}{1 + g_{p11}g_{c11} + g_{p22}g_{c22} + g_{p11}g_{p22}g_{c11}g_{c22}} \tag{9.20}$$

The initial guess value of K_{p12} was calculated, by substituting $Y_{s2(s^*)}$ value as $1/s^*$, $y_{12}(s^*)$ value from the Eq. (9.19) and substitute all remaining guess values and controller settings in Eq. (9.21).

$$\frac{y_{12}(s^*)}{y_{s2}(s^*)} = \frac{g_{p12}g_{c22}}{1 + g_{p11}g_{c11} + g_{p22}g_{c22} + ?g_{p11}g_{p22}g_{c11}g_{c22}} \tag{9.21}$$

The initial guess value for process time constant (τ) is considered as $t_s/8$. Once all the initial guess values k_{pii}, k_{pij}, τ_{ij}, θ_{ij} are obtained, a unit step input excitation is introduced in set point Y_{ms1} and remaining reference signals are kept unchanged; thus we obtain the main response Y_{m11} and interaction response Y_{m21}. Similarly, a unit step input excitation is introduced in set point Y_{ms2} and all the remaining reference signals kept unchanged, and we obtain the main response Y_{m22} and the interaction response Y_{m12}. The optimization problem is formulated to minimize the sum of square of the error of the closed loop responses of the actual and model responses.

$$\text{Minimize} f = \sum_{i=1}^{2}\sum_{j=1}^{2}\sum_{k=1}^{n}\left[y_{mij}\left(t_k\right) - y_{ij}\left(t_k\right)\right]^2 \Delta t \tag{9.22}$$

In the objective function in Eq. (9.22), n is the number of data points in the actual process and model.

In the present case, for the closed loop multivariable system, actual and model are simulated in Simulink. Routine *lsqnonlin* in Matlab is used for solving the optimization problems and routine implements the trust–region–reflective algorithm. The fourth-order Runge-Kutta method with fixed step size solver is used for integration of the differential equations. All computational work for the identification of decentralized multivariable system was done in Intel Xeon 2.6 GHz. To show the significance of the proposed method, a few examples are given with simulation results.

TABLE 9.10   Estimated model parameters of Example 4

ij		$K_{p_{ij}}$	τ_{ij}	θ_{ij}	IAE	ISE
11	G	1.25	1.625	0.1	0.0117	7.602×10^{-05}
	C	0.5001	0.1998	0.2131		
21	G	1.62387	3.125	0.1	0.0086	1.969×10^{-05}
	C	0.4993	0.2477	0.025		
12	G	−3.1261	2.875	0.1	0.0048	1.003×10^{-05}
	C	−0.9973	0.2463	0.0257		
22	G	3.3333	1.625	0.1	0.0126	9.758×10^{-05}
	C	2.3994	0.3848	0.2449		

G: guess values; C: converged values; IAE: integral absolute error; ISE: integral square error

9.5 Simulation Examples

Example 4 Consider the multivariable process transfer function of Niederlinski (1971) as given in Eq. (9.23).

$$G_p = \begin{bmatrix} \dfrac{0.5}{(0.1s+1)^2(0.2s+1)^2} & \dfrac{-1.0}{(0.1s+1)(0.2s+1)^2} \\ \dfrac{1.0}{(0.1s+1)(0.2s+1)^2} & \dfrac{2.4}{(0.1s+1)(0.2s+1)^2(0.5s+1)} \end{bmatrix} \quad (9.23)$$

The higher order TITO process transfer function models are approximated as CSOPTD transfer function models. The decentralized PI controller settings used for identification are as in Eq. (9.24).

$$G_c = \begin{bmatrix} 0.4\left(1+\dfrac{1}{0.9s}\right) & 0 \\ 0 & 0.15\left(1+\dfrac{1}{1.5s}\right) \end{bmatrix} \quad (9.24)$$

The initial guess values of process gain are computed as follows:

The process time delay (θ) is considered as 0.1 s. The initial guess value for the process time constant (τ) is considered as $t_s/8$. When the process gain (K_p) is positive, the upper bound for K_p is 20 times of the guess value and the lower bound is 1/20 times of the guess value. Similarly, for negative values of K_p, the upper bound is 1/20 times of the guess value and the lower bound is 20 times of the guess value. For time constant, the upper bound as 5 times of the guess value

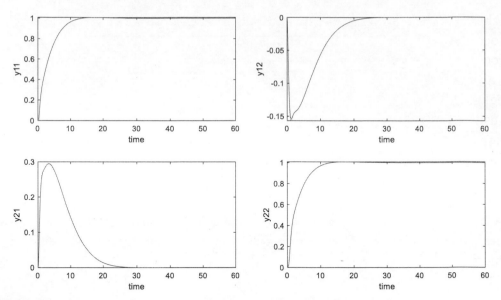

FIGURE 9.8   Comparison between actual process and model of Example 4
Solid line: model; dotted line: actual

and the lower bound as 1/20 times of the guess value; and for time delay, upper bound is 4 times of the guess value and the lower bound is 1/4 times of the guess value. Boundaries in an optimization routine reduce computation time and give quick convergence of the optimal parameters. The guess values, final converged parameters, IAE and ISE, number of iterations and computation time are given in Table 9.10. Fig. 9.8 shows a good match between the closed loop response of actual process and (CSOPTD) model responses with the same diagonal controllers. To get the final converged parameters, the number of iterations is taken as 19 and the computation time for optimization is 9.89 minutes.

Example 5 Let us consider a multivariable SOPTD system considered by Garcia et.al as given in Eq. (9.25).

$$
G_p = \begin{bmatrix}
\dfrac{e^{-0.1s}}{s^2 + 2s + 1} & \dfrac{0.3e^{-0.01s}}{2s^2+3s + 1} & \dfrac{0.3e^{-0.01s}}{2s^2+3s + 1} \\[3mm]
\dfrac{0.3e^{-0.01s}}{2s^2+3s + 1} & \dfrac{e^{-0.1s}}{s^2+2s + 1} & \dfrac{0.3e^{-0.01s}}{2s^2 + 3s + 1} \\[3mm]
\dfrac{0.3e^{-0.01s}}{2s^2+3s + 1} & \dfrac{0.3e^{-0.01s}}{2s^2+3s + 1} & \dfrac{e^{-0.1s}}{s^2+2s + 1}
\end{bmatrix}
\tag{9.25}
$$

The decentralized PI controller parameters $K_{cii} = 4.6979$ and $\tau_{ii} = 2.3612$ are used for identification. The closed loop responses of the actual process showed a higher order model behaviour (initial 0 slope). For calculating the guess values of process gains of main responses ($K_{p_{ii}}$), the $K_p K_{c,des}$ value is assumed as 0.5. The process gains of interaction responses are calculated by considering only two loops at a time, and the interaction coming from the remaining loop is considered as negligible. For example, the guess value of the process gain $K_{p_{12}}$ is computed by considering the interaction specifically coming from the main loop and the third loop was assumed to be negligible. This procedure gives the guess value of process gain of interaction response. Similar procedure is used for calculating the guess value of $K_{p_{21}}$. The guess values of $K_{p_{32}}$ and $K_{p_{31}}$ are calculated by neglecting first and the second loops respectively and considering the interaction coming specifically from the main loop. An identical procedure is used for calculating the guess values of $K_{p_{23}}$ and $K_{p_{13}}$. This procedure was reported by Rajapandiyan and Chidambaram (2012a).

From the closed loop responses, the initial guess value for process time constant (τ) are considered as ts/8, time delay is considered as the same as closed loop time delay. The final converged parameters are obtained from the least square optimization method. The guess values, final converged parameters, IAE, ISE, number of iterations, and computational time are mentioned in Table 9.12.

Fig. 9.9 shows a good match between the closed loop responses of the actual process and the CSOPTD model using the same controller settings. To get final converged parameters, number of iterations are taken as 15 and the computational time for optimization are observed as 2,952.14 s. Table 9.12 shows that IAE and ISE values are low for the present example when compared to the values reported by Rajapandiyan and Chidambaram (2012a) for FOPTD models. Reduced values of IAE and ISE depict the perfect match between actual and model responses of the closed loop.

Summary

To conclude, a closed loop identification of SOPTD model parameters of a TITO multivariable system based on an optimization method using the combined dynamics of step-up and step-down response has been presented. Compared to the conventional step response method, the proposed method (step-up and step-down) shows a good match between the identified model parameters and the actual process. From the three simulation examples, it is shown that the proposed method gives lower ISE values compared to the conventional step response. Simple methods for obtaining initial guess values for the model parameters from

TABLE 9.11　Initial guess values and converged parameters (CSOPTD) for Example 5

ij		$K_{p_{ij}}$	τ_{ij}	θ_{ij}	IAE	ISE
11	G	0.1064	1.125	0.1	0.0085	6.9596×10^{-06}
	C	0.9967	0.998	0.1013		
21	G	0.0343	1.25	0.01	0.0073	5.4078×10^{-06}
	C	0.2868	1.398	0.0025		
31	G	0.0343	1.25	0.01	0.0073	5.4078×10^{-06}
	C	0.2868	1.398	0.0025		
12	G	0.0343	1.25	0.01	0.0073	5.4078×10^{-06}
	C	0.2868	1.398	0.0025		
22	G	0.1064	1.125	0.1	0.0085	6.9596×10^{-06}
	C	0.9967	0.998	0.1013		
32	G	0.0343	1.25	0.01	0.0073	5.4078×10^{-06}
	C	0.2868	1.398	0.0025		
13	G	0.0343	1.25	0.01	0.0073	5.4078×10^{-06}
	C	0.2868	1.398	0.0025		
23	G	0.0343	1.25	0.01	0.0073	5.4078×10^{-06}
	C	0.2868	1.398	0.0025		
33	G	0.1064	1.125	0.1	0.0085	6.9596×10^{-06}
	C	0.9967	0.998	0.1013		

G: guess values; C: converged values; IAE: integral of absolute error;
ISE: integral of square error

TABLE 9.12　Comparison of IAE and ISE values for Examples 4 and 5

Example		IAE	ISE
4	Rajapandiyan	0.0554	2.87×10^{-4}
	Present example	0.0377	2.0332×10^{-4}
5	Rajapandiyan	0.6258	0.0132
	Present example	0.0697	5.3326×10^{-5}

Ex: example number; IAE: integral absolute error; ISE: integral square error

the main and interaction responses are proposed in the chapter for getting a quick and guaranteed convergence. The effects of measurement noise, controller settings and perturbation in the guess values on the identified model are also studied. The maximum computational time for the optimization method is considerably reduced from 5 hour 14 minutes to 11 minutes 11 seconds (maximum

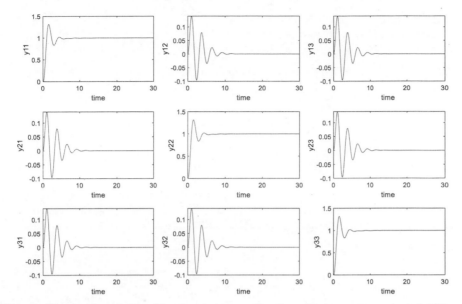

FIGURE 9.9 Comparison of closed loop responses using the actual process and model (Example 5) with the same controllers
Solid line: model; dotted line: actual. Both curves coincide.

computational time) for SOPTD (2×2) model systems; from 1 hour 2 minutes to 2 minutes 20 seconds (maximum computational time) for higher order (2×2) systems; and 9 minutes 14 seconds (maximum computational time) for SOPTD (3×3) model systems.

A method is proposed for the closed loop identification of CSOPTD transfer function of multivariable system based on optimization method. Only step-up input excitation is required. The proposed method was implemented on 2×2, higher order, and 3×3 SOPTD transfer function multivariable systems. This example shows a good match between the actual process and identified model under closed loop using the same controller settings. From the two simulation examples, it is shown that the proposed method gives lower IAE and ISE values when compared with values of IAE and ISE of conventional step response.

Problems

1. Consider the Niederlinski (1971) multivariable transfer function model presented by Viswanathan et al. (2001):

$$G_p(s) = \begin{bmatrix} \dfrac{0.5}{(0.1s+1)^2 (0.2s+1)^2} & \dfrac{-1.0}{(0.1s+1)(0.2s+1)^2} \\ \dfrac{1.0}{(0.1s+1)(0.2s+1)^2} & \dfrac{2.4}{(0.1s+1)(0.2s+1)^2(0.5s+1)} \end{bmatrix}$$

The decentralized PI controllers parameters are $k_{c11} = 0.4$, $\tau_{I11} = 0.4$; $k_{c22} = 0.1$, $\tau_{I22} = 0.3$. Carry out the closed loop optimization method for identifying (a) SOPTD model and (b) CSOPTD model.

2. Consider the 2×2 (SOPTD) distillation column transfer function model matrix (Weischedel and McAvoy, 1980):

$$G(s) = \begin{bmatrix} \dfrac{0.7e^{-s}}{12.35s^2 + 8.26s + 1} & \dfrac{-0.45e^{-s}}{17.04s^2 + 8.69s + 1} \\ \dfrac{0.35e^{-1.28s}}{3.35s^2 + 5.67s + 1} & \dfrac{-0.5e^{-s}}{1.69s^2 + 5.06s + 1} \end{bmatrix}$$

The decentralized PID controller settings used for the MIMO identification were $k_{c11} = 2$, $\tau_{i11} = 2$, $\tau_{d11} = 0.5$; $k_{c22} = -3$, $\tau_{i22} = 2$, $\tau_{d22} = 0.5$. Carry out the closed loop optimization method for identifying (a) SOPTD model and (b) CSOPTD model.

3. For the system:

$$G(s) = \begin{bmatrix} \dfrac{0.58e^{-s}}{(12.35s^2 + 8.26s + 1)} & \dfrac{-0.45e^{-s}}{(17.04s^2 + 8.68s + 1)} \\ \dfrac{0.35e^{-1.28s}}{(3.35s^2 + 5.67s + 1)} & \dfrac{-0.48e^{-s}}{(1.69s^2 + 5.06s + 1)} \end{bmatrix}$$

Find a suitable input out pairing. Design approximates decentralized PI controllers. Perform the closed loop optimization method to identify the model parameters of (a) SOPTD model and (b) CSOPTD model.

4. For the system:

$$G(s) = \begin{bmatrix} \dfrac{0.562e^{-s}}{(7.74s + 1)(7.74s + 1)} & \dfrac{-0.516e^{-1.5s}}{(7.1s + 1)(7.1s + 1)} \\ \dfrac{0.33e^{-1.5s}}{(15.8s + 1)(0.5s + 1)} & \dfrac{-0.394e^{-s}}{(13.8s + 1)(0.4s + 1)} \end{bmatrix}$$

Find a suitable input out pairing. Design approximates decentralized PI controllers. Perform the closed loop optimization method to identify the model parameters of CSOPTD model.

5. For the system:

$$G(s) = \begin{bmatrix} \dfrac{0.471e^{-s}}{(30.7s + 1)(30.7s + 1)} & \dfrac{0.495e^{-2s}}{(28.5s + 1)(28.5s + 1)} \\ \dfrac{0.749e^{-1.7s}}{(57s + 1)(57s + 1)} & \dfrac{-0.832e^{-s}}{(50.5s + 1)(50.5s + 1)} \end{bmatrix}$$

Find a suitable input out pairing. Design approximate decentralized PI controllers. Perform the closed loop optimization method to identify the model parameters of CSOPTD model.

10

Identification of Unstable TITO Systems by Optimization Technique

---

In this chapter, a generalized technique is discussed to obtain the initial guess values for individual transfer function processes of the unstable Two Input Two Output (TITO) multivariable systems. To determine the lower and the upper bounds to be used in the optimization technique, a simple method is explained. Section 10.1 discusses the proposed design method to identify the model parameters of a TITO system under decentralized controller along with the analytical expressions to determine the initial guess values. The applicability of the method is demonstrated with two simulated unstable systems. The method is also extended to unstable TITO system under centralized controllers. Simulation examples show that the proposed method gives a quick convergence with less computational time. For solving the optimization problem, *lsqnonlin* routine in Matlab is used.

10.1 Identification of Systems with Decentralized PI Controllers

The decentralized multivariable system shown in Fig. 3.2 (Chapter 3) is considered. The main loop diagonal transfer function models are unstable FOPTD. The process transfer function matrix $G_p(s)$ and the decentralized controller matrix $G_c(s)$ are given as in Eq. (10.1).

$$G_p\left(s\right) = \begin{bmatrix} g_{11} & g_{12} \\ g_{21} & g_{22} \end{bmatrix}; \qquad G_c\left(s\right) = \begin{bmatrix} g_{c11} & 0 \\ 0 & g_{c22} \end{bmatrix} \tag{10.1}$$

The unstable system can be stabilized by the decentralized control scheme. The controller settings can be selected to obtain a reasonable stable process response. The present example focuses on the identification of the systems having the main loop diagonal transfer functions (g_{11} and g_{22}) as unstable and the off diagonal transfer function (g_{12} and g_{21}) as stable.

In general, the transfer function models used are expressed as in Eq. (10.2).

$$g_{ii}(s) = \frac{k_{pii}e^{-\theta_{ij}s}}{(\tau_{ii}s - 1)}; \; i = 1,2; \quad g_{ij}(s) = \frac{k_{pij}e^{-\theta_{ij}s}}{(\tau_{ij}s + 1)}; \; i \neq j = 1,2 \tag{10.2}$$

The set point y_{r1} is perturbed with the other loops closed and other set points unchanged. From this set point change, the main response y_{11} and the interaction y_{21} is obtained. Similarly, y_{r2} is perturbed to obtain the main response y_{22} and the interaction y_{12}. The initial guess value plays an important role in the optimization technique. For the identification of the model, these response values are used to find the initial guess values for which a straightforward method is suggested. The initial guess values for time delay are considered to be the same as the corresponding closed loop time delay values and the time constant is considered as $t_s/8$ where t_s are the settling time of closed loop responses. The initial guess values for k_{p11} and k_{p22} are obtained from the relation given in Eq. (10.3) (Chidambaram, 1998).

$$k_{Pii}k_{cii} = 2 \tag{10.3}$$

To obtain the guess values of the off-diagonal process gains (k_{p21} and k_{p12}), the Laplace transforms of interaction responses $y_{21}(s^*)$ and $y_{12}(s^*)$ are taken as Eq. (10.4) and Eq. (10.5).

$$y_{21}(s^*) = \int_0^\infty y_{21}(t)e^{-ts^*}dt \tag{10.4}$$

$$y_{12}(s^*) = \int_0^\infty y_{12}(t)e^{-ts^*}dt \tag{10.5}$$

The evaluation of Eq. (10.4) and Eq. (10.5) requires the value of s^*. In the present example, to decrease the time for calculating the integral, the value of

s^* is considered as $8/t_s$ for $t \geq t_s$, e^{-8} tending towards 0, indicating no further change in integral value. To obtain the numerical values of $y_{21}(s^*)$ and $y_{12}(s^*)$ the value of s^* is substituted in Eq. (10.4) and Eq. (10.5). For the cross-loops, the closed loop transfer functions are assumed by considering the interaction from the main loop as the disturbance to that of interacted loop. Some signals are neglected for simplification as shown in Fig. 7.1 (Chapter 7), so as to avoid g_{p12} as in Fig. 7.1a and g_{p21} as in Fig. 7.1b in deriving Eq. (10.6) and Eq. (10.7).

$$y_{21}(s^*) = \frac{\left(1/s^*\right)k_{c11}\left(1+\left(1/\tau_{i11}s^*\right)+\tau_{d11}s^*\right)\frac{k_{p21}e^{-\theta_{21}s^*}}{(\tau_{21}s^*+1)}}{\left[\begin{array}{c}\left(1+k_{c11}\left(1+\left(1/\tau_{i11}s^*\right)+\tau_{d11}s^*\right)\frac{k_{p11}e^{-\theta_{11}s^*}}{(\tau_{11}s^*-1)}\right) \\ \left(1+k_{c22}\left(1+\left(1/\tau_{i22}s^*\right)+\tau_{d22}s^*\right)\frac{k_{p22}e^{-\theta_{22}s^*}}{(\tau_{22}s^*-1)}\right)\end{array}\right]} \tag{10.6}$$

$$y_{12}(s^*) = \frac{\left(1/s^*\right)k_{c22}\left(1+\left(1/\tau_{i22}s^*\right)+\tau_{d22}s^*\right)\frac{k_{p12}e^{-\theta_{12}s^*}}{(\tau_{12}s^*+1)}}{\left[\begin{array}{c}\left(1+k_{c22}\left(1+\left(1/\tau_{i22}s^*\right)+\tau_{d22}s^*\right)\frac{k_{p22}e^{-\theta_{22}s^*}}{(\tau_{22}s^*-1)}\right) \\ \left(1+k_{c11}\left(1+\left(1/\tau_{i11}s^*\right)+\tau_{d11}s^*\right)\frac{k_{p11}e^{-\theta_{11}s^*}}{(\tau_{11}s^*-1)}\right)\end{array}\right]} \tag{10.7}$$

The guess values for k_{p21} and k_{p12} are calculated from Eq. (10.6) and Eq. (10.7). Keeping all loops under closed loop condition, the model set point y_{mr1} is perturbed by a unit step, with all other remaining set points kept unchanged. The main response y_{m11} and interaction response y_{m21} are recorded. Similarly, the same magnitude of step perturbation is applied in set point y_{mr2} to record the main response y_{m22} and interaction response y_{m12}. The recorded responses of the model are compared to the actual process response.

In order to identify the parameters of the system, an optimization problem is formulated. The problem focuses on selecting the model parameters ($k_{p,ij}$, τ_{ij}, and θ_{ij}) to minimize the sum of squared errors between the model and the actual process responses.

$$\text{Minimize } f = \sum_{i=1}^{2}\sum_{j=1}^{2}\sum_{k=1}^{n}[y_{mij}(t_k) - y_{ij}(t_k)]^2 \Delta t \tag{10.8}$$

where n is the number of data points in the response of the model and the process.

Any TITO system has four response curves (two main response curves and two interaction curves) leading to occurrence of four ISE values. The sum of all four ISE values is considered as a scalar objective function as in Eq. (10.8). In this present example, the routine *lsqnonlin* in Matlab is used to solve the optimization problem. The initial guess determines the convergence rate and computational time. The routine depends on the Trust–Region–Reflective algorithm. The present example focuses on proposing a simple method to determine the initial guess values for the model parameters in order to solve the optimization problem with lesser computation time to minimize the scalar function.

In the present case, the values chosen for convergence of the model parameters are TolFun $= 1.0 \times 10^{-6}$, TolX $= 1.0 \times 10^{-3}$ and the maximum number of iterations are 400. To solve the differential equations, fourth-order Runge–Kutta method with fixed step size is used. Simulink is used for the simulation of the closed loop multivariable system.

10.1.1   Simulation Examples

Two TITO transfer function matrix are taken for simulation study.

Example 1 To check the applicability of the proposed method, the system in Eq. (10.9) is considered.

$$G_p(s) = \begin{bmatrix} \dfrac{2.5e^{-1s}}{15s-1} & \dfrac{1e^{-1.5s}}{14s+1} \\[2ex] \dfrac{1e^{-1.5s}}{15s+1} & \dfrac{-4e^{-1s}}{20s-1} \end{bmatrix} \tag{10.9}$$

The decentralized Proportional and Integral (PI) controller settings used in the simulation are:

$$k_{c11} = 3.3459, \ \tau_{i11} = 8.4362; \ k_{c22} = -2.6733, \ \tau_{i22} = 11.0124.$$

To show the enhanced performance of the proposed method, TITO is simulated with the individual transfer function as a FOPTD process. The comparison between the converged parameters and the actual process is depicted. In the first step (Step 1), the decentralized PI controller is used and the process is monitored for a step change in the set point for the process response and the interaction. The response and the interaction curves are analysed to obtain the data for determining the guess value. In the second step (Step 2), the FOPTD model is assumed and the model parameters are obtained by using the initial guess values. The converged parameters are realized by using the least square optimization method. For optimization, *lsqnonlin* routine is used with the sample time of 0.01?s. In the optimization process, $k_p k_{c,des} = 2$ is used for obtaining the guess value of process gain k_{p11} and k_{p22}. The process gain k_{p12} and k_{p21} are calculated by the method given in Chapter 7. For the positive initial guess, the upper bound and the lower bound of the process gain is used as 20 times and 1/20 times of the guess value, respectively, and for the negative initial guess, the upper and the lower bound is used as 1/20 times and 20 times of the guess value, respectively.

For the system under consideration, the main diagonal transfer functions of the process transfer function matrix are unstable. In this case, $k_p k_{c,des} = 2$ is used for obtaining the initial guess value for the process gain because of the finite non-zero initial slope.

The lower bound is taken as $k_p k_{c,des} = 1$ because of the unstable transfer function (Chidambaram, 1998). For unstable FOPTD system (Chidambaram, 1998), we take $k_P k_{c,des} = 6$. Hence, the upper bound for the process gain is taken as $k_p k_{c,des} = 12$. The guess value of the time constant is taken as $t_s/8$ where t_s is the settling time of closed loop response. The upper and the lower bounds for the time constant are fixed as 5 times and 1/20 times of the guess value respectively. For the time delays, the lower bound is fixed as 0.25 times of the guess value and the upper bound is fixed as 4 times of the guess value. The converged model parameters obtained are similar to the actual parameters of system as represented in Eq. (10.9). The closed loop main responses and interactions for a unit step change in the input for the actual model are shown in Fig. 10.1 along with the closed loop main responses and interactions of the identified model.

Fig. 10.1 shows the actual and the model process response. The initial guess value, converged value, IAE, ISE and computational time are given in Table 10.1 The final converged parameters are obtained in 36 iterations in 361 seconds. The effect of perturbation in initial guess value by ±10% on the IAE, ISE, number of iterations and computational time are also listed separately in Table 10.2. Here the IAE and ISE values used are the sum of individual IAE and ISE value of each transfer function. The results suggest that the method proposed is robust even in the presence of uncertainty in guess values. The model parameters are identical to the actual process parameters. In the presence of noise in the system, the guess values will change according to the response and the interaction curve obtained for the process in Step 1. If the initial guess values for the parameters are considered as any random values, then the optimization method does not converge.

To observe the effect of noise, a random noise of standard deviation 0.01 is added in the actual process model. The noisy output is used to identify the model. The guess values are calculated by using the noisy output of the process. Using

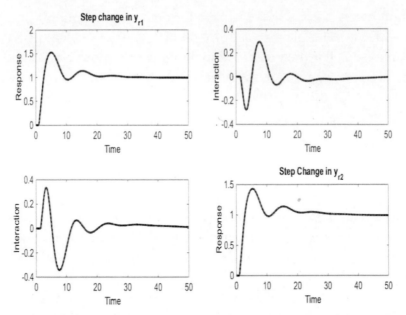

FIGURE 10.1   Closed loop responses of identified model and actual FOPTD (Example 1) Solid line: identified model; dashed line: actual model

the least square optimization method, the final converged values are calculated. The system is simulated for unit step change in the set point and the values of integral time domain performance are calculated. The values indicate the close match between the actual model parameters and the parameters identified. The values along with the computation time and the number of iterations are listed in Table 10.3 for the noise experiment. The computation time and the number of iterations for the optimization in this case are obtained as 335 and 33 respectively. The proposed method is found to be straightforward and easily implementable even in the presence of noise.

Example 2 Consider the unstable transfer function matrix in Eq. (10.10).

$$G_p(s) = \begin{bmatrix} \dfrac{0.3960e^{-0.2s}}{4.572s - 1} & \dfrac{1.7255e^{-0.4s}}{1.807s + 1} \\[2ex] \dfrac{-0.0585e^{-0.2s}}{2.174s + 1} & \dfrac{1.9713e^{-0.4s}}{1.801s - 1} \end{bmatrix} \qquad (10.10)$$

TABLE 10.1 Identified model parameters without noise using PI controllers (Example 1)

Model ij		k_{pij}	τ_{ij}	θ_{ij}	IAE	ISE
11	guess	0.5977	4.0000	1.0000	7.53×10^{-4}	1.29×10^{-9}
	converged	2.5001	15.0006	1.0000		
21	guess	1.2026	6.2500	1.5000	3.29×10^{-4}	5.87×10^{-9}
	converged	1.0001	15.0012	1.4999		
12	guess	1.2626	6.2500	1.5000	1.66×10^{-4}	1.59×10^{-9}
	converged	0.9999	13.9993	1.5001		
22	guess	−0.7481	5.5000	1.0000	1.21×10^{-4}	8.71×10^{-10}
	converged	−3.9991	19.9998	1.0000		

Computation time: 361 s; number of iterations: 36

TABLE 10.2 Effect of the initial guess values on IAE, ISE, computation time, and the number of iterations (Example 1)

Guess variation	N.I.	IAE	ISE	Computation time (s)
+10%	37	6.78×10^{-4}	1.036×10^{-8}	392
−10%	34	9.57×10^{-4}	1.63×10^{-8}	339

TABLE 10.3 Identified model parameters with noise using PI controller (Example 1)

Model ij		K_{pij}	τ_{ij}	θ_{ij}	IAE	ISE
11	guess	0.5977	4.0000	1.0000	0.8009	0.0101
	converged	2.5044	15.0191	0.9997		
21	guess	1.2031	6.2500	1.5000	0.8030	0.0102
	converged	0.9952	14.9173	1.4964		
12	guess	1.2622	6.2500	1.5000	0.8008	0.0101
	converged	1.0018	14.0501	1.5024		
22	guess	−0.7481	5.5000	1.0000	0.8029	0.0102
	converged	−4.0003	20.0008	0.9993		

Computation time: 335 s; number of iterations: 33

The decentralized PI controllers parameters used for identification are given in Eq. (10.11a) and Eq. (10.11b).

$$k_{c11} = 12.1419, \quad \tau_{i11} = 7.1616 \tag{10.11a}$$

$$k_{c22} = 1.3032, \quad \tau_{i22} = 3.4967 \tag{10.11b}$$

TABLE 10.4 Identified model parameters without noise using PI controllers (Example 2)

Model ij		k_{pij}	τ_{ij}	θ_{ij}	IAE	ISE
11	guess	0.1647	5.6250	0.2000	8.01×10^{-4}	7.30×10^{-8}
	converged	0.3960	4.5734	0.2002		
21	guess	−0.2038	5.2500	0.2000	2.88×10^{-4}	5.98×10^{-9}
	converged	−0.0585	2.1739	0.1999		
12	guess	1.7859	5.0000	0.4000	1.30×10^{-3}	6.32×10^{-8}
	converged	1.7251	1.8069	0.3999		
22	guess	1.5347	6.2500	0.4000	6.20×10^{-4}	1.20×10^{-8}
	converged	1.9714	1.8012	0.4001		

Computation time: 353 s; number of iterations: 31

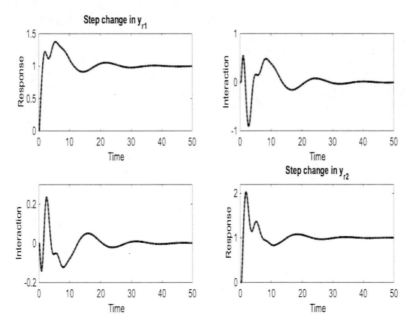

FIGURE 10.2 Closed loop responses of identified model and actual FOPTD (Example 2) Solid line: identified model; dashed line: actual model

Similar procedure as mentioned earlier in this chapter is followed for the system given in Eq. (10.10). The system is simulated using a decentralized PI controller to obtain the response and interaction curve. With the help of these curves, the initial guess values are obtained. Using the least square optimization routine, the converged model parameters are produced. The converged parameters and guess values are listed in Table 10.4. The IAE, ISE, number of iterations, and computation time are also tabulated in Table 10.4. The closed loop main responses and interactions for step unit change in set point for the actual given model and the identified model are shown in Fig. 10.2. The figure shows that the match between the actual model and the identified model is good. The converged parameters and the actual process parameter are found to be identical. The performance of the proposed algorithm in order to identify the model is reasonably acceptable.

The number of iterations and the computation time to obtain the converged parameters are found as 31 and 353 s respectively. Thus this method is time efficient. In the presence of noise, the same procedure can be repeated to identify the process parameters. If the initial guess values for the parameters are considered as any random value, then the optimization method does not converge.

10.2 Identification of Systems Controlled by Centralized Controllers

This section presents a simple and generalized method for obtaining reasonable initial guess values for the identification of unstable FOPTD transfer function model of multivariable unstable systems controlled by centralized PI/PID controllers.

10.2.1 Identification of a Multivariable System

Consider a centralized TITO unstable multivariable system shown in Fig. 3.3 (Chapter 3). The process diagonal transfer function models are unstable FOPTD. $G_p(s)$ and $G_c(s)$ are the process transfer function matrix and the centralized controller matrix respectively.

$$G_p(s) = \begin{bmatrix} g_{11} & g_{12} \\ g_{21} & g_{22} \end{bmatrix} ; \qquad G_c(s) = \begin{bmatrix} g_{c11} & g_{c12} \\ g_{c21} & g_{c22} \end{bmatrix} \qquad (10.12)$$

The controller settings can be selected such that the stable process responses and interaction are obtained. The present example focuses on the identification of the systems having an unstable main loop diagonal transfer functions elements (g_{11} and g_{22}) and a stable off-diagonal transfer function elements (g_{12} and g_{21}). The transfer function models used are expressed as in Eq. (10.13).

$$g_{ii}(s) = \frac{k_{pii}e^{-\theta_{ij}s}}{(\tau_{ii}s - 1)}; \text{ for } i = 1, 2; \; g_{ij}(s) = \frac{k_{pij}e^{-\theta_{ij}s}}{(\tau_{ij}s + 1)}; \text{ for } i \neq j = 1, 2 \qquad (10.13)$$

A unit step change is introduced into set point y_{r1} while the other loop is kept closed with set point unchanged. From this set point change, the main response y_{11} and the interaction y_{21} are obtained. Similar unit step change is introduced in y_{r2} to obtain the main response y_{22} and the interaction y_{12}. The initial guess value plays a vital role in the optimization method. The method proposed by Dhanya Ram and Chidambaram (2016) for stable system is extended to an unstable system. For identification of the model, the process response and interaction values are used to find the initial guess values. The initial guess values for time delay can be taken as the same as the corresponding closed loop time delay values. The guess value for time constant can be taken as $t_s/8$ where t_s are the settling times for the closed loop responses. The relation to obtain the guess value for the process gain of the model is obtained using the system characteristic equation $(I + G_p G_c = 0)$ and applying the Routh Hurwitz method (Gupta, 1995). The initial guess values for process main diagonal gains are obtained from the relation $k_p k_c = 2I$.

Then, the same magnitude of step change is introduced to the model set point y_{mr1} with all other remaining set points kept unchanged and all other loops under closed loop condition. The main response y_{m11} and the interaction response y_{m21} are obtained. Similarly, the same magnitude of step change is introduced in set point y_{mr2} The main response y_{m22} and interaction response y_{m12} are obtained. The closed loop responses are compared with the main and interaction responses of the actual process. The optimization problem is formulated so as to select the model parameters in order to minimize the sum of squared errors between the model and the actual process responses.

$$SSQ = \text{minimize } f = \sum_{i=1}^{2} \sum_{j=1}^{2} \sum_{k=1}^{n} [y_{mij}(t_k) - y_{ij}(t_k)]^2 \Delta t \qquad (10.14)$$

where n is the number of data points in the response of the model and the process. The sum of the ISE values for response and interaction is considered as a scalar objective function, as in Eq. (10.14). The model parameters are to be selected by minimizing the scalar function. The routine *lsqnonlin* in Matlab is used to solve the optimization problem. The initial guess determines the convergence rate and computation time. The routine depends on the Trust–Region–Reflective algorithm. In the present case, the values chosen for convergence of the model parameters are TolFun $= 1.0 \times 10^{-6}$, TolX $= 1.0 \times 10^{-3}$ and the maximum number of iterations are 400. To solve the differential equations, fourth-order Runge–Kutta method with fixed step size is used. Simulink is used for the simulation of the closed loop multivariable system.

10.2.2 Simulation Examples

Example 3 The centralized TITO system in Eq. (10.15) is considered with the L/τ (time delay to time constant) ratio in the range 0.05 to 0. 1.

$$G_p(s) = \begin{bmatrix} \dfrac{2.5e^{-s}}{15s-1} & \dfrac{1e^{-1.5s}}{14s+1} \\ \dfrac{1e^{-1.5s}}{15s+1} & \dfrac{-4e^{-s}}{20s-1} \end{bmatrix} \tag{10.15}$$

To obtain a stable response, centralized PI controller, G_c, is used. The controller parameters are

$$K_c = \begin{bmatrix} 3.1018 & 1.2127 \\ 1.1344 & -2.4783 \end{bmatrix} ; \quad \tau_I = \begin{bmatrix} 7.4656 & 9.9287 \\ 9.3484 & 9.7453 \end{bmatrix} \tag{10.16}$$

The process model is considered as an unstable FOPTD. To obtain the guess values for the process gain, the matrix K_c, as in Eq. (10.16), is substituted in the relation, $K_p K_c = 2I$. For the positive initial guess value of the process gain, the upper bound is taken as 20 times and lower bound is as 1/20 times of the guess value, whereas for the negative initial guess value, the upper bound is taken as 1/20 times and lower bound is taken as 20 times of the guess value. The guess values for time constant are taken as $t_s/8$. The upper bound and lower bound is taken as 5 times and 1/20 times of the guess values, respectively. The guess value for time delay is taken as the same the closed loop time delay. The upper and lower bounds are 4 times and 0.25 times of the guess values, respectively. The least square optimization method is used for obtaining the converged parameters. *lsqnonlin* Matlab routine is used for optimization, with sample time 0.01 s. The closed loop main responses and interactions are shown in Fig. 10.3 for both the actual model and the identified model. The figure shows a good match between the actual model and the identified model. The initial guess value, the converged parameters, IAE, ISE, number of iterations, and computational time are listed in Table 10.5.

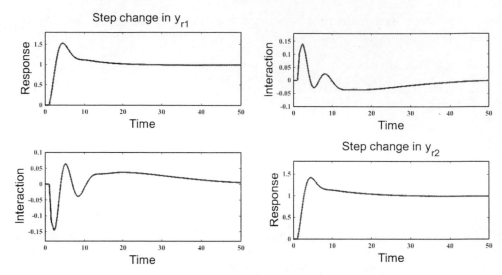

FIGURE 10.3   Closed loop responses of actual FOPTD and identified model (Example 3) Solid line: actual model; dashed line: identified model. Process response curves overlap.

TABLE 10.5   Model identification results without noise using PI centralized controller (Example 3)

Model $G_p(ij)$		K_{pij}	τ_{ij}	θ_{ij}	IAE	ISE
11	Guess	0.5469	3.5000	1.0000	3.39×10^{-10}	2.03×10^{-21}
	Converged	2.5000	15.000	1.0000		
21	Guess	0.2503	7.5000	1.0000	6.34×10^{-10}	7.83×10^{-21}
	Converged	1.0000	15.000	1.5000		
12	Guess	0.2676	6.7500	1.0000	5.46×10^{-10}	7.02×10^{-21}
	Converged	1.0000	14.000	1.5000		
22	Guess	−0.6845	4.3750	1.0000	1.34×10^{-09}	6.75×10^{-20}
	Converged	−4.0000	20.000	1.0000		

Computation time: 274 s; number of iterations: 17

Effect of Change in Guess Value

A heuristic analysis is done to study the effect of the guess values on system identification. A change of +10% and −10% in initial guess value is applied. The effect of variation in the initial guess values on computation time, IAE value and the number of iterations is listed in Table 10.6.

TABLE 10.6 Effect of initial guess values on the number of iterations, IAE, ISE, and computation time (Example 3)

Guess variation	IAE	ISE	N.I.	Computation time (s)
+10%	6.98×10^{-8}	5.00×10^{-17}	13	242
−10%	3.76×10^{-8}	1.45×10^{-17}	19	298

TABLE 10.7 Model identification results with noise using PI centralized controller (Example 3)

Model (ij)		K_{pij}	τ_{ij}	θ_{ij}	IAE	ISE	Time (N.I.)
11	Guess	0.5469	3.7500	1.0000	0.8021	0.0101	400(23)
	Converged	2.5052	15.0253	0.9992			
21	Guess	0.2503	5.6250	1.0000	0.8010	0.0101	
	Converged	0.9916	14.8459	1.5022			
12	Guess	0.2676	5.0000	1.0000	0.8021	0.0101	
	Converged	1.0014	14.0276	1.4981			
22	Guess	−0.6845	4.7500	1.0000	0.8010	0.0101	
	Converged	−4.0037	20.0136	0.9997			

Effect of Measurement Noise

The effect of noise on parameter identification is also studied by including a random noise of SD = 0.01 in the actual system. The initial guess values are obtained from the main response and interactions of noisy output as discussed earlier. The converged parameters, initial guess values, IAE, ISE, number of iterations, and computation time are listed in Table 10.7. The process response to the step change in inputs in the presence of noise is shown in the Fig. 10.4. A close match between the actual model and the identified model, depicting the robust performance of the method, is observed.

Effect of Different Controller Setting

The method is implemented to identify the model using a different PI controller setting as given in Eq. (10.17).

$$K_c = \begin{bmatrix} 1.95 & 0.76 \\ 0.72 & -1.56 \end{bmatrix} \quad \tau_I = \begin{bmatrix} 11.86 & 15.77 \\ 14.85 & 15.48 \end{bmatrix} \tag{10.17}$$

The guess values are obtained from the response and interaction curves for the step change in set point in y_{r1} and y_{r2} respectively. The obtained guess

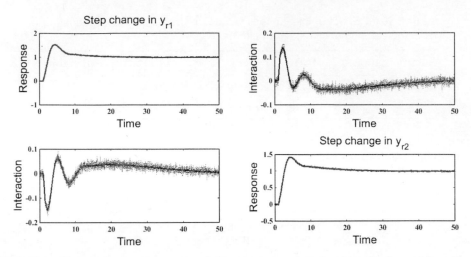

FIGURE 10.4 Closed loop responses of the process and the identified model, with noise (Example 3)
Solid line: actual model; dashed line: identified model. Process response curves overlap.

TABLE 10.8 Model identification results with different controller settings (Example 3)

Model (ij)		K_{pij}	τ_{ij}	θ_{ij}	IAE	ISE	Time (N.I.)
11	Guess	0.4346	7.5000	1.000	5.09×10^{-09}	4.06×10^{-19}	249 (15)
	Converged	2.5000	15.000	1.000			
21	Guess	0.2006	9.0000	1.000	1.03×10^{-08}	1.65×10^{-18}	
	Converged	1.0000	15.000	1.500			
12	Guess	0.2117	11.000	1.000	2.35×10^{-09}	6.69×10^{-20}	
	Converged	1.0000	14.000	1.500			
22	Guess	−0.5433	8.1250	1.000	4.39×10^{-09}	2.76×10^{-19}	
	Converged	−4.0000	20.000	1.000			

values and converged parameters are tabulated in Table 10.8 along with IAE, ISE, computational time and the number of iterations. The value indicates the close match between identified parameters and actual process parameters. The proposed method for the identification of model is found to be unaffected by the values of controller settings used. The responses are shown in Fig 10.5. If arbitrary initial guess values for the model parameters are used, then convergence is not obtained.

Example 4 Diagonally unstable centralized TITO transfer function matrix is considered with the L/τ (time delay to time constant) ratio in the range 0.04 to

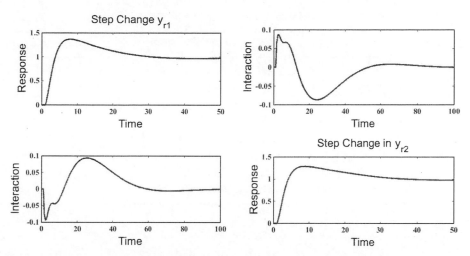

FIGURE 10.5 The closed loop responses of the process and the identified model with different controller settings (Example 3)
Solid line: actual model; dashed line: identified model. Process response curves overlap.

0.22 as given in Eq. (10.18).

$$
G_p(s) = \begin{bmatrix} \dfrac{0.4e^{-0.2s}}{4.5s - 1} & \dfrac{1.7e^{-0.4s}}{1.8s + 1} \\[2ex] \dfrac{-0.06e^{-0.2s}}{2.0s + 1} & \dfrac{2e^{-0.4s}}{1.8s - 1} \end{bmatrix} \tag{10.18}
$$

The centralized PI controller setting used are as in Eq. (10.19).

$$
k_c = \begin{bmatrix} 13.4 & -17.8 \\ 10.5 & 1.44 \end{bmatrix}; \quad \tau_I = \begin{bmatrix} 3.7 & 3.7 \\ 3.5 & 1.8 \end{bmatrix} \tag{10.19}
$$

A method similar to the one implemented in the first example is followed. The system is simulated in Matlab for obtaining the response and interaction curves. The guess value for the process gain of the system, the time delay, and the time constant are selected using the process response and interaction. The upper and lower bounds are selected as mentioned in Example 3. Least square optimization technique is used for obtaining the parameters. The guess value, converged parameters, and IAE and ISE values are listed in Table 10.9. The computation time and the number of iterations are 120 s and 15 respectively. The closed loop response of the actual system and identified model are shown in Fig. 10.6 on the same plot. Both curves are overlap each other which shows good match between the identified model and actual system.

TABLE 10.9   Model identification results without noise using PI centralized controller (Example 4)

Model (ij)		K_{pij}	τ_{ij}	θ_{ij}	IAE	ISE	Time (N.I.)
11	Guess	0.1021	2.6250	0.2000	2.85×10^{-8}	8.48×10^{-17}	120 (15)
	Converged	0.4000	4.5000	0.2000			
21	Guess	−0.0355	3.2500	0.2000	1.95×10^{-8}	3.13×10^{-17}	
	Converged	−0.0600	2.0000	0.2000			
12	Guess	1.2626	2.5000	0.2000	2.85×10^{-8}	6.98×10^{-17}	
	Converged	1.7000	1.8000	0.4000			
22	Guess	0.9505	2.1250	0.2000	1.94×10^{-8}	2.60×10^{-17}	
	Converged	2.0000	1.8000	0.4000			

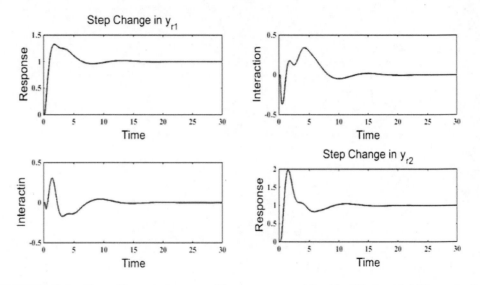

FIGURE 10.6   Closed loop responses of the process and the identified model (Example 4)

The same analysis of the effect of change in the guess values, the effect of measurement noise, and the effect of change in controller settings is studied. It is seen that robust performance is obtained for this example. If arbitrary guess values are used, then no convergence is obtained.

Summary

The closed loop identification technique to determine the FOPTD models of unstable TITO system under decentralized and centralized controllers have been given in this chapter. The technique makes use of the optimization method using step response. To obtain the values of the initial guesses, a simple method relating to the main and interaction responses of the process is presented. The method leads to quick and guaranteed convergence. The method proves to be an efficient way to identify the model. The simulation study shows a good match between the identified model and the actual model. The proposed method is robust for measurement noise also. The proposed method does not depend on the controller setting values used.

Problems

1. For the given system, design decentralized PI controllers and carry out optimization identification method to estimate the FOPTD model parameters.

$$G_p(s) = \begin{bmatrix} \dfrac{0.4e^{-0.3s}}{6s-1} & \dfrac{1.7e^{-0.2s}}{3s+1} \\ \dfrac{-0.06e^{-0.3s}}{4s+1} & \dfrac{2e^{-0.3s}}{3s-1} \end{bmatrix}$$

2. For the given system, design decentralized PI controllers and carry out optimization identification method to estimate the FOPTD model parameters.

$$G_p(s) = \begin{bmatrix} \dfrac{2.5e^{-s}}{20s-1} & \dfrac{1e^{-1.5s}}{16s+1} \\ \dfrac{1e^{-1.5s}}{18s+1} & \dfrac{-4e^{-s}}{25s-1} \end{bmatrix}$$

3. For the given system, design decentralized PI controllers and carry out optimization identification method to estimate the FOPTD model parameters.

$$G_p(s) = \begin{bmatrix} \dfrac{0.4e^{-0.2s}}{6s-1} & \dfrac{1.7e^{-0.4s}}{2s+1} \\ \dfrac{-0.06e^{-0.2s}}{3s+1} & \dfrac{2e^{-0.4s}}{3s-1} \end{bmatrix}$$

4. For the system in Problem (3), design centralized PI controllers and carry out optimization identification method to estimate the FOPTD model parameters.

5. Consider the unstable system where all the subsystems are unstable:

$$K_P G_P = \begin{bmatrix} \dfrac{1.6667e^{-s}}{(1.6667s-1)} & \dfrac{e^{-s}}{(1.6667s-1)} \\ \dfrac{0.8333e^{-s}}{(1.6667s-1)} & \dfrac{1.6667e^{-s}}{(1.6667s-1)} \end{bmatrix}$$

Consider the centralised PI controllers:

$$G_C = \begin{bmatrix} 1.40 + \left(\dfrac{0.05}{s}\right) & -0.84 - \dfrac{0.03}{s} \\ -0.70 - \left(\dfrac{0.03}{s}\right) & 1.40 + \left(\dfrac{0.05}{s}\right) \end{bmatrix}$$

Show by simulation that using the usual step input excitation with the set point, the identification method does not identify the model parameters. Try with the step-up input followed by step-down input excitation in the set points as considered in Chapter 9, and carry out the optimization method of identifying model parameters.

11

Centralized PI Controllers Based on Steady State Gain Matrix

A simple method is given in this chapter to design multivariable Proportional and Integral (PI) and Proportional Integral and Derivative (PID) controllers for stable multivariable systems. The method needs only the Steady State Gain Matrix (SSGM). The method is based on the static decoupler design followed by the SISO PI/PID controllers design and combines the resulting decoupler and the diagonal PI(D) controllers as centralized controllers. The result of the present method is shown to be equivalent to the empirical method proposed by Davison (1976). Three simulation examples have been given in this chapter. The performance of the controllers has been compared with the reported centralized controller based on the multivariable transfer function matrix.

11.1 Introduction

Design of PI controllers for stable MIMO processes is difficult when compared to the SISO processes due to the interaction between the input–output variables. As discussed in Chapter 3, the MIMO processes can be controlled by decentralized or decoupled controllers or by centralized PI controllers. For mild interacting stable MIMO processes, design of decentralized PI controllers based on the diagonal

223

processes (based on proper pairing) is carried out with a suitable detuning method. The detuning step involves decreasing the controller gains by F, multiplying integral by F, and decreasing derivative times by F. Here, F is the detuning factor which may vary from 1.5 to 4, depending on the extent of interactions, dictated by the RGA. For systems with large interactions, the decoupler (D) is designed so as to make the MIMO processes into n SISO processes. The PI controllers (G_c) are designed for the resulting SISO processes. The overall control system is the combined decoupler and the diagonal controllers (DG_c). We can also design, straight away, the centralized PID controllers.

The SSGM of the multivariable system can be obtained more easily than identifying the dynamic model parameters and, hence, it is easier to tune the controllers. In this chapter, derivation of a simple method of designing multivariable PI controllers is explained, based on SSGM of the system.

11.2   Multivariable System

The stable multivariable system (Fig. 3.3 in Chapter 3) is assumed to be of the form given in Eq. (11.1).

$$Y(s) = G_P(s)U(s) \tag{11.1}$$

where Y is the $(n \times 1)$ vector of the output variables, and U is the $(n \times 1)$ vector of the manipulated variable s. $G_P(s)$ is the $n \times n$ transfer function matrix of the system. The centralized controllers are assumed as given in Eq. (11.2).

$$G_c(s) = K_C + (K_I)/s \tag{11.2}$$

where K_C and K_I are matrices of size $n \times n$.

$$K_C = [k_{c,ij}] \tag{11.3}$$

$$K_I = [k_{c,ij}/\tau_{I,ij}], \quad i = 1, \ldots, n \text{ and } j = 1, \ldots, n \tag{11.4}$$

$k_{c,ij}$ is the ij^{th} element of the proportional gain matrix (K_C) and $\tau_{I,ij}$ are the corresponding integral time of the controllers.

11.2.1   Davison Method

Davison (1976) (Section 3.130] has proposed an empirical method of tuning multivariable PI controllers for stable systems, where the matrices K_C and K_I in

Eq. (11.2) are given by Eq. (11.5) and Eq. (11.6).

$$K_C = \delta_1 [G_P(s=0)]^{-1} \tag{11.5}$$

$$K_I = \delta_2 [G_P(s=0)]^{-1} \tag{11.6}$$

Here, $[G_P(s = 0)]^{-1}$ is called the rough tuning parameters. The rough tuning matrix is the inverse of the SSGM. The tuning parameters range is usually 0 to 1 and the recommended values are 0.1 to 0.5 for δ_1 and 0.05 to 0.2 for δ_2. Methods are available to estimate the SSGM of multivariable systems. For stable and integrating system (K_P/s), we can substitute K_P for $G_P(s=0)$.

Consider a nth order transfer function model with equal time constant. It can be shown that the value of $(k_c k_p)_{max} = 8$ for $n = 3$ and $(k_c k_p)_{max} = 1$ for $n = 12$. The design value would be, respectively, 4 and 0.5 for $n = 3$ and $n = 12$. Consider a decentralized control system for a $n \times n$ transfer function matrix of such models. Assuming a detuning factor of 2 for multivariable systems, the controller settings would be 2 and 0.25 respectively for $n = 3$ and $n = 12$. Hence $K_c = \delta_1 K_P^{-1}$ where $\delta_1 = 0.25$ to 2 can be used. The integral time for the decentralized system that can be used is $2\tau_I$ where τ_I is the integral time for the scalar system. Since $k_I (= k_P/\tau_I)$, K_I can be recommended as $\delta_2 K_P^{-1}$ where $\delta_2 = 0.125$ to 1. The range for δ_1 and hence the range for δ_2 depends on the order of the system (or ratio of time delay to time constant of the system) and the interaction of the system. This analysis is for a decentralized control system. Similar analysis can be carried out for a centralized control system after designing a suitable decoupler. This analysis will be presented in the next section.

11.3 Derivation of Controllers Design

Consider a stable MIMO system as given in Fig. 11.2. G_P is the process transfer function matrix. It is assumed that the pairing based on Relative Gain Array (RGA) is carried out and the recommended pairing is already considered as shown in Fig. 11.2. We have to select the decoupler $[D(s)]$ such that the resultant system of the combined process and the decoupler [GpD] becomes a diagonal system (Zhang, 2012) as in Eq. (11.7)

$$G_P(s)D(s) = Q(s) \tag{11.7}$$

where $Q(s)$ is the desired transfer function matrix (diagonal matrix) of the decoupled system. For example, if $D(s)$ is selected such that it is equal to $[G_P(s)]^{-1}$, then $Q(s) = I$. This is an ideal situation. Instead, let us consider Eq. (11.8) and Eq. (11.9).

$$Q_{11} = \exp(-\tau_{d1,d}s)/(\tau_{1,d}s + 1) \qquad (11.8)$$

$$Q_{22} = \exp(-\tau_{d2,d}s)/(\tau_{2,d}s + 1) \qquad (11.9)$$

Q_{12} and Q_{21} are both considered 0. Under this condition, we denote the decoupled system as P. Here, $\tau_{d1,d}$ and $\tau_{1,d}$ are the time delay and time constant of the decoupled system, P_{11}. Similarly, we can define other terms such as $\tau_{d2,d}$ and $\tau_{2,d}$ for P_{22}. The values of time constants ($\tau_{1,d}$, $\tau_{2,d}$) and time delays ($\tau_{d1,d}$, $\tau_{d2,d}$) of the decoupled system are slightly greater than that of the open loop system so as to take into account the interactions among the loops.

Let us consider the static decoupler for the system as in Eq. (11.10).

$$D = [G_P(s = 0)]^{-1} \qquad (11.10)$$

For the resulting decoupled system ($P = G_pD$), suitable PI controllers need to be designed, as in Eq. (11.11).

$$G_{c,d}(s) = (K_{c,d}) + [(K_{I,d})/s] \qquad (11.11)$$

Here, $K_{c,d}$ and $K_{I,d}$ are diagonal matrices. Hence, the overall controller matrix to be implemented on the process (a combination of the decoupler and the diagonal PI controllers) is given by Eq. (11.12).

$$G_c(s) = [G_P(s = 0)]^{-1}[G_{c,d}(s)] \qquad (11.12)$$

Let the controllers ($k_{c,p}$, $k_{I,p}$) be designed for the worst case of a FOPTD model among the diagonal elements of P (i.e., with larger delay, smaller time constant) so that same PI controller setting be used.

$$G_{cd}(s) = [(k_{c,p}) + (k_{I,p}/s)][I] \qquad (11.13)$$

Hence, the overall controller system is given by Eq. (11.14).

$$G_c(s) = [G_P(s = 0)]^{-1}[(k_{c,p}) + (k_{I,p}/s)][I] \qquad (11.14)$$

Here, $G_c(s)$ is the full matrix and is called a centralized controller. Eq. (11.14) can be written as Eq. (11.15).

$$G_c(s) = \delta_1[G_P(s = 0)]^{-1} + (\delta_2/s)[G_P(s = 0)]^{-1} \qquad (11.15)$$

The reason for introducing the new parameters $\delta_1(= k_{c,p})$ and $\delta_2 (= k_{I,p})$ is so that the relation with the Davison method (Section 3.13) can be understood. Eq. (11.15) can be rewritten as Eq. (11.16a) and Eq. (11.16b).

$$G_c = \delta_1[K_P]^{-1} + (\delta_2/s)[K_P]^{-1} \tag{11.16a}$$

$$G_c = K_C + (K_I/s) \tag{11.16b}$$

Where,

$$K_C = \delta_1[K_P]^{-1} \text{ and } K_I = \delta_2[K_P]^{-1}$$

Here, K_C is the centralized controller gain matrix and K_I is the centralized integral gain matrix.

Let us now focus on the tuning of the PI or PID controllers for the resulting SISO–FOPTD systems. For a system, $G(s) = (1/(\tau s + 1))^n$, $k_p k_c$ can vary from ∞ to 1 as n varies from 1 to ∞. Hence, for a scalar system, the minimum of the maximum value of $k_p k_c$ is 1 (Chidambaram, 1998). For a MIMO system, the minimum of maximum is $K_P K_C = I$ (unity matrix). For a SISO system, the design value of k_c is recommended as $0.5 [k_P]^{-1}$. For multivariable systems, due to the interactions in the system, the value of $K_{C,des}$ obtained from the above relation is further reduced. In general, $K_{C,des} = \delta_1[K_P]^{-1}$, where δ_1 varies from 0.1 to 0.3.

For the decoupled scalar system given by Eq. (11.8) or Eq. (11.9), the value of $k_{c,p}$ is given by $0.9(\tau_{1,d}/\tau_{d1,d})$ and $\tau_{I,p}$ as $3\tau_{d1,d}$ (Ziegler and Nichols, 1942). For typical values of model parameters, we can assume the range of δ_1 as 0.1 to 3. The range of values for δ_2 is 0.05 to 1.5. Similarly, for a PID controller, the value of $k_{c,p}$ is given by $1.2(\tau_{1,d}/\tau_{d1,d})$, $\tau_{I,p}$ as $2\tau_{d1,d}$, and $\tau_{D,p}$ as $0.5\tau_{d1,d}$ (Ziegler and Nichols, 1942). For typical values of these parameters, we get the range for δ_1 as 0.1 to 3, for δ_2 as 0.05 to 1.5, and for δ_3 as 0.05 to 0.5.

$$G_C = \delta_1[K_P]^{-1} + (\delta_2/s)[K_P]^{-1} + (\delta_3 s)[K_P]^{-1} \tag{11.17}$$

In case we consider the decoupler as the inverse of the system at ω_c rather than at $\omega = 0$, we get the resulting control structure given by Maciejowski (1989). However, this method requires the transfer function matrix of the process.

11.4 Simulation Studies

11.4.1 Example 1: Wood and Berry Column

The Wood and Berry (WB) distillation column plant is a multivariable stable system that has been studied extensively (Vijaykumar et al., 2012; Zhang, 2012). The transfer function matrix of WB column is given by Eq. (11.18).

$$G_p(s) = \begin{bmatrix} \dfrac{12.8e^{-s}}{16.7s+1} & \dfrac{-18.9e^{-3s}}{21s+1} \\ \dfrac{6.6e^{-7s}}{10.9s+1} & \dfrac{-19.4e^{-3s}}{14.4s+1} \end{bmatrix} \qquad (11.18)$$

The elements of RGA are found to be $\lambda_{11} = \lambda_{22} = 2.01$ and $\lambda_{12} = \lambda_{21} = -1.01$. The centralized controller matrix based on synthesis method (Vijaykumar et. al., 2012) is given by Eq. (11.19).

$$G_C(s) = \begin{bmatrix} 0.1697 + \dfrac{0.0173}{s} & -0.0172 - \dfrac{0.0140}{s} \\ 0.0161 + \dfrac{0.0048}{s} & -0.0723 - \dfrac{0.0096}{s} \end{bmatrix} \qquad (11.19)$$

The inverse of the SSGM is obtained as in Eq. (11.20).

$$K_P^{-1} = \begin{bmatrix} 0.1570 & -0.1529 \\ 0.0534 & -0.1036 \end{bmatrix} \qquad (11.20)$$

For the present example, the PI controllers settings are calculated using different values of the tuning parameter δ_1 and δ_2 and the closed loop performances are evaluated. The settings $\delta_1 = 2.0$ and $\delta_2 = 0.30$ give a better closed loop performance. The resulting centralized control system is given by Eq. (11.21).

$$G_c(s) = \begin{bmatrix} 0.3140 + \dfrac{0.0471}{s} & -0.3058 - \dfrac{0.04587}{s} \\ 0.1068 + \dfrac{0.01602}{s} & -0.2072 - \dfrac{0.03108}{s} \end{bmatrix} \qquad (11.21)$$

The ratio of delay to time constant is small and hence a larger value of controller gain can be allowed However, the interaction is large ($\lambda_{11} = \lambda_{22} = 2.01$) which may reduce the controller gain. Overall, large gain can be used for the controllers ($\delta_1 = 2.0$).

The servo response of the closed loop system for a unit step change in the set point of y_{r1} is evaluated and the response in y_1 and the interaction in y_2 of the closed loop systems are shown in Fig. 11.1 (a). Similarly, the servo response of the closed loop system for a unit step change in the set point of y_{r2} is evaluated and the response in y_2 and the interaction in y_1 of the closed loop system are shown in Fig. 11.1 (b). Fig. 11.1 compares the responses of the present method

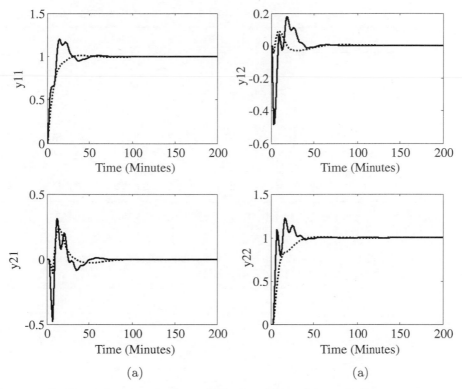

(a)   (a)

FIGURE 11.1 Servo response comparison of Wood and Berry (Example 1)
Solid line: present method with $\delta_1 = 2$ and $\delta_2 = 0.3$; dotted line: synthesis method (Vijaykumar et al., 2012).

with the centralized PI control systems designed based on the synthesis method, using the transfer function matrix (i.e., values of delays, time constants are known), reported by Vijaykumar et al. (2012). Vijaykumar et al. (2012) have shown their method is better than the reported methods. Table 11.1a shows that the sum of the IAE values for the main responses for the servo problem is lower for the present method, whereas the interaction is larger. Fig. 11.2 shows the performances of the two methods for the regulatory problems for a unit step change in the load variable, assuming the disturbances transfer function matrix is the same as the process transfer function matrix as shown in Fig. 3.5 (Chapter 3). An improved performance is obtained for the proposed method. Table 11.2a shows that the sum of the IAE values for the main responses and also the sum of the IAE values for the interactions are also lower for the regulatory problem. If the decouplers are removed, and the same diagonal PI controllers only are used, then the oscillatory responses are obtained for the servo problem as shown in Fig. 11.3.

TABLE 11.1a   IAE values for closed loop system (Example 1)

		IAE values					Sum	
	Method	y_{11}	y_{21}	y_{12}	y_{22}	Overall	Main Action	Interaction
Servo	Proposed	8.103	5.403	4.53	7.866	25.902	15.97	9.93
	Vijaykumar et al. (2012)	8.031	4.046	1.903	11.32	25.3	19.35	5.95
Regulatory	Proposed	55.5	37.32	87.67	89.37	269.86	144.87	124.99
	Vijaykumar et al. (2012)	97.08	48.53	141.6	174.9	462.11	271.98	190.13

Overall: $(y_{11} + y_{21} + y_{12} + y_{22})$; main action: $(y_{11} + y_{22})$; interaction: $(y_{21} + y_{12})$

TABLE 11.1b   IAE values for robustness of the closed loop system: regulatory problem for 10% perturbation in each gain and time delay (Example 1)

		IAE values					Sum	
Uncertainty	Method	y_{11}	y_{21}	y_{12}	y_{22}	Overall	Main Action	Interaction
Process Gain	Proposed	55.25	37.46	86.87	88.78	268.36	144.03	124.33
	Vijaykumar et al. (2012)	97.08	48.54	141.6	174.9	462.12	271.98	190.14
Process Delay	Proposed	57.73	39.77	91.46	93.36	282.32	151.09	131.23
	Vijaykumar et al. (2012)	97.08	48.53	141.6	174.9	462.11	271.98	190.13

The controller settings given by Eq. (11.21) are based on the SSGM and the tuning parameters are selected based on simulation of the closed loop system. The performance of the controllers for perturbation (10% increase of the nominal value) in each gain in the process is studied and the regulatory responses for a unit step change in the load variable are shown in Fig. 11.4. The same values for the tuning parameters ($\delta_1 = 2.0$ and $\delta_2 = 0.30$) are used. The IAE values are given in Table 11.1b. The sum of the IAE values for the main responses and also the sum of IAE values for the interactions are lower for the present method. Similar results are obtained for the variation in each time delay (10% increase of the nominal value) and the regulatory responses are shown in Fig. 11.5. The same values for the tuning parameters ($\delta_1 = 2.0$ and $\delta_2 = 0.30$) are used. As seen from Table 11.1b, a robust performance is obtained for the present method under uncertainty in the model parameters (gain and delay).

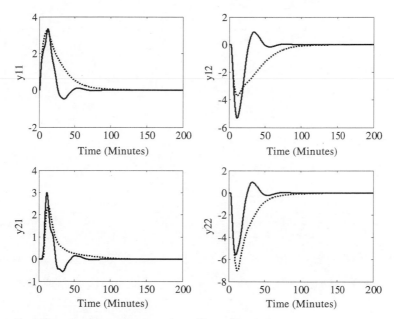

FIGURE 11.2   Regulatory response comparison (Example 1)
Solid line: present method with $\delta_1 = 2$ and $\delta_2 = 0.3$; dotted line: synthesis method (Vijaykumar et al., 2012)

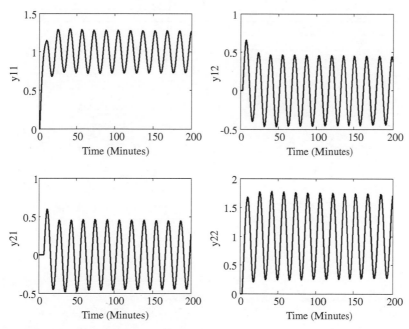

FIGURE 11.3   Servo response using only the diagonal elements of the proposed controller settings without the decouplers (Example 1)

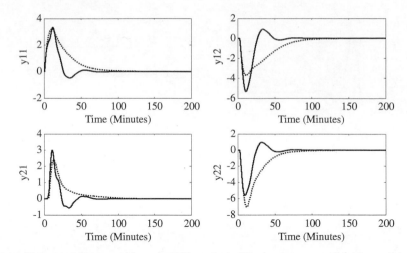

FIGURE 11.4   Robustness comparison (+10% perturbation in process gain) of regulatory response (Example 1)
Solid line: present method with $\delta_1 = 2$ and $\delta_2 = 0.3$; dotted line: synthesis method (Vijaykumar et al., 2012).

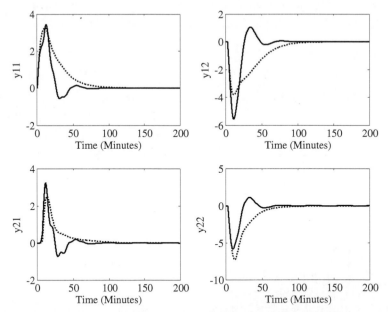

FIGURE 11.5   Robustness comparison (+10% perturbation in process delay) of regulatory response (Example 1)
Solid line: present method with $\delta_1 = 2$ and $\delta_2 = 0.3$; dotted line: synthesis method

11.4.2   Example 2: Industrial-Scale Polymerization (ISP) Reactor

The transfer function matrix for the stable system is given by (Vijaykumar et al., 2012; Chien et. al., 1999) is given in Eq. (11.22).

$$
G(s) = \begin{bmatrix} \dfrac{22.89e^{-0.2s}}{4.572s + 1} & \dfrac{-11.64e^{-0.4s}}{1.807s + 1} \\[3mm] \dfrac{4.689e^{-0.2s}}{2.174s + 1} & \dfrac{5.8e^{-0.4s}}{1.801s + 1} \end{bmatrix} \tag{11.22}
$$

The elements of RGA are found to be $\lambda_{11} = \lambda_{22} = 0.71$ and $\lambda_{12} = \lambda_{21} = 0.29$. The centralized controllers are designed by the synthesis method (Vijaykumar et al., 2012) as given in Eq. (11.23).

$$
G_C(s) = \begin{bmatrix} 0.1644 + \dfrac{0.0424}{s} & 0.1403 + \dfrac{0.0383}{s} \\[3mm] -0.1922 - \dfrac{0.0538}{s} & 0.0843 + \dfrac{0.0764}{s} \end{bmatrix} \tag{11.23}
$$

The inverse of the SSGM is given by Eq. (11.24).

$$
K_p^{-1} = \begin{bmatrix} 0.0310 & 0.0621 \\ -0.0250 & 0.1222 \end{bmatrix} \tag{11.24}
$$

For the present example, the PI controllers settings are calculated using different values of the tuning parameter δ_1 and δ_2 and the closed loop performances are evaluated. In the previous example, it was specified that the range of δ_1 is from 0.1 to 2, but in this example, the tuning parameters giving the best responses are found as to be $\delta_1 = 5$ and $\delta_2 = 1.5$. Since the system is less interactive ($\lambda_{11} = \lambda_{22} = 0.71$ and $\lambda_{12} = \lambda_{21} = 0.29$), and the delay to time constant is less, a larger value of $\delta_1 = 5$ and $\delta_2 = 1$ is permitted here. The resulting centralized PI control system is given by Eq. (11.25).

$$
G_c(s) = \begin{bmatrix} 0.155 + \dfrac{0.0465}{s} & 0.3105 + \dfrac{0.09315}{s} \\[3mm] -0.125 - \dfrac{0.0375}{s} & 0.611 + \dfrac{0.1833}{s} \end{bmatrix} \tag{11.25}
$$

Fig. 11.6 shows the servo responses of the present method along with the synthesis method (Vijaykumar et al., 2012). The present method works well. The IAE values for the responses and the interactions are given in Table 11.2 for both the

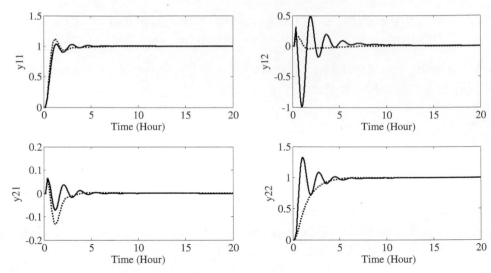

FIGURE 11.6 Servo response comparison (Example 2)
Solid line: present method with $\delta_1 = 5$ and $\delta_2 = 1.5$; dotted line: synthesis method (Vijaykumar et al., 2012)

TABLE 11.2   IAE values (Example 2)

		\multicolumn{7}{c}{IAE values}						
	Method	y_{11}	y_{21}	y_{12}	y_{22}	Overall	Sum Main Action	Interaction
Servo	Proposed	0.7993	0.1078	1.3437	1.068	3.3188	1.8673	1.4515
	Vijaykumar et al. (2012)	0.7594	0.1806	0.2634	1.628	2.8314	2.3874	0.444
Regulatory	Proposed	15.26	3.121	7.717	3.852	29.95	19.112	10.838
	Vijaykumar et al. (2012)	14.29	10.14	7.213	8.001	39.644	22.291	17.353

Overall: $(y_{11} + y_{21} + y_{12} + y_{22})$; main action: $(y_{11} + y_{22})$; interaction: $(y_{21} + y_{12})$

methods. The present method gives lower IAE values for the main responses. The interactions are slightly higher. Fig. 11.7 shows the performance of the two methods for the regulatory problems for a unit step change in the load variable (assuming the disturbances transfer function matrix is the same the process transfer function matrix. As seen from Table 11.2, for regulatory problems, both the main responses and also the interactions of the closed loop system are found to be better for the present method than the controller design proposed by Vijaykumar et al. (2012).

In the first example, the interaction is significant as shown by the RGA (elements of RGA are calculated as $\lambda_{11} = \lambda_{22} = 2.01$ and $\lambda_{12} = \lambda_{21} = -1.01$). The main

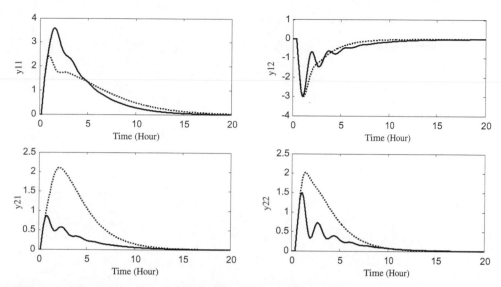

FIGURE 11.7   Regulatory response comparison (Example 2)
Solid line: present method with $\delta 1 = 5$ and $\delta 2 = 1.5$; dotted line: synthesis method (Vijaykumar et al., 2012)

responses of the present method are better than those of the synthesis method which is based on the transfer function matrix. The performances are better for the regulatory problem also. In the second example, the interaction is not significant as shown by the RGA ($\lambda_{11} = \lambda_{22} = 0.71$ and $\lambda_{12} = \lambda_{21} = 0.29$). For Example 2, for the regulatory problems, the main responses as well the interactions are better for the proposed method. For the servo problem, the responses are fewer, for the present method.

11.4.3   Example 3: Ogunnaike et al. Column

The transfer function matrix of a binary ethanol–water system of a pilot plant distillation column proposed by Ogunnaike et. al. (1983) is considered in Eq. (11.26).

$$G_p(s) = \begin{bmatrix} \dfrac{0.66e^{-2.6s}}{6.7s + 1} & \dfrac{-0.61e^{-3.5s}}{8.64s + 1} & \dfrac{-0.0049e^{-s}}{9.06s + 1} \\[2mm] \dfrac{1.11e^{-6.5s}}{3.25s + 1} & \dfrac{-2.36e^{-3s}}{5s + 1} & \dfrac{-0.01e^{-1.2s}}{7.09s + 1} \\[2mm] \dfrac{-34.68e^{-9.2s}}{8.15s + 1} & \dfrac{46.2e^{-9.4s}}{10.9s + 1} & \dfrac{0.87(11.61s + 1)e^{-s}}{(3.89s + 1)(18.8s + 1)} \end{bmatrix} \qquad (11.26)$$

The elements of RGA are found as in Eq. (11.27).

$$\Lambda = \begin{bmatrix} 2.0084 & -0.7220 & -0.2864 \\ -0.6460 & 1.8246 & -0.1786 \\ -0.3624 & -0.1026 & 1.4650 \end{bmatrix} \tag{11.27}$$

The centralized controller matrix based on Xiong et al. method (2007) is given by Eq. (11.28).

$$G_C(s) = \begin{bmatrix} 1.2266\left(1+\dfrac{1}{6.7s}\right) & -0.0716\left(1+\dfrac{1}{3.25s}\right) & 0.0017\left(1+\dfrac{1}{8.15s}\right) \\ 0.5758\left(1+\dfrac{1}{8.64s}\right) & -0.2219\left(1+\dfrac{1}{5.0s}\right) & 4.7035e^{-004}\left(1+\dfrac{1}{10.9s}\right) \\ 61.1085\left(1+\dfrac{1}{9.06s}\right) & 13.9406\left(1+\dfrac{1}{7.09s}\right) & 2.8540\left(1+\dfrac{1}{12.4150s}\right) \end{bmatrix} \tag{11.28}$$

The inverse of the SSGM is obtained as in Eq. (11.29).

$$[G_P(s=0)]^{-1} = \begin{bmatrix} 3.0430 & -0.5820 & 0.0104 \\ 1.1836 & -0.7731 & -0.0022 \\ 58.4481 & 17.8564 & 1.6839 \end{bmatrix} \tag{11.29}$$

For the present work, the PI controllers settings are calculated using different values of the tuning parameter δ_1 and δ_2, and the closed loop performances are evaluated. The settings giving the best responses are obtained for $\delta_1 = 0.5$ and $\delta_2 = 0.125$. Since the delay to time constant is large and the interaction is large for this system, a lesser value of tuning parameters ($\delta_1 = 0.5$ and $\delta_2 = 0.125$) are used. The obtained controller settings are given by:

$$G_c(s) = \begin{bmatrix} 1.5215+\dfrac{0.3804}{s} & -0.291-\dfrac{0.0727}{s} & 0.0052+\dfrac{0.0013}{s} \\ 0.5918+\dfrac{0.1479}{s} & -0.38655-\dfrac{0.0966}{s} & -0.0011-\dfrac{0.0003}{s} \\ 29.2240+\dfrac{7.3060}{s} & 8.9282+\dfrac{2.2320}{s} & 0.84195+\dfrac{0.2105}{s} \end{bmatrix} \tag{11.30}$$

Fig. 11.8 shows the servo responses of the present method with that of Xiong et al. (2007). The IAE values for the responses and the interactions are given in Table 11.3 for both the methods. Table 11.4 shows the sum of the IAE values for the main action ($y_{11} + y_{22} + y_{33}$) and sum of the IAE values for the interaction ($y_{21} + y_{31} +$

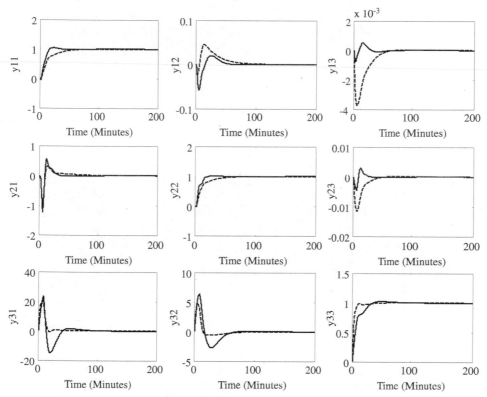

FIGURE 11.8 Servo response comparison (Example 3)
Solid line: present method with $\delta_1 = 0.5$ and $\delta_2 = 0.125$; dashed line: Xiong et al. (2007)

$y_{12} + y_{13} + y_{31} + y_{32}$) for servo and regulatory problems respectively. Fig. 11.9 shows the performance of the two methods for the regulatory problems (assuming the disturbances entering along with the manipulated variables). From Table 11.4, it can be seen that the sum of IAE values of servo response for the main actions is lower for the proposed method when compared with the Xiong method but the interaction is higher. Similar results are obtained for the regulatory problems also. The present method, even though it is based on SSGM, works well. The method is simple to use and the main responses of the closed loop system are found to be better than the method based on the full transfer function matrix.

Summary

Based on the SSGM, a simple method is given in this chapter to tune the centralized PI controllers. The basic idea is to use a static decoupler followed by the design of SISO PI controllers. The derived equations are shown to be equivalent of the

TABLE 11.3   IAE values for the closed loop system (Example 3)

	Method	y_{11}	y_{21}	y_{31}	y_{12}	y_{22}	y_{32}	y_{13}	y_{23}	y_{33}
Servo	Proposed	9.031	10.08	446.2	0.842	8.321	132.6	0.0115	0.0479	9.786
	Xiong et al. (2007)	16.64	11.16	223.5	1.447	15.45	66.04	0.0744	0.1722	3.958
Regulatory	Proposed	5.819	9.902	385.7	5.719	19.63	508.8	0.4294	0.0835	7.494
	Xiong et al. (2007)	7.034	10.51	294.3	4.434	27.81	354	0.0176	0.0490	4.557

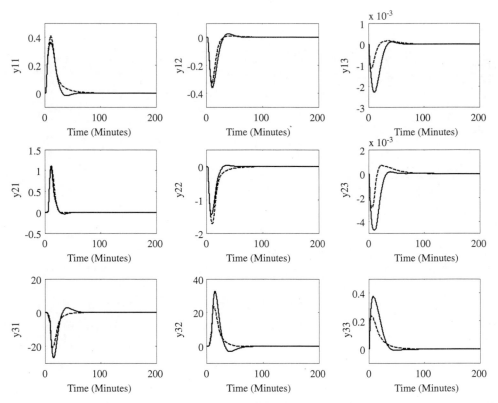

FIGURE 11.9   Regulatory response comparison (Example 3)
Solid line: present method with $\delta_1 = 0.5$ and $\delta_2 = 0.125$; dashed line: Xiong et al. (2007)

empirical method proposed by Davison. The main responses of the proposed controllers are shown to be better than the centralized PI controllers reported in the literature based on the full system transfer function matrix. This is illustrated in this chapter with three simulation examples. For the present method, only the knowledge of the SSGM is needed rather than the full dynamics (gain, time delay, time constant). The obtained main responses are good, signifying that the method can be recommended for interacting multivariable systems.

TABLE 11.4 Total IAE values for the closed loop system (Example 3)

	Method	Main Action	Interaction
Servo	Proposed	27.138	589.7814
	Xiong et al. (2007)	36.048	302.3936
Regulatory	Proposed	32.943	910.6339
	Xiong et al. (2007)	39.401	663.3106

Main Action: $(y_{11} + y_{22} + y_{33})$; interaction: $(y_{21} + y_{31} + y_{12} + y_{32} + y_{13} + y_{23})$

Problems

1. For the given system, design a static decoupler. Derive the transfer function matrix models for the combined process and the static decoupler $[G(s)D(s=0)]$.

$$G(s) = \begin{bmatrix} \dfrac{87.8}{(194s+1)} & \dfrac{-86.4(12.1s+1)}{(194s+1)(15s+1)} \\ \dfrac{108.2}{(194s+1)} & \dfrac{-109.6(17.3s+1)}{(194s+1)(15s+1)} \end{bmatrix}$$

Design PI controllers (g_c) for the diagonal elements of $G(s)D(s=0)$. Get the combined transfer function of $DG_C(s)$. Evaluate the servo responses of the controllers by Matlab and Simulink simulation.

2. For the given system, design a static decoupler. Derive the transfer function matrix models for the combined process and the static decoupler $[G(s)D(s=0)]$.

$$G(s) = \begin{bmatrix} \dfrac{0.5}{(0.1s+1)^2(0.2s+1)^2} & \dfrac{-1}{(0.1s+1)(0.2s+1)^2} \\ \dfrac{-1}{(0.1s+1)(0.2s+1)^2} & \dfrac{2.4}{(0.1s+1)(0.2s+1)^2(0.5s+1)} \end{bmatrix}$$

Design PI controllers for the diagonal elements of G(s)D(s=0). Evaluate the servo responses of the controllers by Matlab and Simulink simulation.

3. For the given system, design a static decoupler. Derive the transfer function matrix models for the combined process and the static decoupler $[G(s)D(s=0)]$.

$$G(s) = \begin{bmatrix} \dfrac{0.66e^{-2.6s}}{(6.7s+1)} & \dfrac{-0.0049e^{-1s}}{(9.06s+1)} \\ \dfrac{-34.7e^{-9.2s}}{(8.15s+1)} & \dfrac{0.87(11.61s+1)e^{-8s}}{(3.89s+1)(18.8s+1)} \end{bmatrix}$$

Design PI controllers for the diagonal elements of $G(s)D(s=0)$. Evaluate the servo responses of the controllers by Matlab and Simulink simulation.

4. Design a static decoupler. Derive the transfer function matrix models for combined the process and the static decoupler $[G(s)D(s = 0)]$ for the system:

$$
G(s) = \begin{bmatrix}
\dfrac{0.66e^{-2.6s}}{6.7s + 1} & \dfrac{-0.61e^{-3.5s}}{8.64s + 1} & \dfrac{-0.0049e^{-s}}{9.06s + 1} \\[3mm]
\dfrac{1.11e^{-6.5s}}{3.25s + 1} & \dfrac{-2.36e^{-3s}}{5s + 1} & \dfrac{-0.01e^{-1.2s}}{7.09s + 1} \\[3mm]
\dfrac{-34.68e^{-9.2s}}{8.15s + 1} & \dfrac{46.2e^{-9.4s}}{10.9s + 1} & \dfrac{0.87(11.61s + 1)e^{-s}}{(3.89s + 1)(18.8s + 1)}
\end{bmatrix}
$$

Design PI controllers for the diagonal elements of $G(s)D(s = 0)$. Evaluate the servo responses of the controllers by Matlab and Simulink simulation.

5. Design multivariable PI controllers by Davison method for problem (1).
Evaluate the servo responses of the controllers designed by Matlab and Simulink simulation.

6. Design multivariable PI controllers by Davison method for problem (2).
Evaluate the servo responses of the controllers designed by Matlab and Simulink simulation.

7. Reddy et al. (1997) compared the methods of designing centralized PI controllers for a MSF desalination plant. Read the paper and give a summary of the work.

8. Wahab et al. (2009) presented a method of designing multivariable PID controllers for an activated sludge process with nitrification and de-nitrification. Read the paper and give a summary of the methods of designing controllers.

9. Zarei (2018) reported the design of MIMO PI controllers by Davison method to control power in Pressurized Water Reactor (PWR) reactors. Please read the paper and a write a summary of the method of designing controllers.

12

SSGM Identification and Control of Unstable Multivariable Systems

In this chapter, a method is presented to identify the Steady State Gain Matrix (SSGM) of an unstable multivariable system under closed loop control. Effects of disturbances and measurement noise on the identification of SSGM are also studied. Davison (1976) method is modified to design single-stage multivariable PI controllers using only the multivariable gain matrix of the system. Since the overshoots in the responses are larger, a two-stage P–PI control system is proposed. Based on the SSGM, a simple proportional controller matrix is designed by the modified Davison (1976) method to stabilize the system. Based on the gain matrix of the stabilized system, diagonal PI controllers are designed.

12.1 Introduction

SISO unstable systems with time delay are difficult to control compared to stable systems. The performance specifications like overshoot, settling time for such systems are larger for unstable systems than that of the stable systems (Padmasree and Chidambaram, 2006). Methods of designing PI/PID controllers for scalar unstable systems are given by pole placement method, synthesis method, IMC method, gain and phase margin method, and optimization method

(Padmasree and Chidambaram, 2006). Conventional PI controllers give a relatively large overshoot for unstable systems. Jacob and Chidambaram (1996) have proposed a two-stage design of controllers for unstable SISO systems. First, the unstable system is stabilized by a simple proportional controller. For the stabilized system, a PI controller is designed. It is shown that an improved performance is obtained. Even for the stable multivariable systems, the method of designing controllers is complicated due to the interactions among the loops. Simple methods of designing centralized PI controllers for stable systems are reviewed by Tanttu and Lieslehto (1991) and Katebi (2012).

Govindakannan and Chidambaram (1997) have applied the method of Tanttu and Lieslehto (1991) to unstable MIMO processes. Decentralized PI controllers are also designed by the detuning method proposed by Luyben and Luyben (1997). The decentralized PI controllers do not stabilize the system if the unstable pole is present in each of the transfer functions of the system. Only centralized PI controllers stabilize such systems. Papastathopoulo and Luyben (1990) have proposed a method of calculating the steady state gain of stable multivariable systems by introducing a step change in one set point at a time.

12.2   Identification of SSGM

The transfer function matrix of the process is denoted as $K_P G_P$ and the multivariable control system is represented by $K_C G_C$. Consider a TITO system Fig. 3.3. A step change is separately given in each of the set points of the closed loop system. For a step change in the set point of y_1, the output variables y_1 and y_2 will reach 1 and 0 respectively at the steady state condition due to the presence of the integral mode in the control system. The corresponding changes in the steady state values in the manipulated variables ($u_{1,1}$ and $u_{2,1}$) are noted as $U_1 = [u_{1,1} \ u_{2,1}]^T$. Similarly, the changes in the steady state values in the manipulated variables for a step change in the set point of y_2 are noted as $U_2 = [u_{1,2} \ u_{2,2}]^T$. For stable systems, the output vector and input vector are related at steady state by $Y = K_P U$. Since a step change is given in y_1 and y_2 separately, $Y = \text{diag}[1]$. Hence, from the noted values of the steady state U vector, the SSGM is calculated as in Eq. (12.1) (Papastathopoulo and Luyben, 1990).

$$K_P = [U_1 \quad U_2]^{-1} \tag{12.1}$$

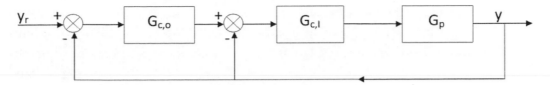

FIGURE 12.1 Block diagram for the two-stage control scheme
G_P: multivariable system; $G_{C,I}$: inner loop stabilizing centralized P controllers; $G_{C,O}$: outer loop decentralized PI controllers; Y: output variables; Y_r: set point values

However, for open loop unstable systems, because of the definition of $k_p G_p$ (say, for FOPTD systems) as $k_p \exp(-\theta s)/(\tau s - 1)$, we have Eq. (12.2).

$$K_P = -[U_1 \quad U_2]^{-1} \tag{12.2}$$

12.3 Multivariable Control System Design

Now let us design centralized PI controllers using only the SSGM of the system. The K_C matrix is given by Eq. (12.3) and Eq. (12.4) (Davison, 1976).

$$K_C = \delta K_P^{-1} \tag{12.3}$$

$$K_I = \varepsilon K_P^{-1} \tag{12.4}$$

For stable multivariable systems, Davison (1976) has recommended the range of values of δ from 0.1 to 0.4 and ε is considered from 0.05 to 0.3. There is no reported work on the application of this method to unstable systems. The Davison method needs to be suitably modified for unstable systems.

For the unstable SISO systems, there is a minimum value of K_C below which the system cannot be stabilized. There is also a maximum value $K_{C,\max}$ above which the closed loop system cannot be stabilized (Luyben and Luyben, 1997; Chidambaram, 1998). The system cannot be stabilized by a proportional controller if the ratio of the delay time to time constant is more than 1.0. For the unstable first order plus delay systems, typical values for these two stability limits $k_{c,\min} = 1$ and $k_{c,\max} = 6$ (Chidambaram, 1998; Venkatashankar and Chidambaram, 1994). Thus, for unstable multivariable systems, the recommended value of δ is from 1.2 to 2. The design value is to be obtained by simulation. Since the objective is to stabilize the system, only rough values of the elements of K_C matrix are required. In the present example, $\delta = 1.6$ is used in Eq. (12.3) to calculate K_C.

To select a suitable value of ε, we can refer to the design formula for the control of pure integrating system (as an approximation of the first order unstable system) (Chidambaram, 1998), $k_c = 10/(t_s k_P)$ and $\tau_I = 0.4\zeta^2 t_s$ where t_s is the settling time and ζ is the damping coefficient. Hence $k_I = 25/(\zeta^2 t_s^2 k_P)$. By selecting suitable values for t_s and ζ, for example with the typical values of $t_s = 10$ to 25 and $\zeta = 1$, we get the range for k_I as $(0.04/k_P)$ to $(0.25/k_P)$.

12.4 Design of Two-Stage Centralized PI Controllers

The overshoot is usually larger for the single-stage multivariable control of unstable systems. To improve the performance, two-stage P–PI control system (Fig. 12.1) can be recommended. Using this SSGM, the process is first stabilized by a simple P controller matrix (Davison, 1976) as given in Eq. (12.3), that is, $K_C = \delta K_P^{-1}$, where δ is considered from 1.0 to 2.0. Since the objective is to stabilize the system, rough values of the elements of K_C matrix are required.

The gain matrix of the closed loop system (M) is given by Eq. (12.5).

$$M = [-I + K_p K_c]^{-1}[K_P K_c] \tag{12.5}$$

For the present method, this matrix (M) is taken to be a diagonal matrix since K_C is considered as δ times the inverse of K_P. For this pure gain system, we can design PI controllers by Davison method. For the stabilized system, the multivariable PI controller system can be designed by the Davison method as in Eq. (12.6) and Eq. (12.7).

$$K_{C,O} = \delta_O \{\text{diag}[k_{p11o} \quad k_{p22,o}]\}^{-1} \tag{12.6}$$

$$K_{I,O} = \varepsilon_O \{\text{diag}[k_{p11,o} \quad k_{p22,o}]\}^{-1} \tag{12.7}$$

For stable systems, δ_O can be considered from 0.1 to 0.4 and ε_O can be considered from 0.05 to 0.25, and suitable values can be obtained by simulation. Since the present example considers only the steady state gain of the system for the design of multivariable controllers, some simulation work or an ad hoc method will be used in the selection of the controller settings.

12.5 Simulation Examples

12.5.1 Example 1

The system matrix considered here is given by Govindakannan and Chidambaram (1997, 2000) as in Eq. (12.8).

$$K_P G_P = \begin{bmatrix} \dfrac{1.6667e^{-s}}{(1.6667s - 1)} & \dfrac{e^{-s}}{(1.6667s - 1)} \\ \dfrac{0.8333e^{-s}}{(1.6667s - 1)} & \dfrac{1.6667e^{-s}}{(1.6667s - 1)} \end{bmatrix} \tag{12.8}$$

Where,

$$K_P = \begin{bmatrix} 1.6667 & 1 \\ 0.8333 & 1.6667 \end{bmatrix} \tag{12.9}$$

The system is assumed to be under closed loop by suitable multivariable PI controllers. In the present example, the multivariable PI settings given by Govindakannan and Chidambaram (1997) are considered as in Eq. (12.10).

$$G_C = \begin{bmatrix} 1.40 + \left(\dfrac{0.05}{s}\right) & -0.84 - \dfrac{0.03}{s} \\ -0.70 - \left(\dfrac{0.03}{s}\right) & 1.40 + \left(\dfrac{0.05}{s}\right) \end{bmatrix} \tag{12.10}$$

For the case study considered, we noted from the steady state portion of the u_1 versus time, u_2 versus time curves and y_1 versus time, y_2 versus time curves (similar to Fig. 12.2 and Fig. 12.4):

$$U = \begin{bmatrix} -0.8569 & 0.5141 \\ 0.4282 & -0.8569 \end{bmatrix} \tag{12.11}$$

The first column in U gives the changes required in u_1 and u_2 at steady state for a step change in y_1 to be achieved for unit change in the set point of y_1. Similarly, the second column of U gives the changes required in u_1 and u_2 at steady state for a step change in the set point of y_2. From Eq. (12.2), we get the process gain matrix as in Eq. (12.12).

$$K_P = \begin{bmatrix} 1.6667 & 0.9999 \\ 0.8328 & 1.6667 \end{bmatrix} \tag{12.12}$$

which is exactly the same as that of the process in Eq. (12.8).

TABLE 12.1 Identified SSGM under load and separately under measurement noise (Example 1)

Load	$U = [U_1 \ U_2]$	$K_P = -U^{-1}$
0.02 in v_1	$\begin{bmatrix} -0.8769 & 0.4942 \\ 0.4282 & -0.8569 \end{bmatrix}$	$\begin{bmatrix} 1.5874 & 0.9155 \\ 0.7933 & 1.6245 \end{bmatrix}$
0.02 in v_2	$\begin{bmatrix} -0.8569 & 0.5141 \\ 0.4082 & -0.8769 \end{bmatrix}$	$\begin{bmatrix} 1.6192 & 0.9403 \\ 0.7537 & 1.5823 \end{bmatrix}$
Noise in y_1	$\begin{bmatrix} -0.855 & 0.515 \\ 0.428 & -0.856 \end{bmatrix}$	$\begin{bmatrix} 1.6736 & 1.0069 \\ 0.8368 & 1.6717 \end{bmatrix}$
Noise in y_2	$\begin{bmatrix} -0.855 & 0.515 \\ 0.43 & -0.855 \end{bmatrix}$	$\begin{bmatrix} 1.6779 & 1.0106 \\ 0.8438 & 1.6779 \end{bmatrix}$

Load variable v_1 entering the system with the input u_1; v_2 entering the system with the input u_2
Noise is a random signal with 0 mean and variance 0.0001.

If the data in U is rounded off to two digits as in Eq. (12.13).

$$U = \begin{bmatrix} -0.86 & 0.51 \\ 0.43 & -0.86 \end{bmatrix} \quad (12.13)$$

then, Eq. (12.2) gives Eq. (12.14).

$$K_P = \begin{bmatrix} 1.6529 & 0.9802 \\ 0.8264 & 1.6529 \end{bmatrix} \quad (12.14)$$

The identified gain matrix is found to be closer to the actual matrix of Eq. (12.8).

$$K_P = \begin{bmatrix} 1.6667 & 1 \\ 0.8333 & 1.6667 \end{bmatrix} \quad (12.15)$$

Effect of measurement noise and, separately, the effect of disturbances entering the process (along with the input variables) are also evaluated. Since the system is under the closed loop control, the effects of load and the disturbances do not affect the gain matrix significantly. The results are given in Table 12.1.

12.6   Design of Centralized PI Controllers for Example 1

Let us consider the identified SSGM as in Eq. (12.16).

$$K_P = \begin{bmatrix} 1.6529 & 0.9802 \\ 0.8264 & 1.6529 \end{bmatrix} \tag{12.16}$$

Now let us design centralized PI controllers using only the SSGM of the system. The K_C matrix is given by (Davison, 1976) as in Eq. (12.3). As stated earlier, for unstable systems, the recommended value of δ is from 1.2 to 2. The design value is to be obtained by simulation. Since the objective is to stabilize the system, only rough values of the elements of K_C matrix are required. In the present example, $\delta = 1.4$ is considered in Eq. (12.3) to calculate K_C. As mentioned earlier, for the present example being considered, typical values of $t_s = 10$ to 25 and $\zeta = 1$ can be assumed, for which we get the range for k_I as $(0.04/k_P)$ to $(0.25/k_P)$. For the present simulation example, we consider ε as 0.07. In general, the value of ε can be from 0.05 to 0.5.

Using the values of $\delta = 1.4$ and $\varepsilon = 0.07$, the calculated values of K_C and K_I from Eq. (12.3) and Eq. (12.4) are given by Eq. (12.17) and Eq. (12.18).

$$K_c = \begin{bmatrix} 1.204 & -0.7140 \\ -0.6019 & 1.2040 \end{bmatrix} \tag{12.17}$$

$$K_I = \begin{bmatrix} 0.0602 & -0.0357 \\ -0.0301 & 0.0602 \end{bmatrix} \tag{12.18}$$

The closed loop performances for the servo and, separately, for the regulatory problems are evaluated. The responses are shown in Fig. 12.2 and Fig. 12.3. The manipulated variable versus time requirement for the servo problem is shown Fig. 12.4. Fig. 12.2 shows that the overshoot is high. The IAE values for the responses are given in Table 12.2 for the servo and the regulatory problems. As stated earlier, the two-stage P–PI controllers reduce the overshoot significantly (Govindakannan and Chidambaram, 2000). However, the method (Govindakannan and Chidambaram, 2000; Tanttu and Lieslehto, 1991) requires the knowledge of the process gain, time constant, and time delay of each of the elements of the transfer function matrix. Let us discuss the method of designing the two-stage multivariable control system based on only the SSGM.

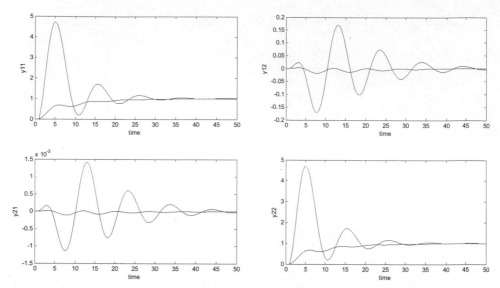

FIGURE 12.2 Responses and interactions for step changes in the set point (Example 1)
y_{11}, y_{21}: response and interaction in y_1 and y_2 for step change in set point of y_1 y_{12}, y_{22}: interaction and response in y_1 and y_2 for step change in set point of y_2
Solid line: two-stage controllers; dashed line: single-stage controllers

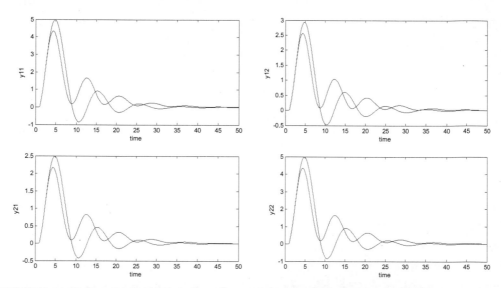

FIGURE 12.3 Responses and interactions for step changes in the load variable (Example 1)
y_{11}, y_{21}: response and interaction in y_1 and y_2 for step change in load v_1; y_{21}, y_{22}: interaction and response in y_1 and y_2 for step change in load v_2; Solid line: two-stage controllers; dashed line: single-stage controllers

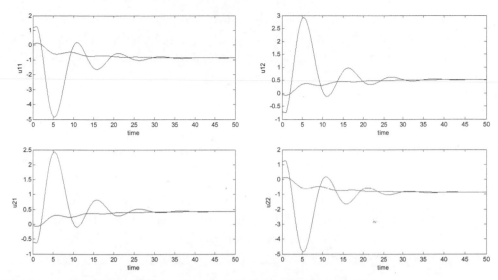

FIGURE 12.4 Manipulated variable versus time requirement for Fig. 12.2
Solid line: two-stage controllers; dashed line: single-stage controllers

12.7 Design of Two-Stage Centralized PI Controllers for Example 1

As stated earlier, the overshoot is larger for a single-stage multivariable controlled system. Let us consider the method of designing the two-stage P–PI controllers for the identified system. Using this SSGM, the process is first stabilized by a simple P controller matrix (Davison, 1976) as given in Eq. (12.19).

$$K_c = \delta K_P^{-1} \tag{12.19}$$

where δ is considered from 1.0 to 2.0. Since the objective is to stabilize the system, rough values of the elements of K_C matrix are required. Since for the design of single-stage PI controllers, $\delta = 1.4$ is assumed, a slightly increased value of δ will be used for proportional controller matrix. The value is varied around this and found by simulation that $\delta = 1.44$ in Eq. (12.3) gives an improved performance. Eq. (12.20) can be got from Eq. (12.3).

$$K_c = \begin{bmatrix} 1.2384 & -0.7344 \\ -0.6191 & 1.2384 \end{bmatrix} \tag{12.20}$$

The gain matrix of the closed loop system (M) is given by Eq. (12.21).

$$M = [-I + K_p K_c]^{-1}[K_P K_c] \tag{12.21}$$

For the present method, this matrix (M) is found to be a diagonal matrix since K_C is considered as δ times the inverse of K_P. We get M as diag [3.273 3.273]. For this pure gain system, we can design PI controllers by Davison method. For the stabilized system, the centralized PI controller system is designed by the Davison method as in Eq. (12.22) and Eq. (12.23).

$$K_{C,O} = \delta_O \{ \text{diag}[k_{p11,o} \quad k_{p22o}] \}^{-1} \tag{12.22}$$

$$K_{I,O} = \varepsilon_O \{ \text{diag}[k_{p11,o} \quad k_{p22o}] \}^{-1} \tag{12.23}$$

As discussed earlier, for the stable systems δ_O is considered from 0.1 to 0.4 and ε_O is assumed from 0.05 to 0.25. The suitable values need be obtained by simulation. In the present example, the values of $\delta_O = 0.25$ and $\varepsilon_O = 0.125$ are found to give a good performance. Hence, we get Eq. (12.24) for the outer loop:

$$K_{C,O} = \text{diag}[0.075 \quad 0.075] \text{ and } K_{I,O} = \text{diag}[0.0375 \quad 0.0375]. \tag{12.24}$$

The closed loop performance is evaluated for the servo problem and, separately, for the regulatory problem. The servo responses are shown in Fig. 12.2. The regulatory responses are shown in Fig. 12.3. The manipulated variable versus time behaviour for the servo problems is shown in Fig. 12.4. The control system designed by the two-stage design method gives improved performance compared to the single-stage method. In both the methods, only the SSGM of the system is used for the design of the controllers. The IAE values for the responses are given in Table 12.2, separately for the servo and the regulatory problems. The load variable is assumed to enter the system along with the manipulated variable. The sum of IAE values for both the servo and regulatory responses are given in Table 12.2. A significant improvement in the performance is obtained for the

TABLE 12.2 IAE values for servo and regulatory problems for single-stage and two-stage methods (Example 1)

Step input in	Single-stage method			Two-stage method		
	y_1	y_2	Sum	y_1	y_2	Sum
$y_{1,r}$	23.58	0.016	23.596	8.180	0.0012	8.18
$y_{2,r}$	2.144	23.37	24.514	0.192	8.268	8.46
L_1	30.07	15.03	45.10	30.46	15.23	45.69
L_2	17.81	30.07	47.88	18.05	30.46	48.51

Total sum for single-stage method = 141.09; total sum for two-stage method = 110.84

TABLE 12.3 IAE values for robust servo problems for single-stage and two-stage methods (Example 1)

Step input in	Single-stage method			Two-stage method		
	y_1	y_2	Sum	y_1	y_2	Sum
$y_{1,r}$	51.86	0.07817	51.93817	8.982	0.00539	8.98739
$y_{2,r}$	10.3	52.11	62.41	0.8796	9.157	10.0366

Sampling time 60 s; +10% perturbation in each time delay in the process

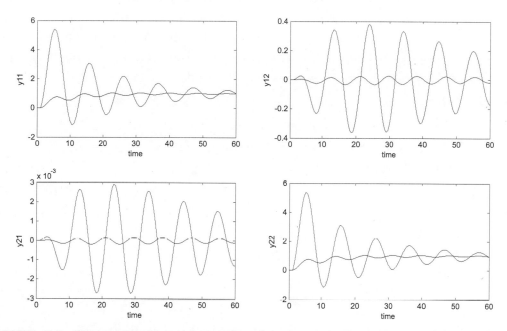

FIGURE 12.5 The responses and interactions for step changes in the set point) for +10% perturbation in delay in the process (Example 1)
y_{11}, y_{21}: response and interactions in y_1 and y_2 for step change in set point of y_1; y_{12}, y_{22}: interaction and response in y_1 and y_2 for step change in set point of y_2; Solid line: two-stage controllers; dashed line: single-stage controllers

two-stage controlled system. The effect of uncertainty in the time delay in the process is also evaluated. A 10% increase in each of the time delay in the diagonal element of the process is given. The same multivariable PI controller settings are used. Robust responses are shown in Fig. 12.5 for a set point change in y_1 and separately for a step change in y_2 (Table 12.3 gives values for the comparison of IAE).

In this example, for the single stage PI controller $\delta = 1.4$ is used and for design of P controller for the inner loop of two-stage controllers, $\delta = 1.44$ is used. The results show improved performance for the two-stage control system for servo problems whereas for the regulatory problem, there is no improvement (Table 12.2). Since the P controller for the inner loop should be larger than (1.1 times) that of the P setting for the single stage PI settings, let us use $\delta = 1.6$ for calculating the P controller settings of the inner loop. The results show improved performance for the two-stage controller for the regulatory control also.

As stated earlier, there is no rigorous method reported for the control of multivariable unstable systems with time delay. Govindakannan and Chidambaram (2000) proposed two-stage P–PI controllers by Tanttu and Lieslehto method which make use of the process gain, time delay, and time constant values of the process. The present method (modified Davison method) requires only the steady state gain of the process. Fig. 12.6 gives the comparison of the present method with the Tanttu and Lieslehto method. In both the methods, the two-stage P–PI control scheme is used.

12.7.1   Design of Two-Stage P Controllers by Tanttu and Lieslehto Method

The transfer function matrix for the process is considered as in Eq. (12.25).

$$K_p G_p = \begin{bmatrix} \dfrac{1.6667e^{-s}}{(1.6667s - 1)} & \dfrac{e^{-s}}{(1.6667s - 1)} \\ \dfrac{0.8333e^{-s}}{(1.6667s - 1)} & \dfrac{1.6667e^{-s}}{(1.6667s - 1)} \end{bmatrix} \tag{12.25}$$

The system in Eq. (12.25) is stabilized by using a centralized multivariable proportional (P) control system. The P controller is designed for each of the scalar transfer functions by the formula in Eq. (12.26) given by De Paor and O'Malley (1989).

$$k_C = [(\tau/L)^{0.5}]/k_p \tag{12.26}$$

The Tanttu and Lieslehto method is used for designing the controller matrix for stabilizing the system as given in Eq. (12.27).

$$K_c = \begin{bmatrix} 1/k_{c,11} & 1/k_{c,12} \\ 1/k_{c,21} & 1/k_{c,22} \end{bmatrix}^{-1} \tag{12.27}$$

Here $k_{c,ij}$ is the proportional controller designed for $(K_pG_P)_{ij}$. By using Eq. (12.26), we get Eq. (12.28).

$$K_c = \begin{bmatrix} 1.1065 & -0.6639 \\ -0.5532 & 1.1065 \end{bmatrix} \tag{12.28}$$

Using the controller settings in Eq. (12.28), the main and the interaction responses of the system for a set point change in y_1 and y_2 are noted. The responses are approximated to a FOPTD model. To identify the model, $k_{P,ij}$ is taken as the final steady state value, time constant is taken as $1/8$ times the settling time and the time delay for each of the elements of the FOPTD model is noted as the time delay of the response. Thus, the identified FOPTD model is as given in Eq. (12.29).

$$G(s) = \begin{bmatrix} \dfrac{4.436e^{-s}}{(2.375s+1)} & \dfrac{0.000261e^{-4.5s}}{(3.563s+1)} \\ \dfrac{-0.00033e^{-4.5s}}{(2.813s+1)} & \dfrac{4.437e^{-s}}{(2.625s+1)} \end{bmatrix} \tag{12.29}$$

For the stabilized system in Eq. (12.29), a centralized PI controller is designed.

12.7.2 PI Controllers for the Outer Loop by Tanttu and Lieslehto Method

Let us now design the outer loop PI controllers. Let $k_{c,ij}$ and $k_{I,ij} = k_{c,ij}/\tau_{Iij}$ be the controller settings for the individual transfer functions.

$$K_{C,O} = \begin{bmatrix} \dfrac{1}{k_{co,11}} & \dfrac{1}{k_{co,12}} \\ \dfrac{1}{k_{co,21}} & \dfrac{1}{k_{co,22}} \end{bmatrix}^{-1}$$

and

$$K_{I,O} = \begin{bmatrix} \dfrac{1}{k_{IO,11}} & \dfrac{1}{k_{IO,12}} \\ \dfrac{1}{k_{IO,21}} & \dfrac{1}{k_{IO,22}} \end{bmatrix}^{-1} \tag{12.30}$$

where the PI settings can be calculated using the IMC method (Seborg et al., 1996 and $k_{ci,j} = \tau_{pi,j}/[(\tau_{cij} + \theta_{ij})k_{pi,j}]$, where $\tau_{cij} = \lambda\tau_{pij}$ and $\tau_{Iij} = \tau_{pij}$. Here, $\tau_{p,ij}$ is the time constant, θ_{ij} is the time delay, k_{pij} is the gain of the inner loop controlled

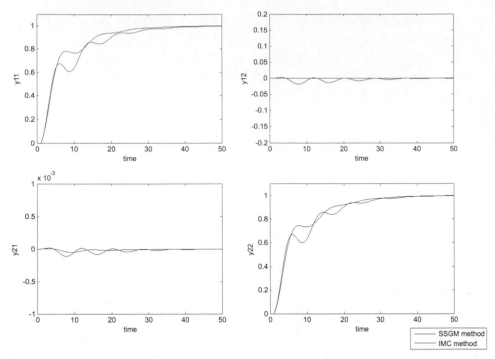

FIGURE 12.6   Comparison of the two-stage controllers of the present method with that of the IMC method
y_{11}, y_{21}: response and interactions in y_1 and y_2 for step change in set point of y_1; y_{12}, y_{22}: interaction and response in y_1 and y_2 for step change in set point of y_2; Solid line: present method; dashed line: IMC method (Tanttu and Lieslehto method)

system, and $\tau_{c,ij}$ is the desired closed loop time constant for the closed loop system. The same value of the tuning parameter is used for each of the transfer functions.

The PI controller settings for the outer loop are calculated for different values of λ (tuning factor). For $\lambda = 2.5$, the controller settings are obtained as in Eq. (12.31).

$$G_C = \begin{bmatrix} 0.0772 + \dfrac{0.0325}{s} & 0 \\ 0 & 0.0782 + \dfrac{0.0298}{s} \end{bmatrix} \tag{12.31}$$

The closed loop response for $\lambda = 2.5$ is found to be good.

Both the methods give similar performance in the main responses. The present method needs only the knowledge of the steady state gains. The interactions are lower for the Tanttu and Lieslehto method.

12.8 Robustness of the Control Systems

Among the various methods available for the study of the robustness analysis in multivariable systems (Morari and Zafiriou, 1989), the method based on the inverse of maximum singular value is easy to use and to compare the different control system stabilities (Maciejowski, 1989; Vijaykumar et al., 2012). To analyse the stability robustness of a proposed control system graphically, several uncertainty models are considered. First, for a process multiplicative input uncertainty represented as $G(s)[I + \Delta_I(s)]$, the closed loop system is stable if (Maciejowski, 1989):

$$\|\Delta_I(j\omega)\| < \frac{1}{\bar{\sigma}}\left\{[I + G_C(j\omega)G(j\omega)]^{-1} G_C(j\omega) G(j\omega)\right\} \qquad (12.32)$$

where, $\bar{\sigma}$ is maximum singular value.

This can be derived from the characteristic equation of:

$$
\begin{aligned}
T(s) &= \det\left\{I + G(s)[I + \Delta_I(s)]G_C(s)\right\} & (12.33a)\\
&= \det\left[I + G(s)G_C(s) + G(s)\Delta_I(s)G_C(s)\right]\\
&= det\left[I + G(s)G_C(s)\right]det\left\{I + [I + G(s)G_C(s)]^{-1} G(s)\Delta_I(s)G_C(s)\right\}\\
&= \det\left[I + G(s)G_C(s)\right]det\left\{I + G_C(s)[I + G(s)G_C(s)]^{-1} G(s)\Delta_I(s)\right\}\\
&= \det\left[I + G(s)G_C(s)\right]det\left\{I + [I + G_C(s)G(s)]^{-1} G_C(s)G(s)\Delta_I(s)\right\}
\end{aligned}
$$

$$(12.33b)$$

We can also use multiplicative output uncertainty $[I + \Delta_o(s)]G(s)$. The closed loop system is stable if:

$$\|\Delta_O(j\omega)\| < \frac{1}{\bar{\sigma}}\left\{[I + G(j\omega)G_C(j\omega)]^{-1} G(j\omega)G_C(j\omega)\right\} \qquad (12.34)$$

The frequency plot obtained for the right hand side of Eq. (12.32) or Eq. (12.34) indicates stability bounds of the closed loop system. The area under the curve represents stability of the system. More area under the curve indicates high stability of the system. By using this plot, it is easy to compare the stability of different controllers. The controller which gives maximum area under the curve is more stable.

As stated earlier, the frequency plot obtained for the right-hand side of Eq. (12.32) or Eq. (12.34) indicates stability bounds of the closed loop system. Fig. 12.7

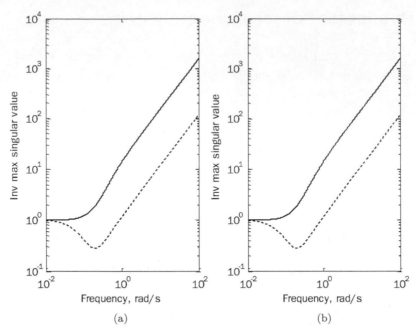

FIGURE 12.7 Stability region of output and input uncertainties for the controlled system (Example 1)
(a): output uncertainties; (b): input uncertainties
Solid line: two-stage controller; dashed line: single-stage controller

shows stability bounds for Example 1. In this figure, the region below the curve represents the stability region and above the curve represents the instability region. From the figure we can see that two-stage controllers have more stability region than single-stage controllers. This means that two-stage controllers have more robust stability. We can see that the peak value is reduced for two-stage controllers which implies that the control system can quickly track set point changes while suppressing low frequency disturbances.

12.8.1　Example 2

This example is reported by Jevtovic and Matausek (2010) and is suitably modified to get a FOPTD system or a SOPTD system transfer function model as given in Eq. (12.35).

$$
K_P G_P = \begin{bmatrix} \dfrac{e^{-0.8s}}{(2s+1)\,(4s-1)} & \dfrac{e^{-s}}{(s+1)\,(4s-1)} \\[2ex] \dfrac{-3e^{-0.1s}}{(4s-1)} & \dfrac{e^{-2s}}{(4s-1)} \end{bmatrix} \tag{12.35}
$$

The system is assumed to be under a closed loop with a suitable multivariable PI controller. In the present example, the multivariable PI settings in Eq. (12.36) are considered.

$$K_C G_C = \begin{bmatrix} 0.35 + \left(\dfrac{0.007}{s}\right) & -0.35 - \left(\dfrac{0.007}{s}\right) \\ 1.0 + \left(\dfrac{0.02}{s}\right) & 0.35 + \left(\dfrac{0.007}{s}\right) \end{bmatrix} \qquad (12.36)$$

12.9 Identification of SSGM for Example 2

As discussed earlier, the closed loop identification for the SSGM of the system is carried out by simulation. For the step change in the set point of y_1, the steady state values of $u_{1,1}$ and $u_{2,1}$ are noted as $U_1 = [0.2499 \quad -0.7502]^T$. Similarly for the step change in the set point of y_2, the steady state values of $u_{1,2}$ and $u_{2,2}$ are noted as $U_2 = [0.2499 \quad -0.2498]^T$. Hence, the SSGM of the open loop system is calculated from Eq. (12.37) as in Eq. (12.38).

$$K_P = -[U_1 \quad U_2]^{-1} \qquad (12.37)$$

$$= \begin{bmatrix} 1 & 1 \\ -3 & 1 \end{bmatrix} \qquad (12.38)$$

The identified gain matrix matches with that of the actual system given in Eq. (12.35).

12.9.1 Single-Stage Multivariable Controllers for Example 2

As discussed earlier, based on the modified Davison method and using the values of $\delta = 1.4$ and $\varepsilon = 0.028$, the controller matrices K_C and K_I are calculated as in Eq. (12.39) and Eq. (12.40).

$$K_C = \begin{bmatrix} 0.35 & -0.35 \\ 1.05 & 0.35 \end{bmatrix} \qquad (12.39)$$

$$K_I = \begin{bmatrix} 0.007 & -0.007 \\ 0.021 & 0.007 \end{bmatrix} \qquad (12.40)$$

The closed loop performances for the servo and, separately, for the regulatory problems are evaluated and are shown respectively in Fig. 12.8 and Fig. 12.9. The

TABLE 12.4   IAE values for servo and regulatory problems for single-stage and two-stage methods (Example 2)

Step input in	Single-stage method			Two-stage method		
	y_1	y_2	Sum	y_1	y_2	Sum
$y_{1,r}$	43.89	30.34	74.23	10.0	3.292	13.29
$y_{2,r}$	3.991	45.94	49.93	0.4224	10.42	10.84
L_1	43.01	133.0	176.01	21.04	52.8	73.84
L_2	35.71	35.76	71.47	17.61	17.72	35.33

Total sum for single-stage method = 371.64; total sum for two-stage method = 133.30

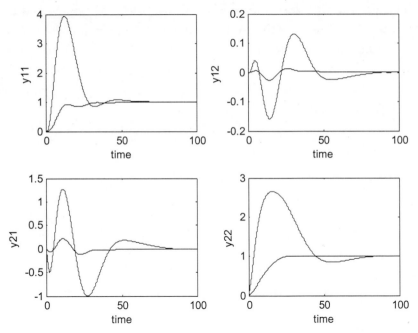

FIGURE 12.8   Responses and interactions for step changes in the set point (Example 2) y_{11}, y_{21}: response and interaction in y_1 and y_2 for step change in set point of y_1 y_{21}, y_{22}: interaction and response in y_1 and y_2 for step change in set point of y_2
Solid line: two-stage controllers; dashed line: single-stage controllers

manipulated variable versus time requirement for the servo problem is shown in Fig. 12.10. Fig. 12.8 shows that the overshoot is high. The IAE values for the responses are given in Table 12.4 separately for the servo and the regulatory problems. The sum of the IAE values for the combined servo and regulatory problems are also given in Table 12.4.

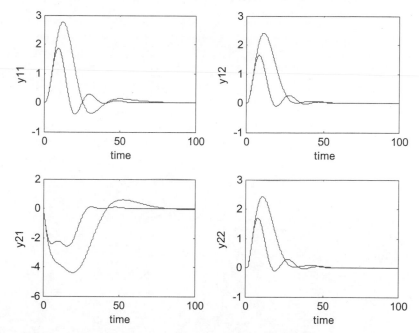

FIGURE 12.9 Responses and interactions for step changes in the load (Example 2)
y_{11}, y_{21}: response and interaction in y_1 and y_2 for step change in load v_1
y_{12}, y_{22}: interaction and response in y_1 and y_2 for step change in load v_2
Solid line: two-stage controllers; dashed line: single-stage controllers

12.9.2 Two-Stage Centralized PI Controllers for Example 2

Let us consider the method of designing two-stage P–PI controllers for the identified system. Using the SSGM, the process is first stabilized by a simple P controller matrix by the modified Davison method in Eq. (12.3), where δ is considered to range from 1.2 to 2. Since the objective is to stabilize the system, only rough values of the elements of the K_C matrix are required. In the present example, $\delta = 1.6$ is considered. Thus Eq. (12.3) gives Eq. (12.41).

$$K_C = \begin{bmatrix} 0.4 & -0.4 \\ 1.2 & 0.4 \end{bmatrix} \tag{12.41}$$

Let us now discuss the method of designing multivariable PI controller for the stabilized system.

The gain matrix of the closed loop system (M) is given by Eq. (12.42).

$$M = [-I + K_p K_c]^{-1} [K_P K_c] \tag{12.42}$$

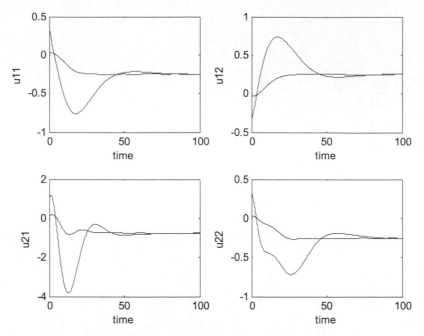

FIGURE 12.10 Manipulated variable versus time requirements for Fig. 12.8
Solid line: two-stage controllers; dashed line: single-stage controllers

As discussed earlier, the matrix (M) becomes a diagonal matrix. Since K_C is considered as δ times the inverse of K_P, we get $M = \mathrm{diag}[2.6667\ \ 2.6667]$. For this pure gain system, we can design PI controllers by the Davison method.

For the stabilized system, a centralized PI controller system is designed by the Davison method as in Eq. (12.43) and Eq. (12.44).

$$K_{C,O} = \delta_O \{\mathrm{diag}[k_{p11,o}\ \ k_{p22o}]\}^{-1} \tag{12.43}$$

$$K_{I,O} = \varepsilon_O \{\mathrm{diag}[k_{p11,o}\ \ k_{p22o}]\}^{-1} \tag{12.44}$$

where δ_O is considered to range from 0.1 to 0.4 and ε_O from 0.05 to 0.25. Suitable values are obtained by simulation. As discussed earlier, the values of $\delta_O = 0.2$ and $\varepsilon_O = 0.1$ are considered. Here $k_{pij,o}$ is the gain of the inner loop closed system, i.e., the elements of M. Hence, as discussed earlier, we get the outer loop as in Eq. (12.45).

$$K_{C,O} = \mathrm{diag}[0.075\ \ 0.075]; \quad K_{I,O} = \mathrm{diag}[0.0375\ \ 0.0375] \tag{12.45}$$

The closed loop performance is evaluated for the servo problem and, separately, for the regulatory problem. The servo responses are shown in Fig. 12.8. The regulatory

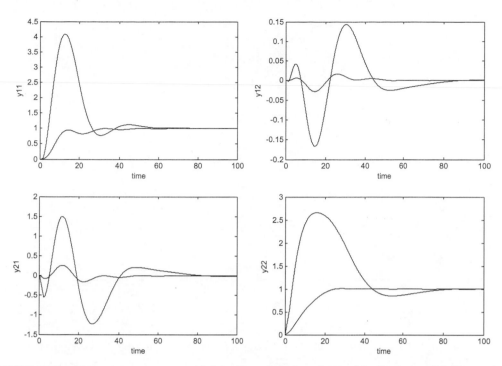

FIGURE 12.11 Robust responses and interactions for step changes in the set point (Example 2) +10% perturbation is given in each time delay in the process y_{11}, y_{21}: response and interactions in y_1 and y_2 for step change in set point of y_1 y_{12}, y_{22}: interaction and response in y_1 and y_2 for step change in set point of y_2
Solid line: two-stage controllers; dashed line: single-stage controllers

responses are shown in Fig. 12.9. The time variation of the manipulated variable for the servo problem is shown in Fig. 12.10.

The two-stage method gives a better performance than that of the single-stage method. In both the methods, only the SSGM of the system is used. The IAE values for the responses are given in Table 12.4 separately for the servo the regulatory problems. The sum of IAE values for both the servo and the regulatory problems are also given in Table 12.4. A significant improvement is obtained for the two-stage controlled system.

The overshoot for the single-stage controller for the servo problem can also be reduced by using the set point weighted PI controllers (Padmasree and Chidambaram, 2006). However, the use of set point weighted PI controllers cannot change the performance of the regulatory problems. The present two-stage controllers can improve both performance of the servo and the regulatory problems as shown in the Example 2. The effect of uncertainty in the time delay in the process is also evaluated. A 10% increase in all the time delay in the diagonal element of the

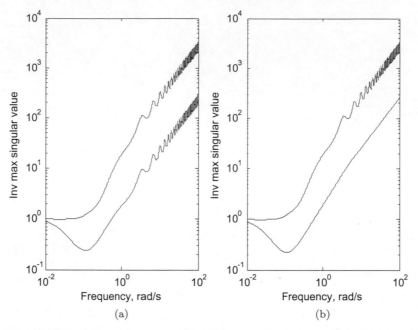

FIGURE 12.12   Stability region of output and input uncertainties for a controlled system (Example 2)
(a): output uncertainties; (b): input uncertainties
Solid line: two-stage controller; dashed line: single-stage controller

TABLE 12.5   IAE values for robust servo problems for single-stage and two-stage methods (Example 2)

Step input in	Single-stage method			Two-stage method		
	y_1	y_2	Sum	y_1	y_2	Sum
$y_{1,r}$	45.17	35.09	80.26	10.05	3.893	13.943
$y_{2,r}$	4.086	45.92	50.006	0.4376	10.34	10.7776

+10% perturbation is given in each time delay in the process

process is given. The same multivariable PI controller settings are used. The robust responses are shown in Fig. 12.11 for the set point changes in y_1 and y_2. Table 12.5 gives comparisons of IAE values.

As stated earlier, the frequency plot obtained for the right hand side of Eq. (12.32) or Eq. (12.34) indicates stability bounds of the closed loop system. Fig. 12.12 shows stability bounds for the problem in Example 2. In this figure, the region below the curve represents the stability region and above the curve represents instability region. From the figure we can see that two-stage controllers have more stability region than the single-stage controllers. This means that two-stage controllers have more robust stability. We can see that the peak value is reduced for two-stage controllers, which implies that the control system can quickly track set point changes while suppressing low frequency disturbances.

Summary

In this chapter, the method of Papastathopoulo and Luyben to identify the SSGM of stable systems is extended to unstable multivariable systems. The Davison method is modified to design a single-stage PI controller matrix using only the SSGM. A two-stage P–PI centralized controller matrix is also designed. The centralized P controller matrix is first designed by the modified Davison method. For the stabilized system, a decentralized PI controller matrix is designed by the Davison method. A systematic method is given in the chapter for the selection of the tuning parameters. Two simulation examples are given. The performance of two-stage multivariable controllers is shown to be better than the single-stage multivariable controllers. The proposed method gives a robust performance for the uncertainty in the model parameters of the system.

Problems

1. For the system,

$$G(s) = \begin{bmatrix} \dfrac{-1.6e^{-s}}{(2s-1)} & \dfrac{0.6e^{-1.5s}}{(25s+1)} \\ \dfrac{0.07e^{-1.5s}}{(3s+1)} & \dfrac{-1.7e^{-s}}{(2.2s-1)} \end{bmatrix}$$

given the diagonal PI controllers for the diagonal elements of $G(s)$ are $k_{c,11} = -1,1974$; $\tau_{I,11} = 2.805$; $\tau_{D,11} = 0.7012$, and $k_{c,22} = -1,1974$; $\tau_{I,22} = 2.805$; $\tau_{D,22} = 0.7012$, obtain the SSGM of the system for step changes in the set point as discussed in Section 12.2. Design centralized PI controllers by the Davison method. Evaluate the servo responses of the controllers by using Matlab and Simulink.

2. For the system,

$$G(s) = \begin{bmatrix} \dfrac{(2s+1)e^{-s}}{(2s-1)} & \dfrac{0.6e^{-1.5s}}{(2.5s+1)} \\ \dfrac{0.07e^{-1.5s}}{(3s+1)} & \dfrac{(15s+1)e^{-s}}{(3s-1)(8s-1)} \end{bmatrix}$$

given the diagonal PI controllers for the diagonal elements of $G(s)$ are $k_{c,11} = -1.5497$; $\tau_{I,11} = -3.175$; $\tau_{D,11} = 0.7937$ and $k_{c,22} = -1,4879$; $\tau_{I,22} = 3.1755$; $\tau_{D,22} = 0.7937$, obtain the SSGM of the system for step changes in the set point as discussed in Section 12.2. Design centralized PI controllers by the Davison method. Evaluate the servo responses of the controllers by using Matlab and Simulink.

3. For the system,

$$(s) = \begin{bmatrix} \dfrac{1.6667e^{-s}}{(1.6667s - 1)} & \dfrac{0.3e^{-s}}{(2.5s + 1)} \\ \dfrac{0.8333e^{-s}}{(1.6667s - 1))} & \dfrac{0.5e^{-s}}{(0.5s + 1)} \end{bmatrix}$$

Given the diagonal PI controllers for the diagonal elements of G(s) are $k_{c,11} = 1.2029$; $\tau_{I,11} = 3.8$; $\tau_{D,11} = 0.608$ and $k_{c,22} = -0.4458$; $\tau_{I,22} = 3.8$; $\tau_{D,22} = 0.608$, obtain the SSGM of the system for step changes in the set point as discussed in Section 12.2. Design centralized PI controllers by the Davison method. Evaluate the servo responses of the controllers by using Matlab and Simulink.

13

Control of Stable Non-square MIMO Systems

In this chapter, the simple centralized controller tuning methods – Davison method, and Tanttu and Lieslehto method – are extended to non-square systems with the Right Half-Plane (RHP) zeros. The proposed methods are applied by simulation on two examples – coupled pilot plant distillation columns and a crude distillation unit. The performances of the square and non-square controllers are compared.

Genetic algorithm (GA) optimization technique is applied for tuning of centralized and decentralized controllers for linear non-square multivariable systems with RHP zeros. Using Ziegler–Nichols (ZN) method, all the loops are initially tuned independently. To obtain the range of controller parameters, a detuning factor is used and this range is used to improve the range of GA search. Simulation studies are applied on a non-square (3 input, 2 output) coupled distillation column. ISE values show that centralized controllers designed by the GA technique give good performance when compared to the decentralized controllers designed by the GA method and other analytical methods. The method of designing decentralized PI controllers for non-square systems by relay auto-tuning method is also proposed in this chapter.

13.1   Introduction

Processes with unequal number of inputs and outputs often arise in the chemical process industry. Such non-square systems may have either more outputs than inputs or more inputs than outputs. Non-square systems with more outputs than inputs are generally not desirable as all the outputs cannot be maintained at the set points since they are over specified. The control objective in this case is to minimize the sum of square errors of the outputs with the given (fewer) inputs. For these systems, robust performance (with no offset) is impossible to achieve due to the presence of an inevitable permanent offset that results in at least one of the outputs.

More frequently encountered in the chemical industry are non-square systems with more inputs than outputs. Here, better control can be achieved by redesigning the controller, eliminating the steady state offsets. Examples of non-square systems are mixing tank process (Reeves and Arkun, 1989), 2×3 system, Shell standard control problem (Prett and Morari, 1987), 5×7 system, crude distillation unit (Muske et al., 1991), 4×5 system, and so on. A common approach towards the control of non-square processes is to first 'square up' or 'square down' the system through the addition or removal of appropriate inputs (manipulated variables) or outputs (controlled variables) to obtain a square system matrix. Then the multivariable control design methods can be employed to achieve design specifications. But adding or removing variable is not desirable. Adding unnecessary outputs to be measured can be costly, while deleting inputs leaves fewer variables to be automatically manipulated in achieving the desired control. Similarly, reducing the number of measured outputs decreases the amount of feedback information available to the system, and arbitrarily adding new manipulated inputs can incur unnecessary cost. Hence, if superior performance can be achieved by the original non-square system, it is preferable to squaring the system (Loh and Chiu, 1997).

In multi input and multi output systems, interaction and location of transmission 0 are important. The system with one or more Right Half-Plane Transmission (RHPT) zeros is called a non-minimum phase system. These RHP zeros impose limitations on stability and controllability of the system. They affect both the amplitude and phase angle. The extra phase lag that is added by the RHP zero contributes to the instability and makes the control difficult. So, the controller design for having positive zeros is of greater concern in the present example. A few methods are available to design multivariable square systems for non-minimum phase systems but they involve complicated control strategies and lengthy calculations. Though several rigorous methods are available for designing multivariable PI controllers (Huang et al., 2003), simple methods are preferable. Simple tuning methods are available to design the multivariable centralized controllers for minimum phase system such as (i) Davison method (1976), and (ii) Tanttu and Lieslehto method (1991). Dinesh et al. (2003) have shown that the Davison method gives good performance for square non-minimum phase system. Shaji and Chidambaram (1996) have used the method for square-down crude distillation, a 4×4 system. In this chapter, Section 13.1 describes these two simple methods extended to non-square systems with RHP zeros.

There are two design structures for a non-square system. One of them is the centralized controller and the other is the decentralized controller. The centralized controller uses feedback from all the measured outputs to manipulate each input, whereas the decentralized controller pairs one output with one (or more) inputs and implements feedback-based control. Even though decentralized controller can lead to a high degree of interactions, it is popular in industries due to the many reasons. Decentralized controllers are easy to implement, they are easy for the operator to understand, and the operator can easily retune the controllers to take into account the variations in the process conditions.

Vlachos et al. (1994) have proposed a simple GA for designing decentralized PI controllers for square systems. GAs are search and optimization procedures that are motivated by the principles of natural genetics and natural selection. The powerful capacity of GA in locating a global optimal solution is used in the design of PI controllers. In this method, different performance criteria are combined into a single objective function using the penalty function method. The range of values for the proportional and integral term of the controller is given randomly. The disadvantage of this method is that a large number of computations are involved, which is time consuming. Sadasivarao and Chidambaram (2006) have shown that an approximate initial guess for controller settings used in the GA is able to converge faster than the method proposed by Vlachos et al. (1994). In Section 13.6, these approximate ranges are used to design centralized and decentralized PI controllers for non-square systems.

13.2   Davison Method, and Tanttu and Lieshleto Method for Stable Systems

13.2.1   Davison Method

Davison (1976) has proposed a centralized multivariable PI controller tuning method for stable square systems. Here the proportional and integral gain matrices are given by Eq. (13.1) and Eq. (13.2).

$$K_C = \delta[G(s = 0)]^{-1} \tag{13.1}$$

$$K_I = \varepsilon[G(s = 0)]^{-1} \tag{13.2}$$

where $[G(s = 0)]^{-1}$ is called the rough tuning matrix, and δ and ε are the fine-tuning parameters which generally range from 0 to 1. In the present example, this method is extended to a non-square system. As inverse does not exist for non-square system, Moore–Penrose pseudo-inverse is used. For matrix A, Moore–Penrose pseudo-inverse is given in Eq. (13.3).

$$A^\dagger = A^H(A * A^H)^{-1} \tag{13.3}$$

A^H is the Hermitian matrix of A. So, for a non-square system, PID controller gains are as given in Eq. (13.4).

$$K_c = \delta[G(s = 0)]^\dagger$$
$$K_I = \varepsilon[G(s = 0)]^\dagger \tag{13.4}$$
$$K_D = \gamma[G(s = 0)]^\dagger$$

13.2.2   Calculation of Tuning Parameters

The system is not stable for the entire range (0 to 1) of tuning parameters in the Davison method. In the present method, the tuning parameters (δ, ε, γ) are calculated as follows:

First the characteristic equation, $\det(I + KG)$, is obtained. For a 2×2 system, characteristic equation can be obtained easily but for higher-order systems, software like *Symbolic Math* (Matlab toolbox) have to be used. Then, Routh stability criterion is applied to find the range of the tuning parameters for which the system is stable. The system is then simulated by tuning around these values and fine-tuning parameters are chosen based on performance.

Usually for design of controllers for SISO systems with a positive 0, only the stable invertible portion is considered. That is, the numerator dynamics due to positive 0 is ignored. For square MIMO systems, system positive 0 should be separated and this should not be considered for the controller design. In the Davison method, only the steady state gain matrix is considered. As such, no change or modification is required for the Davison method for the system with a positive 0. In the present example, we try to check how the Davison method works for such systems.

13.2.3 Tanttu and Lieslehto (TL) Method

Morari and Zafiriou (1989) have discussed a method for the design of PI controller known as internal model control (IMC). Tanttu and Lieslehto (1976) have developed a multivariable PI controller tuning method based on IMC method. First, PID controller $(k_{c,ij})$ for each of the scalar transfer functions $(g_{p,ij})$ of the process is designed based on the IMC method. For a FOPTD system, this is given in Eq. (13.5a) and Eq. (13.5b).

$$k_{ij}k_{cij} = (2\tau_{ij} + L_{ij})/2\lambda \tag{13.5a}$$

$$\tau_{Iij} = (\tau_{ij} + 0.5L_{ij}) \tag{13.5b}$$

Here k_{ij} and L_{ij} are the gain and time delay of an element in the process model for the ith output and j^{th} input; and k_{cij} and τ_{Iij} are the proportional gain and integral time constant of the IMC controller of the ij^{th} loop. In this method, the same filter constant is used for each of the scalar systems so that there is only one tuning parameter (λ). Then the multivariable PI controllers can be designed as in Eq. (13.6), Eq. (13.7), and Eq. (13.8).

$$R_C = \begin{bmatrix} 1/k_{c11} & 1/k_{c12} & 1/k_{c1n} \\ 1/k_{c21} & 1/k_{c22} & 1/k_{c2n} \\ \dots & \dots & \dots \\ 1/k_{cm1} & 1/k_{cm2} & 1/k_{cmn} \end{bmatrix} \tag{13.6}$$

$$R_I = [k_{Iij}] \tag{13.7}$$

$$R_D = \begin{bmatrix} 1/k_{D11} & 1/k_{D12} & /k_{D1n} \\ 1/k_{D21} & 1/k_{D22} & 1/k_{D2n} \\ \dots & \dots & \dots \\ 1/k_{Dm1} & 1/k_{Dm2} & 1/k_{Dmn} \end{bmatrix} \tag{13.8}$$

Where,

$$k_{Iij} = (k_{cij}/\tau_{Iij}) \tag{13.9a}$$

$$k_{Dij} = (k_{cij}*\tau_{Dij}) \tag{13.9b}$$

For a square system,

$$K_c = [R_c]^{-1} \tag{13.10a}$$

$$K_I = [R_I]^{-1} \tag{13.10b}$$

$$K_D = [R_D]^{-1} \tag{13.10c}$$

This method is extended to a non-square system taking the pseudo-inverse.

$$K_c = [R_c]^{\dagger} \tag{13.11a}$$

$$K_I = [R_I]^{\dagger} \tag{13.11b}$$

$$K_D = [R_D]^{\dagger} \tag{13.11c}$$

The values of k_{cij}, k_{Iij} and k_{Dij} are calculated from equations given in Morari and Zafiriou (1989). Tuning parameter (λ) has to be calculated by trial-and-error-method. The lower limit of λ is given in literature for individual elements (Morari and Zafirio, 1989). This limit is taken as the initial value and the final value of the tuning parameter is obtained by trial-and-error method by simulation.

13.2.4   Robust Decentralized Controller

The independent procedure for robust decentralized controllers proposed by Hovd and Skogestad (1993) was extended by Loh and Chiu (1997) to non-square systems with more inputs than outputs. The controller design equation is given in Eq. (13.12).

$$C = Q[I - G_M Q]^{-1} \tag{13.12}$$

$$= [G_{M-}]^{\dagger} F \tag{13.13}$$

Where,

Q is a non-square IMC controller, G_M^- is the minimum phase part of G_M and $F = \text{diag}\{f_I\}_{i \sim k}$ is a low-pass diagonal filter with a steady state gain of 1.

13.3 Simulation Studies

13.3.1 Example 1: Control of Distillation Column

Levien and Morari (1987) have discussed the example of two coupled distillation column in their work. They have considered a square system (3×3). In the present example, a system having non-minimum phase is considered with 2 outputs and 3 inputs. Here the outputs are a mole fraction of ethanol in the distillate (y_1) and a mole fraction of water in the bottom (y_2), and manipulated variables are distillate flow rate (u_1), steam flow rate (u_2), and product fraction from the side column (u_3). The system transfer function is given as in Eq. (13.14).

$$G(s) = \begin{bmatrix} \dfrac{0.052e^{-8s}}{19.8s+1} & \dfrac{-0.03(1-15.8s)}{108s^2+63s+1} & \dfrac{0.012(1-47s)}{181s^2+29s+1} \\ \dfrac{0.0725}{890s^2+64s+1} & \dfrac{-0.0029(1-560s)}{293s^2+51s+1} & \dfrac{0.0078}{42.3s+1} \end{bmatrix} \tag{13.14}$$

13.3.1.1 *Davison Method for Distillation Columns*

From the transfer function matrix in Eq. (13.14), SSGM is given by Eq. (13.15).

$$G(0) = \begin{bmatrix} 0.052 & -0.03 & 0.012 \\ 0.0725 & -0.009 & 0.0078 \end{bmatrix} \tag{13.15}$$

The pseudo-inverse for the matrix in Eq. (13.15) is calculated and substituted in Eq. (13.4) to obtain the proportional and integral gains of PI controllers. The system is stable over the range $\delta = 0.7$ to 1 and $\varepsilon = 0$ to 0.1. Tuning parameters are obtained by tuning the controller around these values. The overall controller is obtained with $\delta = 1$ and $\varepsilon = 0.03$. The final PI controller matrix is obtained as in Eq. (13.16).

$$G_c = \begin{bmatrix} -2.0845 + \dfrac{-0.0625}{s} & 15.1612 + \dfrac{0.4548}{s} \\ -33.0046 + \dfrac{-0.9901}{s} & 23.9571 + \dfrac{0.7187}{s} \\ 7.1044 + \dfrac{0.213}{s} & -3.8095 + \dfrac{-0.1142}{s} \end{bmatrix} \tag{13.16}$$

13.3.1.2 *Tanttu and Lieslehto Method for Distillation Columns*

A centralized PI controller is designed for the two coupled distillation columns using the Tanttu and Lieslehto method as given in Section 13.2.3. Individual IMC settings for the two coupled distillation column systems, with process model $G(s)$, are found and arranged in matrix form given by Eq. (13.17a) and Eq. (13.17b).

$$
Rc = \begin{bmatrix} \dfrac{\lambda}{457.69} & \dfrac{-31.6+\lambda}{2032.2} & \dfrac{98+\lambda}{2416.7} \\[2mm] \dfrac{\lambda}{882.7} & \dfrac{1120+\lambda}{17586.2} & \dfrac{\lambda}{5423.07} \end{bmatrix} \tag{13.17a}
$$

$$
R_I = \begin{bmatrix} \dfrac{\lambda}{19.23} & \dfrac{-31.6+\lambda}{32.25} & \dfrac{98+\lambda}{83.3} \\[2mm] \dfrac{\lambda}{13.79} & \dfrac{1120+\lambda}{344.82} & \dfrac{\lambda}{128.205} \end{bmatrix} \tag{13.17b}
$$

Here, λ is the tuning parameter. The initial the value of $\lambda = 13$ ($\lambda > 1.7*$ time delay) is used. The final controller settings are found as discussed in Section 13.2.2. The recommended value for the tuning parameter is $\lambda = 15$. The final PI controller is obtained as in Eq. (13.18).

$$
G_c = \begin{bmatrix} 9.6833 + \dfrac{0.3145}{s} & 3.4640 + \dfrac{0.1084}{s} \\[2mm] -3.1519 + \dfrac{-0.1221}{s} & 14.5772 + \dfrac{0.2666}{s} \\[2mm] 14.0520 + \dfrac{0.5107}{s} & 0.1198 + \dfrac{0.0390}{s} \end{bmatrix} \tag{13.18}
$$

In the controller matrix by Tanttu and Lieslehto method, the signs of two of the controllers settings are different from the Davison method as shown in Eq. (13.16) and Eq. (13.18). Since system zeros are not defined for non-square systems, it is not clear how to separate the positive system zero from the non-square transfer function matrix. The presence of such zeros may cause the change of sign in some of the individual controllers. Further research is required in this area.

13.3.1.3 *Decentralized Controller Method for Distillation Columns*

The Loh and Chiu (1997) method is applied to coupled pilot plant distillation columns and the decentralized controller is designed. The pairing of the manipulated and controlled variables is obtained using block relative gain (BRG) (Reeves and Arkun, 1989). The first output variable y_1 is paired with u_2 and y_2 is paired with u_1 and u_3. This pairing leads to less interaction. From the pairings, the block diagonal model (Reeves and Arkun, 1989) is obtained and given by Eq.

(13.19).

$$G_M(s) = \begin{bmatrix} \dfrac{-0.03(1 - 15.8s)}{108s^2 + 63s + 1} & 0 & 0 \\[3mm] 0 & \dfrac{0.0725}{890s^2 + 64s + 1} & \dfrac{0.0078}{42.3s + 1} \end{bmatrix} \tag{13.19}$$

The IMC controller Q is designed by substituting G_M in Eq. (13.12). Here, a first-order filter is used. Simulations are carried out for different values of the tuning parameter (Loh and Chiu, 1997) and the best value is obtained as $\varepsilon_1 = \varepsilon_2 = 28$. With this tuning parameter value, the final controller is given by Eq. (13.20). From the controller transfer function matrix (Q) it is clear that it is not of the conventional PI/PID form.

$$Q = \begin{bmatrix} \dfrac{108s^2 + 63s + 1}{-0.474s^2 - 1.314s - 0.03} & 0 \\[5mm] 0 & \dfrac{115453s^4 + 13761s^3 + 586s^2 + 10.7s + 0.0725}{1349s^5 + 241.4s^4 + 280.18s^3 + 22.64s^2 + 0.6112s + 0.0054} \\[5mm] 0 & \dfrac{261345s^5 + 43765s^4 + 2827s^3 + 88s^2 + 1.3s + 0.0078}{1349s^5 + 241.4s^4 + 280.18s^3 + 22.64s^2 + 0.6112s + 0.0054} \end{bmatrix}$$

$$\tag{13.20}$$

13.4 Simulation Results

Simulation was carried out for both the servo and the regulatory problems using Simulink. Results are compared using ISE values for both the methods (Table 13.1 and Table 13.2). The load transfer function matrix for the disturbances is not

TABLE 13.1 ISE values for the servo problem for centralized controller: two coupled distillation columns example

Method	Step Change in	ISE values y_1	y_2	Sum of ISE
Davison	y_1	33.04	12.41	45.45
	y_2	3.11	11.46	14.57
Tanttu and Lieslehto	y_1	34.04	44.04	78.07
	y_2	0.4532	52.765	53.218
Loh and Chiu	y_1	44.99	9.91	54.90
	y_2	19.64	28.34	47.98

TABLE 13.2 ISE values of the regulatory problem for two coupled distillation columns: example for disturbances at the input

Method	Step in	ISE values		Sum of ISE	IAE values		Sum of IAE
		y_1	y_2		y_1	y_2	
Davison	v_1	0.0319	0.0465	0.0784	1.79	3.58	4.37
	v_2	0.01	0.0056	0.156	1.18	0.84	2.02
	v_3	0.0075	0.0046	0.0127	1.045	0.847	1.893
Tanttu and Lieslehto	v_1	0.0481	0.0216	0.0698	2.151	2.076	4.227
	v_2	0.0080	0.0248	0.0334	1.254	3.065	4.319
	v_3	0.0092	0.0043	0.0136	1.069	1.227	2.296
Loh and Chiu	v_1	0.041	0.107	0.148	2.67	4.67	7.34
	v_2	0.036	0.021	0.057	2.75	1.54	4.39
	v_3	0.0065	0.0045	0.011	1.03	0.658	1.688

available. For the regulatory problem, the load transfer function matrix is assumed as the process transfer function matrix (i.e., load enters along with the manipulated variable). The Davison method gives the lowest ISE values compared to the other two methods for both servo and regulatory problems. The decentralized controller gives lower ISE values compared to the Tanttu and Lieslehto method for the servo problem, whereas for the regulatory problem the Tanttu and Lieslehto method gives lower ISE values compared to the decentralized controller (Fig. 13.1 and Fig. 13.2). For step change in y_1, the decentralized controller gives sluggish response compared to the centralized controller and settling time is more in the case of the Tanttu and Lieslehto method. For step change in y_2, the Tanttu and Lieslehto method gives sluggish response and the decentralized controller gives more interactions compared to centralized controllers.

13.4.1   Robustness Studies for Coupled Pilot Plant Distillation Column

Robustness studies were carried out for the closed loop system of Eq. (13.4) by increasing the individual element gain by 10%. The same controller setting as previously obtained was used. The performance of the three methods for the perturbed system is shown in Fig. 13.3. The performance of the centralized controllers is similar to that of the system without changing the gain value. The sum of ISE for the perfect parameter system and the perturbed system is compared in Table 13.3. The Davison method gives the ISE values for the perturbed system close to that of the nominal system. It gives a more robust performance compared to the other two methods.

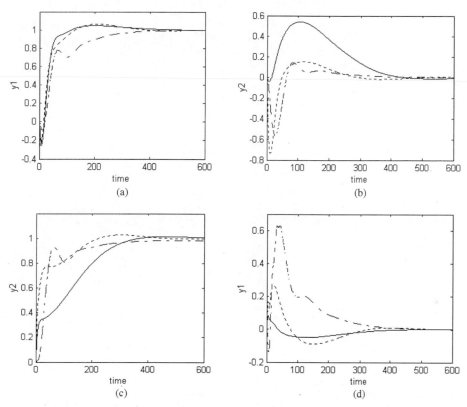

FIGURE 13.1 Performance comparison of the three methods for unit step changes in $y1$ or $y2$ for coupled pilot plant distillation columns (a) Response, (b) Interaction for step change in y_1; (c) Response, (d) Interaction for step change in y_2
Solid line: Tanttu and Lieslehto method; dotted line: Davison method; dash–dot line: decentralized controller

13.5 Simulation Example 2: Crude Distillation Process

The crude distillation unit lies at the front end of a refinery. This unit performs the initial distillation of the crude oil into fractions with different boiling range. The crude is pumped in from storage tanks and, after desalination, it is preheated using the crude tower products and overhead streams. The crude is then partially vaporized in two parallel fuel gas-fired heaters. The vapour and liquid from the heaters enter the flash zone at the bottom of the crude column. Muske et al. (1991) have considered the crude distillation unit at Cosmo Oil's Sakai Refinery. They have given the transfer function for three general crudes and average crude. Crude 2 is considered in this example and its transfer function is given in Eq. (13.21). In this example, controlled variables are naphtha/kerosene cut point (y_1), kerosene/LGO cut point (y_2), LGO/ HGO cut point (y_3) and measured over

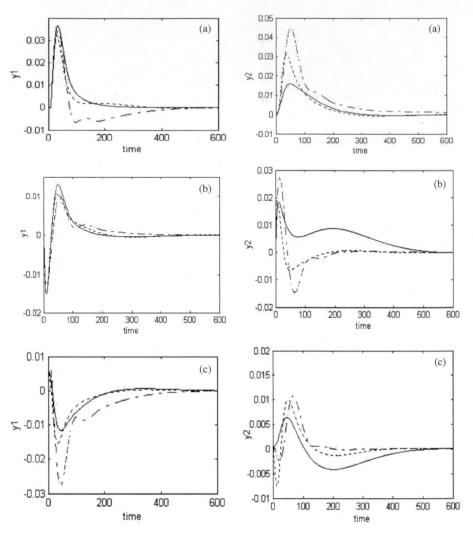

FIGURE 13.2 Performance comparison of three methods for unit step changes in load variables
$v1/v2/v3$ for coupled pilot plant distillation columns
Step changes in (a) v_1 (b) v_2 and (c) v_3
Solid line: Tanttu and Lieslehto method; dotted line: Davison method; dash–dot line: decentralized
controller

flash (y_4). Manipulated variables are top temperature (u_1), kerosene yield (u_2),
LGO yield (u_3), HGO yield (u_4) and heater outlet temperature (u_5). Here LGO
and HGO mean the light gas oil and heavy ga soil respectively. For this, process,
centralized PID controllers are designed using Davison as well as Tanttu and
Lieslehto methods.

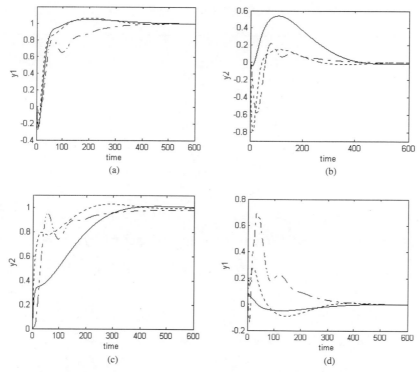

FIGURE 13.3 Performance comparison of the two methods for unit step change in y_1 or y_2 for coupled pilot plant distillation column with 10% deviation in each scalar system gain (a) Response, (b) Interaction for step change in y_1; (c) Response, (d) Interaction for step change in y_2
Solid line: Tanttu and Lieslehto method; dotted line: Davison method; dash–dot line: decentralized controller

$$
G(s) = \begin{bmatrix}
\dfrac{3.8(16s+1)}{140s^2+14s+1} & \dfrac{2.9e^{-6s}}{10s+1} & 0 & 0 \\[3mm]
\dfrac{3.9(4.5s+1)}{96s^2+17s+1} & \dfrac{6.3}{20s+1} & 0 & 0 \\[3mm]
\dfrac{3.8(16s+1)}{140s^2+14s+1} & \dfrac{6.1(12s+1)e^{-s}}{337s^2+34s+1} & \dfrac{3.4e^{-2s}}{6.9s+1} & 0 \\[3mm]
\dfrac{-0.73-16s+1)e^{-4s}}{150s^2+20s+1} & \dfrac{-1.53(3.1s+1)}{5.1s^2+7.1s+1} & \dfrac{-1.3(7.6s+1)}{4.7s^2+7.1s+1} & \dfrac{-0.6e^{-s}}{2s+1}
\end{bmatrix}
$$

$$
\begin{bmatrix}
\dfrac{-0.73-16s+1)e^{-4s}}{150s^2+20s+1} \\[3mm]
\dfrac{16se^{-2s}}{(5s+1)(14s+1)} \\[3mm]
\dfrac{22se^{-2s}}{(5s+1)(10s+1)} \\[3mm]
\dfrac{0.32-9.1s+1)e^{-s}}{12s^2+15s+1}
\end{bmatrix}
\tag{13.21}
$$

TABLE 13.3   Sum of ISE values for robustness comparison for coupled pilot plant distillation column

Method	Step change in	Sum of ISE Perfect parameter	Sum of ISE +10% chance in time constant
Davison	y_1	45.45	44.2
	y_2	14.57	13.23
Tanttu and Lieslehto	y_1	78.07	72.8
	y_2	53.218	47.96
Loh and Chiu	y_1	54.90	55.32
	y_2	47.98	48.64

13.5.1   Davison Method for Crude Distillation Process

For the transfer function matrix given in Eq. (13.21), SSGM is given in Eq. (13.22).

$$G(0) = \begin{bmatrix} 3.8 & 2.9 & 0 & 0 & -0.73 \\ 3.9 & 6.3 & 0 & 0 & 0 \\ 3.8 & 3.1 & 3.4 & 0 & 0 \\ -1.62 & -1.5 & -1.3 & -0.6 & 0.32 \end{bmatrix} \tag{13.22}$$

For this matrix, the pseudo-inverse is calculated and substituted in Eq. (13.11) to get proportional, integral and derivative gains. As discussed in Section 13.2.3, tuning parameters are calculated. Simulation is carried out for the designed controller and the tuning parameter. The values used are $\delta = 1$, $\varepsilon = 0.22$, $\gamma = 0.9$. A controller with these tuning parameter values gives better performance and minimum ISE values. The final controller transfer function is given in Eq. (13.23).

$$G_c = \begin{bmatrix} 0.443 + \dfrac{0.133}{s} + 0.399s & -0.201 - \dfrac{0.060}{s} - 0.181s \\ -0.274 - \dfrac{0.082}{s} - 0.247s & 0.283 + \dfrac{0.085}{s} + 0.255s \\ -0.003 - \dfrac{0.0009}{s} - 0.003s & -0.284 - \dfrac{0.085}{s} - 0.256s \\ -0.586 - \dfrac{0.176}{s} - 0.527s & -0.491 - \dfrac{0.147}{s} - 0.442s \\ -0.153 - \dfrac{0.046}{s} - 0.138s & 0.078 + \dfrac{0.023}{s} + 0.007s \end{bmatrix}$$

$$
\begin{bmatrix}
0.443 + \dfrac{0.133}{s} + 0.399s & 0.061 + \dfrac{0.018}{s} + 0.055s \\[2mm]
-0.015 - \dfrac{0.005}{s} - 0.014s & -0.037 - \dfrac{0.011}{s} - 0.033s \\[2mm]
0.294 + \dfrac{0.088}{s} + 0.265s & -0.0004 - \dfrac{0.00012}{s} - 0.0003s \\[2mm]
-0.294 - \dfrac{0.088}{s} - 0.265s & -1.647 - \dfrac{0.494}{s} - 1.482s \\[2mm]
0.064 + \dfrac{0.019}{s} + 0.058s & 0.166 + \dfrac{0.049}{s} + 0.149s
\end{bmatrix} \tag{13.23}
$$

13.5.2 Tanttu and Lieslehto Method

For the crude distillation problem, a centralized PID controller is designed using the Tanttu and Lieslehto method as discussed in Section 13.2.3. The initial value of the tuning parameter is $\lambda = 11$ ($\lambda > 1.7*$time delay). Simulation studies are carried out for the system with different tuning parameter values and the recommended value is $\lambda = 30$. The controller with this tuning parameter value gives better performance and lower ISE values. The final PID controller matrix with this λ value is given by Eq. (13.24).

$$
G_c = \begin{bmatrix}
0.143 + \dfrac{0.011}{s} + 0.689s & -0.071 - \dfrac{0.004}{s} + 0.36s \\[2mm]
-0.072 - \dfrac{0.005}{s} - 0.453s & 0.129 + \dfrac{0.007}{s} - 0.519s \\[2mm]
-0.037 - \dfrac{0.00002}{s} - 0.075s & -0.032 - \dfrac{0.009}{s} - 0.556s \\[2mm]
-0.005 - \dfrac{0.021}{s} & -0.032 - \dfrac{0.017}{s} \\[2mm]
-0.191 - \dfrac{0.008}{s} - 1.959s & 0.171 + \dfrac{0.004}{s} + 1.309s \\[2mm]
0.021 + \dfrac{0.002}{s} & 0.027 + \dfrac{0.004}{s} \\[2mm]
-0.019 - \dfrac{0.002}{s} - 0.202s & -0.025 - \dfrac{0.004}{s} \\[2mm]
0.068 + \dfrac{0.009}{s} - 0.215s & -0.013 - \dfrac{0.00007}{s} - 0.121s \\[2mm]
-0.043 - \dfrac{0.021}{s} & -0.118 - \dfrac{0.055}{s} \\[2mm]
0.024 + \dfrac{0.002}{s} & 0.032 + \dfrac{0.005}{s}
\end{bmatrix} \tag{13.24}
$$

TABLE 13.4   ISE values of the servo problem for centralized controller crude distillation column example

Method	Step In	ISE values				Sum of ISE
		y_1	y_2	y_3	y_4	
Davison	y_1	3.194	0.134	0.792	0.066	4.185
	y_2	1.033	7.128	1.758	0.264	10.18
	y_3	0.061	0.071	5.031	0.239	5.403
	y_4	0.074	0.088	0.142	3.153	3.457
Tanttu and Lieslehto	y_1	13.57	0.769	1.262	0.022	15.62
	y_2	0.254	18.0	1.264	0.016	19.54
	y_3	0.078	0.068	17.28	0.915	17.82
	y_4	0.377	0.366	0.376	16.57	17.69

The Tanttu and Lieslehto method is based on the IMC method which gives only PI controller, as in Eq. (13.24), for the FOPTD transfer function models. The IMC method gives PID controller settings for higher-order systems whereas in the Davison method we can calculate K_D as well as get PID controllers too, as in Eq. (13.23).

13.5.3   Simulation Results

Crude distillation is a large-scale problem with 4 outputs and 5 inputs. For this example, centralized controllers are designed using two proposed methods. Simulations are carried out on the two designed controllers for both the servo and regulatory problems. These are compared based on the ISE values given in Tables 13.4 and 13.5 for the servo and regulatory problems, respectively. ISE values show that the Davison method gives a good response and lower ISE values compared to the Tanttu and Lieslehto method. ISE values of the Tanttu and Lieslehto method are about 2–3 times compared to the Davison method for the servo problem. For the regulatory problem, still larger ISE values are obtained in the Tanttu and Lieslehto method. Simulation results are given in Fig. 13.4 for the servo problem for step changes in all the controlled variables, one at a time. The settling time for the Tanttu and Lieslehto method is more than the settling time for the Davison method. The Tanttu and Lieslehto method gives sluggish response.

Robustness Studies for the Crude Distillation Process

Robustness studies are carried out for the crude distillation process by increasing the individual element process gain by 10% using the same controller settings. The

TABLE 13.5 ISE values for the regulatory problem for the crude distillation column example with disturbances at the input

Method	Step in	ISE values				Sum of ISE
		y_1	y_2	y_3	y_4	
Davison	v_1	17.23	17.20	15.02	2.70	52.17
	v_2	10.22	31.57	23.16	2.22	67.17
	v_3	0.098	0.108	13.69	2.143	16.04
	v_4	0.021	0.025	0.038	0.371	0.457
	v_5	0.796	2.23	4.428	0.079	6.534
Tanttu and	v_1	204.1	133.9	125.4	27.7	491.2
Lieslehto	v_2	8.98	392.3	324.8	22.43	827.5
	v_3	0.128	0.200	148.1	21.98	170.4
	v_4	0.134	0.131	0.131	5.239	5.636
	v_5	6.121	3.148	7.211	1.108	17.58

sum of ISE values is given in Table 13.6 for both the perfect parameter and the perturbed systems for step change in set points. From the table it is clear that the Davison method gives ISE values closer to that of perfect parameter system and hence this method is more robust than the Tanttu and Lieslehto method.

13.5.4　Comparison of Controllers for Square and Non-square Crude Distillation Process

The crude distillation process transfer function model is given in Eq. (13.21). The columns represent manipulated variables and the rows represent controlled variables. From the transfer function matrix, it is clear that the fourth manipulated variable is affecting only the fourth output variable. Hence the fourth manipulated variable is kept constant so that the transfer function matrix becomes square (4×4) system. For this square transfer function matrix, a centralized controller is designed using the Davison method. The controller is given by Eq. (13.25).

$$G_c = \begin{bmatrix} 2.038 + \dfrac{1.223}{s} + 1.834s & -1.53 - \dfrac{0.918}{s} - 1.377s \\[2ex] -1.261 - \dfrac{0.757}{s} - 1.135s & 1.106 + \dfrac{0..663}{s} + 0.995s \\[2ex] -0.014 - \dfrac{0.0862}{s} - 0.012s & -0.2741 - \dfrac{0.164}{s} - 0.246s \\[2ex] 4.228 + \dfrac{2.536}{s} + 3.805s & -3.572 - \dfrac{2.143}{s} - 3.215s \end{bmatrix}$$

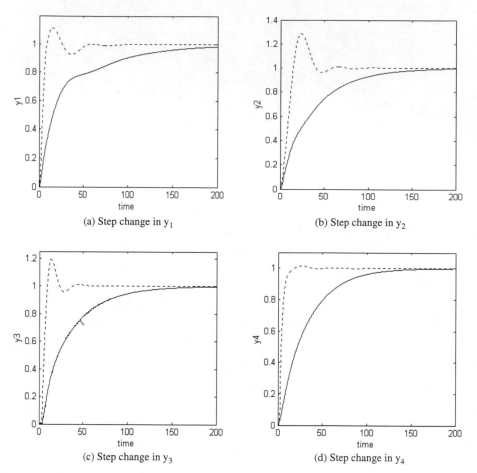

FIGURE 13.4 Performance comparison of Davison method, and Tanttu and Lieslehto method for unit step changes in y_1, y_2, y_3 or y_4 for crude distillation process
Solid line: Tanttu and Lieslehto method; dotted line: Davison method

$$\begin{bmatrix} 1.778 + \dfrac{0.002}{s} + 1.6s & 4.651 + \dfrac{2.790}{s} + 4.18s \\[2ex] -1.100 - \dfrac{0.002}{s} - 0.990s & -2.878 - \dfrac{1.727}{s} - 2.590s \\[2ex] 0.281 + \dfrac{0.009}{s} + 0.2535s & -0.032 - \dfrac{0.019}{s} - 0.0293s \\[2ex] 4.882 + \dfrac{2.929}{s} + 4.394s & 12.77 + \dfrac{7.662}{s} + 11.49s \end{bmatrix} \qquad (13.25)$$

The performance of the square controller is compared in Fig. 13.5 with the non-square controller designed by Eq. (13.23) previously. The square controller gives oscillatory response and has a large settling time. The sum of ISE values for the square and non-square controllers for the servo problem is also compared

TABLE 13.6 Comparison of ISE values for perfect parameter system and perturbed system for crude distillation process

Method	Step in	Sum of ISE for Perfect Parameter System	+10% Change in Process Gain
Davison	y_1	4.185	4.03
	y_2	10.18	9.999
	y_3	5.403	5.32
	y_4	3.457	3.27
Tanttu and Lieslehto	y_1	15.62	14.262
	y_2	19.54	17.994
	y_3	17.82	16.18
	y_4	17.69	16.23

TABLE 13.7 ISE values for square and non-square systems

Step change in	Sum of ISE For Non-square Systems	For Square Systems
y_1	4.185	8.93
y_2	10.18	19.78
y_3	5.403	15.16
y_4	3.457	37.69

in Table 13.7. For square controller, ISE values are two times higher than the non-square controller for step changes in y_1 or y_2 or y_3 and three times higher for step change in y_4.

13.6 Optimization Method for Controller Design for Non-square Stable Systems with RHP Zeros

13.6.1 Genetic Algorithm

A simple Genetic Algorithm (GA) as presented by Chipperfield et al. (1994) is used in this example. The binary alphabet and grey coding were used for encoding the controller parameters. A generation gap of 0.9, linear rank-based fitness assignment with a selective pressure of 2, proportionate selection with stochastic universal sampling, single-point crossover and fitness-based reinsertion is used as chosen by Vlachos et al. (1999). The size of the population chosen is 80 so that the search space is attacked at many points simultaneously, thus resulting in a faster convergence to the global optimum. The initially selected population is

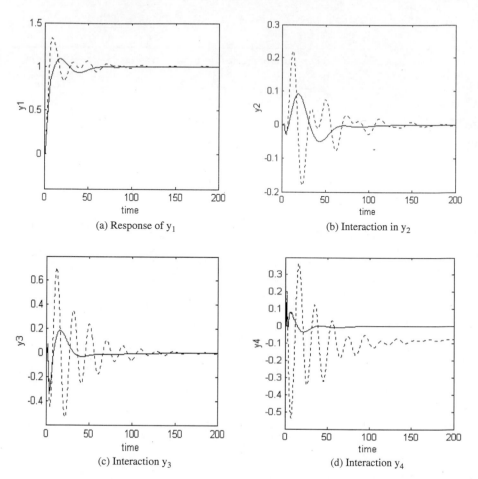

(a) Response of y_1

(b) Interaction in y_2

(c) Interaction y_3

(d) Interaction y_4

FIGURE 13.5   The performance comparison of the square controller and non-square controller for step change in y_1 for crude distillation example
Solid line: non-square controller; dotted line: square controller

left to evolve for 50 generations. A 20-bit string element is used for the encoding of each of the controller parameters. The crossover and mutation probabilities are chosen to be 0.45 and 0.01, respectively.

In the present case, Ziegler–Nichols tuning method is used for the systems having stability limit. Direct synthesis tuning method is used for the systems having no stability limit. For the open loop transfer function models (first order and second order), which have no stability limit on K_c, the ranges of K_c and T_I are selected by extending $K_{c,des}$, 5 times and $T_{I,des}$, 10 times in both directions (i.e., lower limit of proportional gain is taken as $K_{c,des}/5$ and upper limit as $5*K_{c,des}$). If the individual transfer function models have a stability limit, then the lower limit of proportional gain is taken as 0 and the upper limit as $K_{c,des}/1.5$; the lower limit of T_I is taken as $T_{I,des}/2$ and no upper limit is placed on T_I so that the system will not go to an unstable region. The parameters are encoded as K and K/T in the algorithm. Given the range of the controller parameters, the GA will generate initial guesses randomly in that range and evaluate the objective function. In this way, the unstable region is eliminated from the region of search. This approximate range for the GA is able to reduce the number of computations and the algorithm converges faster.

13.6.2 Objective Function

Objective function (sum of deviation from performance criteria) is the mathematical way of representing the desired response. The method of Vlachos et al. (1999) is followed to represent the objective function. The function for the ith output and the jth set point pattern can be expressed as in Eq. (13.26).

$$J_{ij}(K_c, T_I) = \int_0^{t_{max}} [\max\{f_{ij}^{(l)}(t) - y_i(t), 0\} + \max\{y_i(t) - f_{ij}^{(u)}(t), 0\}]dt \quad (13.26)$$

Objective K_c and T_I are parameters (proportional gain and integral time) of the n number of PI controllers associated with the n loops of the multivariable system; $f_{ij}^{(l)}$ and $f_{ij}^{(u)}$ are the user defined lower and upper boundaries of the region representing the performance objectives for the ith output and the jth set point pattern respectively. A unit step change is given to one of the set points (j^{th} set point) by keeping the remaining set points constant and the performances of all outputs are evaluated using the equation. The results are weighted and added together to form a single performance. The procedure is repeated for the other set points (j varies from 1 to n) and the final objective function is represented as in Eq. (13.27).

$$J_0(K_c, T_I) = \max_{1 \leq j \leq n} \left\{ \sum_{i=1}^{n} w_{ij} J_{ij}(K_c T_I) \right\} \quad (13.27)$$

J_{ij} denotes the objective function for the ith output and the jth set point; w_{ij} denotes the weighing factor of the objective function for the ith output and the jth set point. The higher the weighing factor, the more important is the corresponding component of the objective function. In case of the centralized control system, K_c and T_I are the parameters of the $n \times n$ size of PI controller matrix. In the present case, w_{ij} is taken as 1 when $i = j$ and is taken as 0.25 when $i \neq j$, to give more importance to set point tracking objectives.

13.6.3   Controllers for Non-square Systems with RHP Zeros

In the multi-input and multi-output system, interaction and location of transmission zero are important (Ogunnaike and Ray, 1994). The optimization method (GA) is applied to design the centralized and decentralized controllers for non-square systems with RHP zeros.

13.6.4   Design Example

The proposed ranges for the GA are applied to a non-square coupled distillation column to design both the centralized and decentralized PI controllers. These two design settings are compared with the method suggested by Sarma and Chidambaram (2005) to design the PI controller settings for a non-square system. A coupled distillation column (3 input, 2 output) studied by Levien and Morari (1987) is described by Eq. (13.14).

Here the outputs are a mole fraction of ethanol in distillate (y_1) and a mole fraction of water in bottoms (y_2); the manipulated variables are distillate flow rate (u_1), steam flow rate (u_2), and product fraction from the side column (u_3).

The performance objectives are chosen as follows:

Peak overshoot : $\leq 20\%$

Settling time : ≤ 50 minutes

Loop coupling : $\leq 50\ \%, 0 \leq t \leq 75$ minutes

$\leq 5\%, 25 \leq t \leq 125$ minutes

$$G(s) = \begin{bmatrix} \dfrac{0.052e^{-8s}}{19.8s+1} & \dfrac{-0.03(1-15.8s)}{108s^2+63s+1} & \dfrac{-0.012(1-47s)}{181s^2+29s+1} \\ \dfrac{0.0725}{890s^2+64s+1} & \dfrac{-0.0029(1-560s)}{293s^2+51s+1} & \dfrac{0.0078}{42.3s+1} \end{bmatrix} \quad (13.28)$$

In the Vlachos et al. (1999) method, the range of controller parameters is chosen randomly (proportional term 0.5 to 50 and integral term 0.1 to 100), the algorithm takes nearly 40 generations to achieve the desired objective function, whereas the proposed GA is able to converge faster and even first generation itself the desired objective function is reached (Fig. 13.6) and the optimal controller settings are achieved.

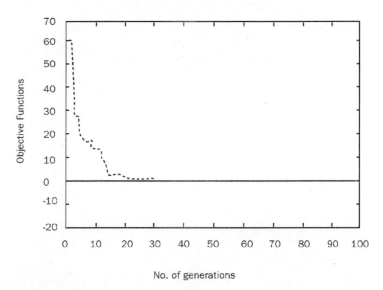

FIGURE 13.6 Convergence of GA: objective function versus number of generations
Solid line: proposed GA; dashed line: Vlachos method

13.6.4.1 *Centralized Controllers*

A centralized controller is designed for the coupled distillation column system by using the simple GA. For convenience in representing the ranges, the parameters are specified as $K_{c,ij}$ and $K_{c,ij}/T_{I,ii}$ in the algorithm. Since 6 PI controllers are used for the centralized control system for the example as in Eq. (13.28), 12 parameters are to be estimated. The approximate ranges for the controller parameters are calculated and used in the GA to get the optimum controller parameters. The system response is shown in Fig. 13.7 for unit step change in set point. The optimum controller parameters are given in the following matrix as proportional and integral gain form:

$$G_c = \begin{bmatrix} 7.5748 + \dfrac{4.1371}{s} & 10.8965 + \dfrac{9.9103}{s} \\[2mm] 4.4523 + \dfrac{1.2653}{s} & 33.5059 + \dfrac{21.8233}{s} \\[2mm] -34.505 - \dfrac{2.5336}{s} & 36.725 + \dfrac{1.5431}{s} \end{bmatrix} \qquad (13.29)$$

13.6.4.2 *Decentralized Controller*

In this decentralized controller, first the pairing has to be done using Block Relative Gain (BRG). BRG (Reeves and Arkun, 1989) has to be calculated for

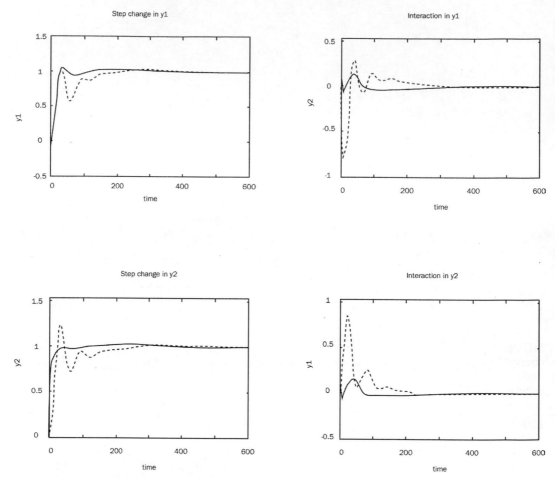

FIGURE 13.7 Performance comparisons of centralized and decentralized controllers: unit step change in y_1 and y_2 (servo problem)
Solid line: centralized controller; dashed line: decentralized controller

all possible pairings. BRG is equal to unity for the best pairing and for lower interaction. The first output variable y_1 is paired with u_2 and y_2 is paired with u_1 and u_3. This pairing leads to a lower interaction and a unity BRG. The optimal closed loop response for the decentralized controller is shown in Fig. 13.8. For convenience in representing the ranges, the parameters are encoded as $K_{c,ij}$ and $K_{c,ij}/\tau_{I,ii}$ in the algorithm. Since 3 PI controllers are used for the decentralized controllers for the system given by Eq. (13.28), 6 parameters are to be estimated. The approximate ranges for the controller parameters are calculated and used in the GA to get the optimum controller parameters. The resulting decentralized controller parameters designed by the simple GA (Chipperfield, 1994) are shown as

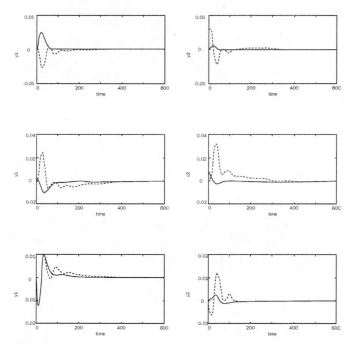

FIGURE 13.8 Performance comparisons of centralized and decentralized controllers for unit step change in load variables v_1, v_2, and v_3
The ith row figures are for change in load variables v_i
Solid line: centralized controller; dashed line: decentralized controller

following:

$$
G_c = \begin{bmatrix} -23.8095 - \dfrac{0.8275}{s} & 0 \\ 0 & 26.88 - \dfrac{0.3798}{s} \\ 0 & 0.1 - \dfrac{0.1787}{s} \end{bmatrix} \tag{13.30}
$$

The optimal closed loop response for the decentralized controller is shown in Fig. 13.8.

13.6.5 Comparison Criterion of Controller Performance

There are three ways the performance of the controller system can be measured: Integral of Absolute Error (IAE), Integral of Squared Error (ISE), and Integral of Time Weighted Absolute Error (ITAE). In multivariable systems, different criteria have been considered to compare the performance of the controllers. Stephanopoulos (1984) has discussed that ISE should be used to suppress large errors, while ISE should be used to suppress small errors, and for errors that persist for long times, ITAE criterion should be used.

$$IAE = \int_0^\infty |y_s(t) - y(t)| dt \tag{13.31}$$

$$ISE = \int_0^\infty [y_s(t) - y(t)]^2 dt \tag{13.32}$$

In a multivariable system, most of the controller settings give comparable responses. To distinguish the responses, IAE or ISE values should be compared. While comparing two controller settings, the ISE values in response may be lower, but at the same time, it may be higher in interaction. In this case, the criterion preferred is one which considers the minimization of the sum of the IAE or ISE values in both the response and the interaction.

13.6.6   Comparison of Controller Performance for Coupled Distillation Column

The performances of centralized and decentralized controllers designed by optimization method (GA) were compared along with the analytical controllers designed by Sarma and Chidambaram (2005). The results have shown that the centralized controllers designed by the optimization technique performed better than the decentralized controllers designed by the optimization technique and previously reported analytical methods. Fig. 13.7 compares the response of y_1 and y_2 for unit step change in set point y_1 and y_2 respectively. It also compares the interaction in y_2 and y_1 for unit step change in set point y_1 and y_2 respectively. It is clear that the interaction is lower in the centralized controller performance compared to the decentralized controller. Also, the settling time is lower for the centralized controller performance as compared to the decentralized controller. It is also shown that decentralized controllers give a larger overshoot than centralized controllers. The performances of the designed controllers were also compared for change in load variables. It is assumed that the load transfer function matrix is the same as the process transfer function matrix. Fig. 13.8 compares the output variable y_1 and y_2 for unit step change in load variable v_1, v_2 and v_3 respectively. In the regulatory problem also, the centralized controllers gave a better response, that is, less overshoot compared to the decentralized controllers.

ISE values for the centralized controller and the decentralized controller are given in Table 13.8a along with those values from Sarma and Chidambaram (2005) for the servo problem, and in Table 13.8b for the regulatory problem. For the

TABLE 13.8a ISE values for regulatory problem: GA optimization technique

Sl. no.	Method	Step in	ISE values		Sum of ISE
			y_1	y_2	
1.	Centralized	y_1	7.5381	0.18	7.72
	controller (GA)	y_2	0.8468	2.25	3.09
2.	Decentralized	y_1	14.225	10.68	24.91
	controller (GA)	y_2	6.3149	9.42	15.74
3.	D method	y_1	33.04	12.41	45.45
	centralized	y_2	3.11	11.46	14.57
4.	LC method	y_1	44.99	9.91	54.90
	decentralized	y_2	19.64	28.34	47.98

D: Davison; LC: Loh and Chiu; reference for serial nos. 3 and 4: Sarma and Chidambaram

TABLE 13.8b ISE values for servo problem: GA optimization technique

Sl. no.	Method	Step in	ISE values		Sum of ISE
			y_1	y_2	
1.	Decentralized	v_1	0.0202	0.0168	0.037
	controller (GA)	v_2	0.0172	0.0404	0.057
		v_3	0.0048	0.0045	0.09
2.	Centralized	v_1	0.0147	0.0005	0.015
	controller (GA)	v_2	0.0040	0.0012	0.005
		v_3	0.0041	0.0002	0.004
3.	D method	v_1	0.0319	0.0465	0.078
	centralized	v_2	0.01	0.0056	0.156
		v_3	0.0075	0.0046	0.013
4.	LC method	v_1	0.041	0.107	0.148
	decentralized	v_2	0.036	0.021	0.057
		v_3	0.0065	0.0045	0.011

D: Davison: LC: Loh and Chiu

servo problem, the sum of ISE values is lower for the centralized controller when compared to the decentralized controller for the Davison method as compared to the Loh and Chiu method. It is noted that for the regulatory problem also, centralized controller gives lower ISE values when compared to the others.

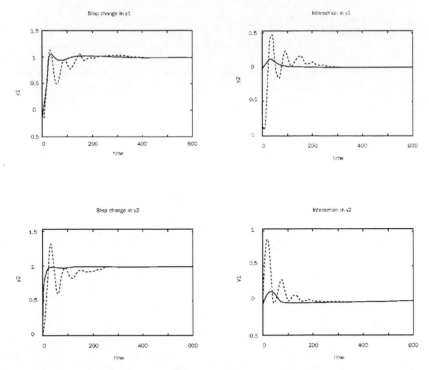

FIGURE 13.9 Controller performance of perturbed system for unit step change y_1 and y_2
Solid line: centralized controller; dashed line: decentralized controller

13.6.7   Robustness Studies

Robustness studies were carried out for this system by increasing the individual element gain of the transfer function model by 10%. The same controller settings as previously obtained were used. Performance of the centralized controller and the decentralized controller for the perturbed system is shown in Fig. 13.9. From the figure, it is clear that the performance of the centralized controllers and the decentralized controllers is similar to the corresponding performance of the perfect parameter system.

13.7   Auto-Tuning of PI Controllers for a Non-square System

In a control system, by using the relay autotuning method, the ultimate periods (P_u) and ultimate gains (K_u) are determined. Using the tuning formula of Ziegler–Nichols continuous cycling technique, decentralized PI controllers are designed. Good reviews are given by Yu (2006), Chidambaram and Vivek (2014)

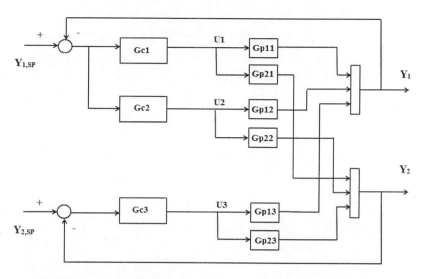

FIGURE 13.10 Decentralized control of a 2×3 non-square system

and Chidambaram and Nikita (2018). The simultaneous relay tuning method for stable square MIMO systems is presented by Chidambaram and Nikita (2018). The reasons for considering the higher order harmonics (HOH) in analysing the relay oscillations are established and shown in the following paragraph.

The SISO system consists of one critical point whereas the MIMO systems consists of many critical points. Therefore, the relay can generate the limit cycles and it is represented as stability limits as in Palmor et al. (1995). The decentralized controllers are exchanged by relays. The oscillations have the same period and are close to the lower frequency of the SISO system. The proofs for this are given by Friman and Waller (1994) and Atherton (1975). Also, the relative magnitude (h_1, h_2) is constant as in Palmor et al. (1995).

The general block diagram of a decentralized control for a non-square 2×3 MIMO system is as shown in Fig. 13.10. The relay method is one of the most commonly used methods for identification of critical points. In the MIMO system, every controller is changed by relays during the identification of critical points.

The relay feedback loop provides sustained oscillations if the set point is not changed. The relay test consists of two steps. Initially, the relay feedback test is performed by using simultaneous loop tuning method, and the ultimate frequency and amplitude from the relay response are measured. Then, the appropriate tuning procedure is selected and incorporated to find the PI controller parameters.

In the relay feedback test, the HOH (Higher Order Harmonics) are not filtered by all the system in relay output. This assumption will be valid when the system consists of the characteristics of low-pass filter. A proper method is needed to find the K_u value when HOH are present in the output response. Otherwise, they will affect the valuation of ultimate gain. The technique suggested by Nikita and Chidambaram (2018) is extended here to non-square MIMO systems (Kalpana and Chidambaram, 2020).

A triangular waveform is represented by Eq. (13.33).

$$y(t*) = a^* \left[1 + \frac{1}{3^2} + \frac{1}{5^2} + \frac{1}{7^2} + \cdots + \frac{1}{N^2} + \cdots \right] \qquad (13.33)$$

A rectangular waveform is represented by Eq. (13.34).

$$y(t*) = a^* \left[1 - \frac{1}{3} + \frac{1}{5} - \frac{1}{7} + \cdots + \frac{1}{N} + \cdots \right] \qquad (13.34)$$

Here, the value of $y(t*)$ is obtained at time $t* = 0.5\pi/\omega_u$ from the relay response and a suitable N value is used. The value of a^* is calculated by solving the given equation and it is not the amplitude value which is directly observed from the response.

Generally, the system filters out the HOH based on τ_d/τ ratio. Typically, a value of $N = 3$ (greater than 1) is recommended, when there is an absence of initial dynamics. To improve the ultimate values, a suitable number of HOH terms can be selected based on the output waveform.

13.7.1   Design Procedure

For the MIMO non-square process, decentralized PID controllers are designed by determining the critical points of the system (Nikita and Chidambaram, 2018). Relays are placed instead of controllers to determine the critical points and to get the continuous oscillations in the process output. The relay response shape obtained in the output is not completely sinusoidal, which proves the presence of HOH. Suitable HOH terms are used for estimating the critical points.

For a stable system, the ZN method is used to design all controllers and it is listed in Table 13.9.

TABLE 13.9 Zeigler–Nichols tuning rules for P, PI, and PID controllers

Controller	$K_{c,des}$	τ_I	τ_D
P	$0.5K_{u,\max}$	–	–
PI	$0.45K_{u,\max}$	$0.833P_u$	–
PID	$0.6K_{u,\max}$	$0.5P_u$	$0.125P_u$

TABLE 13.10 Relay feedback test results

h_1	h_2	a_1	a_2	P_{u1}	P_{u2}	ω_{u1}	ω_{u2}	$t_1{}^*$	$t_2{}^*$	$y_1(t*)$	$y_2(t*)$
0.3	0.35	0.62	0.39	20	11	0.3142	0.5712	4.999	2.749	0.5	0.35

For $N = 5, 5$; $a_1* = 0.45$; $a_2* = 0.315$

13.7.2 Simulation Study of a Non-square MIMO System

An example presented of mixing tank with 3 inputs and 2 outputs is considered by adding a delay of 30%. The transfer function matrix is given by Eq. (13.35).

$$G(s) = \begin{bmatrix} \dfrac{4e^{-6s}}{20s+1} & \dfrac{4e^{-6s}}{20s+1} & \dfrac{4e^{-6s}}{20s+1} \\[2mm] \dfrac{3e^{-3s}}{10s+1} & \dfrac{-3e^{-3s}}{10s+1} & \dfrac{5e^{-3s}}{10s+1} \end{bmatrix} \tag{13.35}$$

The block diagonal model and the pairings are found using the BRG presented in Reeves and Arkun (1989). For this example, y_1 is paired with u_1 and u_2; and y_2 is paired with u_3. Based on the design procedure, decentralized PI controllers are designed and the results are reviewed in Table 13.10.

Fig. 13.11 shows the relay response of the process. From the relay response, it is clearly observed that the plant model, represented in Eq. (13.35), offers inadequate filtering. In Fig. 13.11, a value of N=3,3 is selected founded from relay response initial dynamics. The PI controller settings are listed in Table 13.11. The servo response and the regulatory response of the process are shown in Fig. 13.12 and Fig. 13.13 respectively.

The qualitative comparison of the improved method over the conventional method for servo and regulatory operation is analysed by using time–integral performance analysis which is listed in Tables 13.12 and 13.13 respectively. From the Tables 13.12 and 13.13, it is observed that, the performance of the improved method is better than the conventional method.

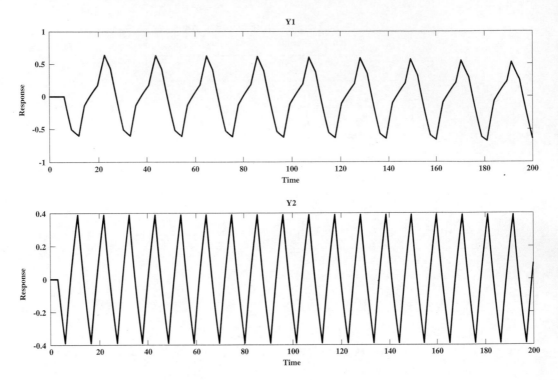

FIGURE 13.11   Relay response in y_1 and y_2

TABLE 13.11   PI controller parameters from relay test

| | Conventional Method | | Improved Method $(N = 3, 3)$ | |
	K_u	$K_{c,des}$	K_u	$K_{c,des}$
$g_{c,11}$	0.6161	0.277245	0.84883	0.38197
$g_{c,22}$	0.6161	0.277245	0.84883	0.38197
$g_{c,32}$	0.9794	0.44073	1.21261	0.54568

$\tau_{I1} = 16.666; \tau_{I2} = 9.1666$

TABLE 13.12   Performance analysis for servo response

	Response y_1	Interaction y_2	Interaction y_1	Response y_2
CM	9.644	0.0131	4.687	2.699
IM	9.329	0	4.54	2.612

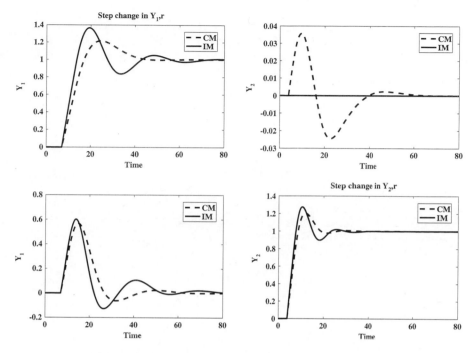

FIGURE 13.12 Servo response and its interaction for $N = 3, 3$

TABLE 13.13 Performance analysis for regulatory response

Step change	Method	y_1	y_2
d_1	CM	8.425	8.387
	IM	6.604	7.236
d_2	CM	76.77	8.868
	IM	55.52	7.236
d_3	CM	6.96	23.53
	IM	5.802	20.1

13.8 Stability Criterion of Modified Inverse Nyquist Array on a Non-square Process

The idea of applying Inverse Nyquist Array (INA) to analyse characteristics of a non-square MIMO system was first put forward by Kurniawan et al. (2018). INA is a method which was proposed by Rosenbrock in 1969 to indicate the degree of interaction among loops and to evaluate the stability of a square MIMO system by utilizing the Nyquist plot. However, this method has not been

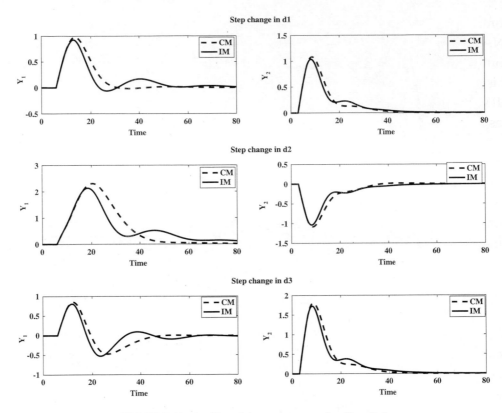

FIGURE 13.13   Regulatory response for $N = 3, 3$

applied to evaluate the stability of non-square MIMO system due to limitations of the inverse. Non-square matrix is a complex matrix which cannot be inverted directly. Pseudo-inverse method is used to get the inverse of the non-square matrix. Subsequently, the result of INA evaluation was compared with the dynamic response given by Simulink to find the accuracy of the INA method. The result of this comparison showed that the stability of the simple non-square MIMO system can be evaluated by INA theory if the system is diagonally dominant.

In addition, Rosenbrock's theory can indicate the degree of interaction of every loop. In 1972, Munro developed the INA theory to the next level. The degree of interaction can be seen just by drawing a Gershgorin band (Munro, 1972) at every frequency of the diagonal loops and he made additional constraint to make the stability theory more accurate (Munro, 1972). After that, Postlethwaite (1977) made a generalized theory for square MIMO stability based on the INA criterion without using Gershgorin band approach. This generalized theory became the INA stability criterion that was applied on the square MIMO system and many researchers focused on developing the INA theory to design compensator and controller. Lourtie (1985) designed a compensator for non-square systems by using INA. Another research using INA was presented by Ivezic. He created a robust controller for a milling circuit based on INA (Ivezic and Petrovi, 2003). However, this INA method has not been applied to evaluate the stability of non-square MIMO systems directly due to limitations of the inverse. Therefore, pseudo-inverse has been applied to the non-square matrix to get its inverse.

Revelo et al. (2020) presented a hybrid centralized control scheme for a non-square multivariable process. The proposed approach combines the Smith predictor, gain-scheduling methodology, and Davison method with the Particle Swarm Optimization (PSO) algorithm, all of which are combined to solve a non-square control system problem that compensates for the multiple and different time delays and process nonlinearities.

Summary

To conclude, the simple centralized controller-tuning methods such as the Davison method, and the Tanttu and Lieslehto method, were extended to non-square systems with RHP zeros. In this chapter, the proposed methods were applied to two examples – coupled pilot plant distillation columns and a crude distillation unit. Simulations were carried out for both servo and regulatory problems. ISE values were given. For the crude distillation process, the two centralized controller methods were compared. The Davison method gave better performance and less settling time than the Tanttu and Lieslehto method. The Tanttu and Lieslehto method gave a sluggish response. For the servo problem, the ISE values of the Tanttu and Lieslehto method were approximately 2–3 times compared to the Davison method. The performances of the square and non-square controllers were compared. Improved performance was observed in the non-square controller.

Also, in this chapter, a GA optimization technique was used to design centralized and decentralized controller for linear non-square (3 input, 2 output) MIMO system with RHP zero. A simulation study was given for a non-square coupled distillation column. The simulation results indicated that in the presence of uncertainty, the two types of controller structure achieved a robust performance. By comparing the ISE values, centralized controller designed by the optimization method gave the lowest ISE values for both the servo and regulatory problems when compared to the decentralized controller and other analytical methods. In addition, the inclusion of approximate ranges for the PI control parameters into the GA improved the convergence of the algorithm significantly. The qualitative and quantitative analysis of relay tuning of 2×3 MIMO systems was performed over the conventional method and compared with the modified method. From the analysis, it was observed that the modified method performs better than the conventional method.

Problems

1. Design centralized PI controllers by (a) the Davison method (b) the Tanttu and Lieslehto method for the given system:

$$
G(s) = \begin{bmatrix} \dfrac{0.5e^{-0.2s}}{(3s+1)} & \dfrac{0.07e^{-0.3s}}{(2.5s+1)} & \dfrac{0.04e^{-0.03s}}{(2.8s+1)} \\ \dfrac{0.004e^{-0.4s}}{1.5s+1} & \dfrac{-0.003e^{-0.2s}}{(s+1)} & \dfrac{0.001e^{-0.4s}}{(1.6s+1)} \end{bmatrix}
$$

2. Design centralized PI control system by (a) the Davison method and (b) the Tanttu and Lieslehto method for the given system:

$$
K_P G_P = \begin{bmatrix} \dfrac{1.7e^{-0.5s}}{(2s+1)} & \dfrac{0.4e^{-s}}{(3s+1)} & \dfrac{e^{-0.5s}}{(4s+1)} \\ \dfrac{0.8e^{-0.4s}}{(3s+1)} & \dfrac{0.2e^{-0.5s}}{(2s+1)} & \dfrac{1.8e^{-0.7s}}{(3s+1)} \end{bmatrix}
$$

3. Revelo et al. (2020) reported a study titled 'Non-square Multivariable Chemical Processes: A Hybrid Centralized Control Proposal'. Give a write-up of the system.

14

Control of Unstable Non-square Systems

For unstable non-square multivariable systems, the Davison method (Davison, 1976) is modified to design single-stage multivariable PI controllers using only the SSGM of the system. Since the overshoots in the closed loop responses are larger, a two-stage P–PI control system is proposed. Based on the SSGM, a simple proportional controller matrix is first designed by the modified Davison (1976) method to stabilize the system. Based on the gain matrix of the stabilized system, diagonal PI controllers are designed. The performance of the two-stage control system (inner loop centralized P controllers and outer loop diagonal PI controllers) is evaluated and compared with the single-stage multivariable control system. Simulation results on two examples show the effectiveness of the proposed methods for both the servo and regulatory problems. A method is presented to identify the SSGM of a non-square multivariable unstable system under closed loop control. Effects of disturbances and measurement noise on the identification of SSGM are also studied.

14.1   Introduction

Control of unstable systems with time delay is a challenging task when compared to the stable systems. Non-square process has a greater number of inputs than outputs or a greater number of outputs than the inputs. Non-square multivariable

301

systems with a greater number of inputs than outputs arise in chemical industries. Examples include vacuum distillation column unit with five inputs and four outputs (Treiber, 1984), a mixing tank with three inputs and one output (Reeves and Arkun, 1989) and a distillation column with four measured impurities and three manipulated variables (Treiber, 1984).

Studies on decentralized control of stable non-square multivariable systems have been reported by (Loh and Chiu, 1997). Ganesh and Chidambaram (2010) have reported the design of centralized multivariable controllers by an optimization method such as the GA. Delay compensator for non-square systems have been reported by Rao and Chidambaram (2006b) and Chen and Ding (2011). Xu and Shin (2011) have presented the design of self-tuning fuzzy logic controller for stable non-square systems. Sarma and Chidambaram (2005) have extended the method of Davison and the method of Tanttu and Lieslehto (1991) to non-square stable systems. Sarma and Chidambaram (2005) have presented that these simple methods for controller design based on SSGM give a good performance for the non-square stable multivariable systems also. The steps in the design of centralized PI controllers applying only the SSGM of the square system are as follows:

The K_C and K_I matrices are given by Davison (1976) as in Eq. (14.1) and Eq. (14.2).

$$K_c = \delta K_P^{-1} \tag{14.1}$$

$$K_I = \varepsilon K_P^{-1} \tag{14.2}$$

The Davison method requires only the SSGM of the system. For stable systems, the value of δ is less than 1 and ranges from 0.05 to 0.9 and ε ranges from 0.05 to 0.5. For stable multivariable non-square systems, Sarma and Chidambaram (2005) have suggested the range of from 0.1 to 0.4 and from 0.05 to 0.3. For non-square systems, pseudo-inverse of the matrices is to be used as in Eq. (14.3) and Eq. (14.4).

$$K_c = \delta K_P^{\dagger} \tag{14.3}$$

$$K_I = \varepsilon K_P^{\dagger} \tag{14.4}$$

For the present purpose, the Davison method has to be suitably modified for unstable non-square systems. Dhanya Ram et al. (2015) extended the method of Davison for unstable square systems. They have reported a value of $\delta > 1$ to be used (range of from 1 to 2) and the range of from 0.05 to 0.5. Dhanya Ram et al. (2015) also proposed a two-stage method of designing controllers for unstable square systems which gives a significantly improved performance over the single-stage method. In the present chapter, the method proposed by Dhanya Ram et al. (2015) is extended to unstable non-square systems. The Papastathopoulo and Luyben (1990) method is suitably adapted to identify the SSGM of an unstable multivariable non-square system in the closed loop condition.

14.2 Design of SSGM Based Multivariable Control

14.2.1 Design of Centralized PI Controllers

For SISO unstable systems, there is a minimum and maximum value of k_c ($k_{c,\min}$ and $k_{c,\max}$ respectively) below and above which the system cannot be stabilized. For unstable FOPTD systems, stability limits values are $k_p k_{c,\min} = 1$ and $k_p k_{c,\max} = 6$. Thus, the recommended design value of is from 1.2 to 2 for unstable systems. The design value has to be achieved by simulation. The reported settings can be referred from the pole placement method also. The range $k_p k_{c,\min}$ to $k_p k_{c,\max}$ is from 1.2 to 2.0 for unstable systems. The unstable system has to be stabilized; hence only rough values of K_c matrix are needed. For the present case, $\delta = 1.6$ is considered in Eq. (14.3) to calculate K_c.

For designing K_I, the value of ε has to be obtained. The design formula for the control of pure integrating system (as an approximation of the first order unstable system) is given as $k_c = 10/(t_s k_p)$ and $\tau_I = 0.4\zeta^2 t_s$. Thus $k_I = 25/(\zeta^2 t_s^2 k_p)$. For the example considered, values of $t_s = 10$ to 25 and $\zeta = 1$ are assumed. Using this formula, the range for k_I is obtained as $(0.04/k_p)$ to $(0.25/k_p)$. For the present simulation example, the value for ε is considered as 0.05. In general, the value of ε can be from 0.05 to 0.5. Let us explain the suggested method with two examples.

14.3 Simulation Studies

14.3.1 Example 1

The system matrix considered here is from Govindakannan and Chidambaram (1997) suitably modified given by Eq. (14.5).

$$K_P G_P = \begin{bmatrix} \dfrac{1.6667e^{-s}}{(1.6667s - 1)} & \dfrac{0.42e^{-s}}{(1.6667s - 1)} & \dfrac{e^{-s}}{(1.6667s - 1)} \\ \dfrac{0.8333e^{-s}}{(1.6667s - 1)} & \dfrac{0.21e^{-s}}{(1.6667s - 1)} & \dfrac{1.6667e^{-s}}{(1.6667s - 1)} \end{bmatrix} \qquad (14.5)$$

Where,

$$K_P = \begin{bmatrix} 1.6667 & 0.42 & 1 \\ 0.8333 & 0.21 & 1.6667 \end{bmatrix} \tag{14.6}$$

14.3.1.1 *Single-Stage Multivariable Controllers for Non-square Systems (Example 1)*

Using the values of $\delta = 1.4$ and $\varepsilon = 0.05$, the calculated values of K_c and K_I are given by Eq. (14.7) and Eq. (14.8).

$$K_c = \begin{bmatrix} 1.1283 & -0.6769 \\ 0.2843 & -0.1705 \\ -0.5999 & 1.1999 \end{bmatrix} \tag{14.7}$$

$$K_I = \begin{bmatrix} 0.0403 & -0.0242 \\ 0.0102 & -0.0061 \\ -0.0214 & 0.0429 \end{bmatrix} \tag{14.8}$$

The closed loop performances for the servo and, separately, for the regulatory problems are evaluated and the responses are shown in Fig. 14.1 and Fig. 14.2. Fig. 14.1 shows that the overshoot is high. The manipulated variable versus time requirement for the servo problem is shown in Fig. 14.3. The IAE values for the responses are given in Table 14.1 separately for the servo and the regulatory problems. As stated earlier, the two-stage P–PI controllers reduce the overshoot significantly (Govindakannan and Chidambaram, 2000). However, the method requires the knowledge of the process gain, time constant, and time delay of each of the elements of the transfer function matrix. Let us discuss the method of designing the two-stage multivariable control system based on only the SSGM.

14.3.2　Two-Stage Centralized PI Controllers for Example 1

Consider the method of designing the two-stages P–PI controllers (Fig. 12.1, Chapter 12). First, centralized multivariable Proportional controllers are used to stabilize the system. For the stabilized system, an outer loop diagonal of the PI control system is designed. In the present method, the system is first stabilized by a simple P controller matrix (Davison, 1976) using the knowledge of SSGM as given

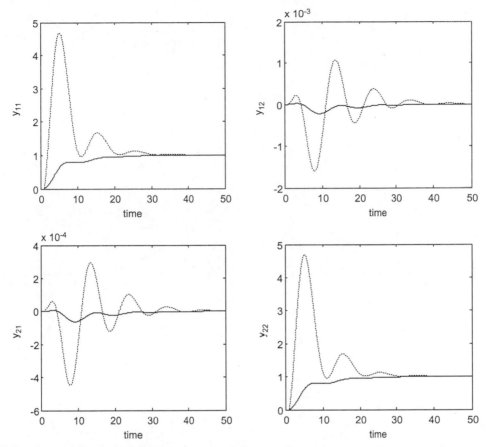

FIGURE 14.1 Responses and interactions for step changes in the set point for Example 1, Eq. (14.5)
y_{11}, y_{21}: response and interaction in y_1 and y_2 for step change in set point of y_1; y_{12}, y_{22} interaction
and response in y_1 and y_2 for step change in set point of y_2
Solid line: two-stage controllers; dashed line: single-stage controllers

in Eq. (14.9).

$$K_c = \delta K_P^{-1} \tag{14.9}$$

where δ is considered to range from 1.0 to 2.0.

Since the objective is to stabilize the system, only rough values of the elements
of K_C matrix are required. For non-square systems, the pseudo-inverse of K_P is
to be used in Eq. (14.3). In Eq. (14.1) K_P^{-1} can be replaced by $K_P^{\dagger}$ denoting the
pseudo-inverse of the matrix K_P. For the design of single-stage PI controllers, $\delta =$
1.4 was assumed; the value of δ is varied around this. By simulation it is found
that $\delta = 1.3$ in Eq. (14.9) gives an improved performance. Using Eq. (14.9), we get

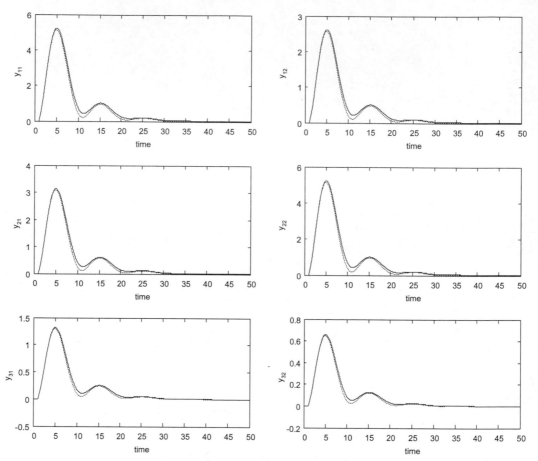

FIGURE 14.2   Responses and interactions for step changes in the load for Example 1, Eq. (14.5)
y_{11} y_{21}: response and interaction in y_1 and y_2 for step change in load v_1
y_{21}, y_{22}: interaction and response in y_1 and y_2 for step change in load v_2
y_{31}, y_{32}: interaction and response in y_1 and y_2 for step change in load v_3
Solid line: two-stage controllers; dashed line: single-stage controllers

Eq. (14.10).

$$K_c = \begin{bmatrix} 1.0477 & -0.6286 \\ 0.2640 & -0.1583 \\ -0.5571 & 1.1142 \end{bmatrix} \tag{14.10}$$

The gain matrix of the closed loop system (M) is given by Eq. (14.11).

$$M = [-I + K_p K_c]^{-1}[K_P K_c] \tag{14.11}$$

For the proposed methodology, the gain matrix (M) is seen to be a diagonal matrix as K_C is considered as δ times the inverse of K_P. M is obtained as

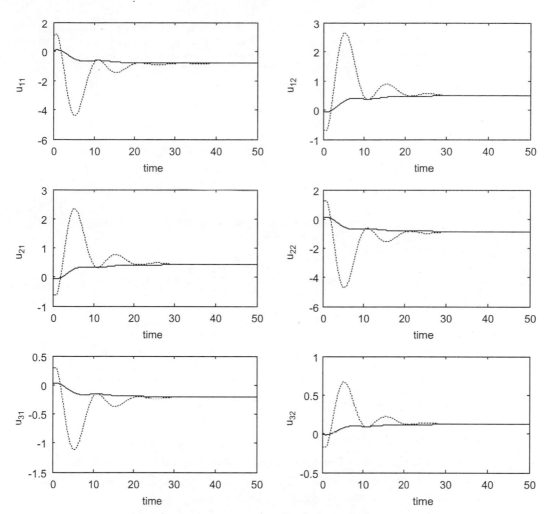

FIGURE 14.3 Manipulated variable versus time requirement for Fig. 14.1
u_{11}, u_{21}, u_{31} represent behaviour for step change in the set point of y_1
u_{12}, u_{22}, u_{32} represent behaviour for step change in the set point of y_2
Solid line: two-stage controllers; dashed line: single-stage controllers

diag[4.333   4.333]. For this pure gain system, PI controllers are designed. In the present example, the values of $\delta_O = 0.3$ and $\varepsilon_O = 0.15$ give a good performance. The outer loop controller parameters are given in Eq. (14.12).

$$K_{C,O} = \text{diag}[0.07 \quad 0.07] \text{ and } K_{I,O} = \text{diag}[0.035 \quad 0.035] \tag{14.12}$$

The closed loop performance is evaluated for the servo problem and, separately, for the regulatory problem. The servo responses are shown in Fig. 14.1. The regulatory responses are shown in Fig. 14.2. The manipulated variable versus time behaviour

TABLE 14.1 IAE values for servo and regulatory problems for single-stage and two-stage methods (Example 1)

Step input in	Single-stage method			Two-stage method		
	y_1	y_2	Sum	y_1	y_2	Sum
$y_{1,r}$	23.05	0.00393	23.054	6.589	0.000606	6.590
$y_{2,r}$	0.014	23.04	23.054	0.00216	6.59	6.592
L_1	33.32	16.66	49.980	36.6	18.3	54.9
L_2	8.396	4.198	12.594	9.223	4.611	13.834
L_3	19.99	33.320	53.310	21.95	36.59	58.54

TABLE 14.2 IAE values for servo and regulatory problems for single-stage and two-stage methods for uncertainty in unstable time constant (Example 1)

Step input In	Single-stage method			Two-stage method		
	y_1	y_2	Sum	y_1	y_2	Sum
$y_{1,r}$	32.44	0.0112	32.451	6.615	0.00095	6.616
$y_{2,r}$	0.03932	32.44	32.479	0.0034	6.616	6.619
$y_{1,r}$	23.06	0.00268	23.063	6.591	0.00061	6.592
$y_{2,r}$	0.00955	23.06	23.070	0.00216	6.592	6.594

two rows are for -10% uncertainty in τ in the system; last two rows are for $+10\%$ uncertainty in τ.

for the servo problems is shown in Fig. 14.3. It can be clearly seen that the control system designed by the two-stage P–PI controller design method offers better performances than the single-stage PI controller design method. The knowledge of only the SSGM of the process is needed for the design of the controllers. Table 14.1 gives the IAE values and the sum of IAE values for the servo and the regulatory problems separately. The load variable is assumed to enter the system along with the manipulated variable. A significant improvement in the performance is obtained for the two-stage controlled system. The robustness of the proposed control systems is evaluated by perturbing the unstable time constant by -10% (1.6667 to 1.5) whereas the controller settings are not changed. Fig. 14.4 shows the performances for both the single-stage multivariable PI control system and the two-stage multivariable PI control system. The proposed control system is robust. The IAE values are given in Table 14.2. Similar robust response is obtained for $+10\%$ uncertainty in the unstable time constant. The IAE values are given in Table 14.2.

In this example, for the single stage PI controller $\delta = 1.4$ is used and for design of P controller for the inner loop of two stage controllers, $\delta = 1.3$ is used. The results

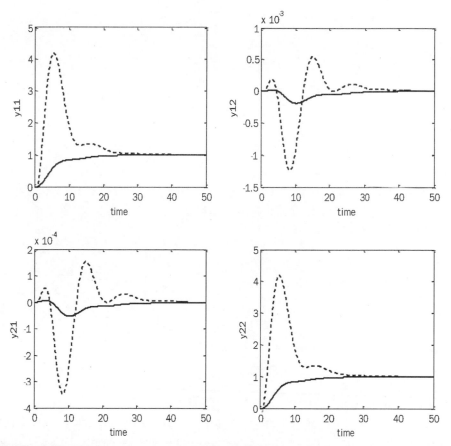

FIGURE 14.4 Responses and interactions for step changes in the set point for -10% uncertainty in time constant (Example 1)

y_{11}, y_{21}: response and interaction in y_1 and y_2 for step change in set point of y_1

y_{21}, y_{22}: interaction and response in y_1 and y_2 for step change in set point of y_2

Solid line: two-stage controllers; dashed line: single-stage controllers

show improved performance for the two stage control system for servo problems whereas for the regulatory problem there is no improvement (Table 14.2). Since the P controller for the inner loop should be larger than (1.1 times) that of the P setting for the single stage PI settings, let us use $\delta = 1.6$ for calculating the P controller settings of the inner loop. The results show improved performance for the two stage controller for the regulatory problem also.

14.3.3 Example 2 for Non-square systems

The case study considered here is the one presented by Jevtovic and Matausek (2010) which is suitably modified to get a FOPTD or SOPTD transfer function models as given in Eq. (14.13).

$$K_p G_p = \begin{bmatrix} \dfrac{e^{-0.8s}}{(2s+1)\,(4s-1)} & \dfrac{0.5e^{-0.8s}}{(4s-1)} & \dfrac{e^{-s}}{(s+1)\,(4s-1)} \\[2ex] \dfrac{-3e^{-0.1s}}{(4s-1)} & \dfrac{1.5e^{-0.2s}}{(4s-1)} & \dfrac{e^{-2s}}{(4s-1)} \end{bmatrix} \tag{14.13}$$

14.3.3.1   *Single-Stage Multivariable Controllers for Example 2*

As discussed earlier, based on the modified Davison method and using the values of $\delta = 1.4$ and $\varepsilon = 0.03$, the controller matrices K_c and K_I are calculated from Eq. (14.3) and Eq. (14.4) as in Eq. (14.14) and Eq. (14.15).

$$K_c = \begin{bmatrix} 0.4577 & -0.2961 \\ 0.4308 & 0.2153 \\ 0.7269 & 0.1884 \end{bmatrix} \tag{14.14}$$

$$K_I = \begin{bmatrix} 0.0098 & 0.0063 \\ 0.0092 & 0.0044 \\ 0.0156 & 0.0040 \end{bmatrix} \tag{14.15}$$

The closed loop performances are noted for the servo and, separately, for the regulatory problem. The servo performances for unit step change in set point are shown in Fig. 14.5. The regulatory responses are presented in Fig. 14.6. The manipulated variable versus time requirement for the servo problem is shown in Fig. 14.7. It is seen from Fig. 14.6 that the overshoot is high for controllers designed using single-stage. Table 14.3 gives the IAE values for the responses for the servo and the regulatory problems. The sum of the IAE values for the combined servo and regulatory problems are also given in Table 14.3.

14.3.3.2   *Two-Stage Centralized PI Controllers for Example 2*

Consider the methodology of designing two-stage P–PI controllers for an unstable non-square system. The modified Davison method which uses only the knowledge of the SSGM of the process is applied first for stabilizing the system by means of a simple P controller matrix. The P controller matrix is given by Eq. (14.16).

$$K_c = \delta K_P^\dagger \tag{14.16}$$

where δ is considered to range from 1.0 to 2. As stated earlier, $K_P^\dagger$ is the pseudo-inverse of K_P. Since the objective is to stabilize the system, only rough

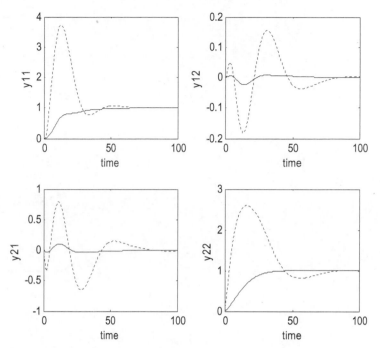

FIGURE 14.5 Responses and interactions for step changes in the set point (Example 2)
y_{11}, y_{21}: response and interaction in y_1 and y_2 for step change in set point of y_1
y_{21}, y_{22}: interaction and response in y_1 and y_2 for step change in set point of y_2
Solid line: two-stage controllers; dashed line: single-stage controllers

TABLE 14.3 IAE values for servo and regulatory problems for single-stage and two-stage methods (Example 2)

Step input In	Single-stage method			Two-stage method		
	y_1	y_2	Sum	y_1	y_2	Sum
$y_{1,r}$	43.42	20.68	64.10	13.39	1.927	15.317
$y_{2,r}$	4.926	44.63	49.556	0.4614	13.52	13.981
L_1	44.3	127.8	172.10	22.64	67.95	90.59
L_2	17.86	58.27	76.13	11.16	33.50	44.66
L_3	34.52	35.14	69.66	22.31	22.32	44.63

sum for single-stage method = 431.546; total sum for two-stage method = 209.178

values of the elements of the K_C matrix are required. For the present example, $\delta = 1.6$ is considered. Eq. (14.16) gives Eq. (14.17).

$$K_c = \begin{bmatrix} 0.5230 & -0.3384 \\ 0.4923 & 0.2461 \\ 0.8307 & 0.2154 \end{bmatrix} \tag{14.17}$$

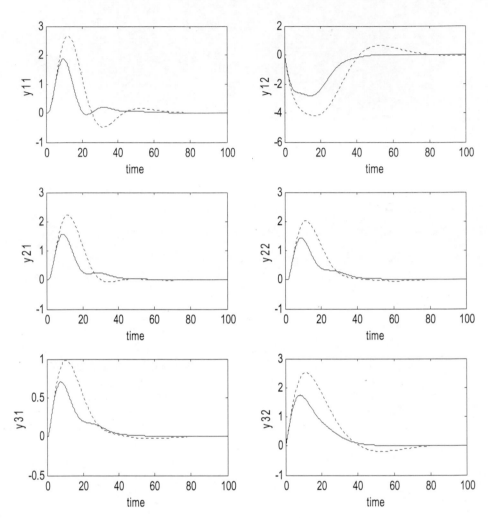

FIGURE 14.6 Responses and interactions for step changes in the load (Example 2)
y_{11}, y_{21}: response and interaction in y_1 and y_2 for step change in load v_1
y_{21}, y_{22}: interaction and response in y_1 and y_2 for step change in load v_2
y_{31}, y_{32}: interaction and response in y_1 and y_2 for step change in load v_3
Solid line: two-stage controllers; dashed line: single-stage controllers

Now multivariable PI controllers have to be designed for the stabilized system. The closed loop system (M) gain matrix is given by Eq. (14.18).

$$M = [-I + K_p K_c]^{-1}[K_p\ K_c] \tag{14.18}$$

The matrix (M) becomes a diagonal matrix since K_C is considered as δ times the inverse of K_P. The gain matrix is obtained as $M = \text{diag}[2.6667\ 2.6667]$. For this pure gain system, PI controllers are designed by the Davison method. For the

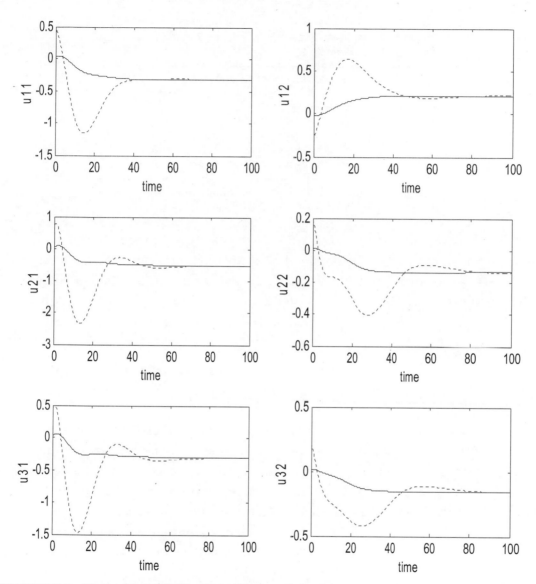

FIGURE 14.7   Manipulated variable versus time requirement for Fig. 14.5
u_{11}, u_{21}, u_{31} represent the behaviour for step change in the set point of y_1
u_{12}, u_{22}, u_{32} represent the behaviour for step change in the set point of y_2
Solid line: two-stage controllers; dashed line: single-stage controllers

stabilized system, a centralized PI controller system is designed by the Davison method as in Eq. (14.19) and Eq. (14.20).

$$K_{c,o} = \delta_O \{\text{diag}[k_{p11,o} \quad k_{p22,o}]\}^{-1} \tag{14.19}$$

$$K_{I,O} = \varepsilon_O \{\text{diag}[k_{p11,o} \quad k_{p22,o}]\}^{-1} \tag{14.20}$$

TABLE 14.4 IAE values for servo and regulatory problems for single-stage and two-stage methods for uncertainty in unstable time constant (Example 2)

Step input In	Single-stage method			Two-stage method		
	y_1	y_2	Sum	y_1	y_2	Sum
$y_{1,r}$	42.78	21.96	64.74	13.39	1.932	15.322
$y_{2,r}$	5.163	42.49	47.65	0.4668	13.83	14.297
$y_{1,r}$	45.64	19.88	65.52	13.39	1.932	15.322
$y_{2,r}$	4.774	46.58	51.35	0.4668	13.85	14.317

The first two rows are for -10% uncertainty in τ in the system; the last two rows are for $+10\%$ uncertainty in τ.

where δ_O is considered to range from 0.1 to 0.4 and ε_O ranges from 0.05 to 0.2. The suitable values are to be obtained by simulation. The values of $\delta_O = 0.15$ and $\varepsilon_O = 0.075$ are considered. Here $k_{pij,o}$ is the gain of the inner loop closed system, that is, the elements of M. For the outer loop we consider Eq. (14.21).

$$K_{c,o} = \text{diag}[0.056 \quad 0.056] \text{ and } K_{I,O} = \text{diag}[0.028 \quad 0.028] \tag{14.21}$$

The servo and regulatory closed loop responses are evaluated separately. The servo performance for unit step change in set point is shown in Fig. 14.5. The regulatory responses are presented in Fig. 14.6. Fig. 14.7 shows the time variation of the manipulated variable for the servo problem. It can be clearly seen that the control system designed by the two-stage method offers better performance than the single-stage method for the given multivariable unstable non-square system. Only the knowledge of SSGM of the system is used for the design of controllers. Table 14.3 gives separately the IAE values and the sum of IAE values for the responses of the servo and regulatory problems. The robustness of the proposed control system is evaluated by perturbing the unstable time constant -10%, keeping the controller settings the same. The IAE values are given in Table 14.4. Similarly, robust response is obtained for $+10\%$ uncertainty in the unstable time constant. The IAE values are given in Table 14.4. The proposed control system is robust.

14.4   Identification of SSGM

The system considered is an open loop unstable process, hence the SSGM is to be obtained only under closed loop condition (with PI controlled system). The closed loop system is excited by giving a step change separately in each of the set points.

When a step change is given in the set point of y_1 at steady state conditions, the output variables y_1 and y_2 will reach 1 and 0 respectively. The resulting changes in the steady state values of the manipulated variables ($u_{1,1}$, $u_{2,1}$, and $u_{3,1}$) are defined as $U_1 = [u_{1,1} \; u_{2,1} \; u_{3,1}]^T$. Similarly, when a step change is given in the set point of y_2, the changes in the steady state values are specified as $U_2 = [u_{1,2} \; u_{2,2} \; u_{3,2}]^T$. Since Y is an Identity matrix $\begin{bmatrix} 1 & 0 \\ 0 & 1 \end{bmatrix}$ and $Y = K_P U$, the SSGM is computed (Papastathopoulo and Luyben, 1990) as given in Eq. (14.22).

$$K_P = [U_1 \quad U_2]^{-1} \tag{14.22}$$

For open loop unstable systems because of the definition of $k_p G_p$, we have Eq. (14.23).

$$K_P = -[U_1 \quad U_2]^{-1} \tag{14.23}$$

By using suitable multivariable PI controllers, the system is assumed to be under closed loop stability. For the present example, the multivariable PI settings given by Govindakannan and Chidambaram (2000) are considered as in Eq. (14.24).

$$G_C = \begin{bmatrix} 1.1283 + \dfrac{0.0402}{s} & -0.6769 - \dfrac{0.0242}{s} \\ 0.2843 + \dfrac{0.0102}{s} & -0.1705 - \dfrac{0.0061}{s} \\ -0.5999 - \dfrac{0.0214}{s} & 1.1999 + \dfrac{0.0429}{s} \end{bmatrix} \tag{14.24}$$

Let us consider Example 1. The system is simulated under closed loop identification for the SSGM. The system is excited by giving a step change in the set point of y_1, and the steady state values of $u_{1,1}$, $u_{2,1}$ and $u_{3,1}$ are noted as $U_1 = [-0.8078 \; 0.2028 \; 0.4298]^T$. Similarly, a step change in the set point of y_2 is introduced and the steady state values of $u_{1,2}$, $u_{2,2}$ and $u_{3,2}$ are noted as $U_2 = [0.4806 - 0.1407 - 0.8597]^T$. Thus the SSGM of the open loop system is calculated from Eq. (14.25).

$$K_P = -[U_1 \quad U_2]^\dagger \tag{14.25}$$

For non-square systems, pseudo-inverse of the matrix is to be used as in Eq. (14.26).

$$K_P = \begin{bmatrix} -1.6598 & -0.3558 & -0.9776 \\ -0.8475 & -0.1435 & -1.6604 \end{bmatrix} \tag{14.26}$$

TABLE 14.5 Identified SSGM (K_p) under load and separately under measurement noise (Example 1)

Load	$U = [U_1 \ U_2]$	$K_P = -U^\dagger$
0.02 in v_1	$\begin{bmatrix} -0.8267 & 0.4617 \\ -0.2072 & 0.1363 \\ 0.4298 & -0.8597 \end{bmatrix}$	$\begin{bmatrix} 1.5982 & 0.3427 & 0.9127 \\ 0.8076 & 0.1347 & 1.6183 \end{bmatrix}$
0.02 in v_2	$\begin{bmatrix} -0.8126 & 0.4760 \\ -0.2039 & 0.1396 \\ 0.4298 & -0.8597 \end{bmatrix}$	$\begin{bmatrix} 1.6575 & 0.3545 & 0.9753 \\ 0.8372 & 0.1407 & 1.6496 \end{bmatrix}$
0.02 in v_3	$\begin{bmatrix} -0.8079 & 0.4805 \\ -0.2022 & 0.1414 \\ 0.4099 & -0.8798 \end{bmatrix}$	$\begin{bmatrix} 1.6307 & 0.3492 & 0.9467 \\ 0.7681 & 0.1269 & 1.5765 \end{bmatrix}$
Noise in y_1	$\begin{bmatrix} -0.8137 & 0.4747 \\ -0.2044 & 0.1391 \\ 0.4330 & -0.8566 \end{bmatrix}$	$\begin{bmatrix} 1.6598 & 0.3558 & 0.9776 \\ 0.8475 & 0.1435 & 1.6604 \end{bmatrix}$
Noise in y_2	$\begin{bmatrix} -0.8035 & 0.4849 \\ -0.2019 & 0.1417 \\ 0.4235 & -0.8661 \end{bmatrix}$	$\begin{bmatrix} 1.6810 & 0.3611 & 1.0002 \\ 0.8303 & 0.1410 & 1.6425 \end{bmatrix}$

Load variable v_1 entering the system with the input u_1; v_2 entering the system with the input u_2; v_3 entering the system with the input u_3; noise is a random signal with 0 mean and variance 0.0001; † refers to pseudo-inverse.

The influence of measurement noise and the influence of disturbances entering the process (along with the input variables) are also assessed separately. The effect of load and the disturbances do not affect the gain matrix significantly as the system is under closed loop control. The results are given in Table 14.5.

Summary

To conclude, Davison method is modified to design a single-stage PI controller matrix using the knowledge of the SSGM of unstable non-square multivariable systems. Since the overshoots are high, a two-stage P–PI centralized controller matrix is designed. First the system is stabilized by the Centralized P controller matrix designed by the modified Davison method. For this stabilized system, a decentralized PI controller matrix is designed by the Davison method. A systematic method for the selection of the tuning parameters is presented. The two-stage multivariable controllers are found to be better than that of the single-stage multivariable controllers. The method proposed by Papastathopoulo and Luyben

(1990) to identify the steady state gain matrix (SSGM) of stable systems is extended to the unstable non-square multivariable systems.

Problems

1. For the given unstable non-square system, design centralized PI controllers by Davison method, and Tanttu and Lieslehto method. Compare the closed loop servo response.

$$
K_c G_c =
\begin{bmatrix}
\dfrac{e^{-0.8s}}{(2s+1)(6s-1)} & \dfrac{e^{-0.8s}}{(6s-1)} & \dfrac{e^{-s}}{(s+1)(6s-1)} \\[3mm]
\dfrac{-3e^{-0.1s}}{(6s-1)} & \dfrac{2e^{-0.2s}}{(6s-1)} & \dfrac{e^{-2s}}{(6s-1)}
\end{bmatrix}
$$

2. For the given unstable non-square system, design centralized PI controllers by Davison method, and Tanttu and Lieslehto method. Compare the closed loop servo response.

$$
K_P G_P =
\begin{bmatrix}
\dfrac{1.5e^{-s}}{(2s-1)} & \dfrac{0.4e^{-s}}{(2s-1)} & \dfrac{e^{-s}}{(2s-1)} \\[3mm]
\dfrac{0.8e^{-s}}{(2s-1)} & \dfrac{0.25c^{-s}}{(2s-1)} & \dfrac{1.5e^{-s}}{(2s-1)}
\end{bmatrix}
$$

3. For the given unstable non-square system, design centralized PI controllers by Davison method, and Tanttu and Lieslehto method. Compare the closed loop servo response.

$$
K_P G_P =
\begin{bmatrix}
\dfrac{1.7e^{-s}}{(4s-1)} & \dfrac{0.4e^{-s}}{(4s-1)} & \dfrac{e^{-s}}{(4s-1)} \\[3mm]
\dfrac{0.8e^{-s}}{(4s-1)} & \dfrac{0.21e^{-s}}{(4s-1)} & \dfrac{1.7e^{-s}}{(4s-1)}
\end{bmatrix}
$$

15

Trends in Control of Multivariable Systems

In this chapter, some of the methods of designing multivariable control system are reviewed. These methods include Gain and Phase Margin (GPM) method, Internal Model Control (IMC) method, synthesis method, PI controllers with no proportional kick, analytical methods, method of inequalities, goal attainment method, Effective Transfer Function (ETF) method, and Model Reference Controller (MRC) design method.

15.1 Gain and Phase Margin Method

The gain and phase margins (GPMs) are typical loop specifications associated with the frequency response (Franklin et al., 1989). The GPMs have always served as important measures of robustness (Kaya, 2004, Lee, 2004). It is known from classical control that phase margin is related to the damping of the system and can therefore also serve as a performance measure (Franklin et al., 1989). The controller design methods to satisfy GPM criteria are not new, and have been widely used (Ho et al., 1995; Hu et. al, 2011). Simple formulae are given by Maghade and Patre (2012) to design a PI/PID controller to meet user-defined Gain Margin (GM) and Phase Margin (PM) specifications.

Using definitions of GPM, the set of equations from Eq. (15.1) to Eq. (15.4) can be written.

$$\arg[l_{ii}(j\omega_{pii})k_{ii}(j\omega_{pii})] = -\pi \tag{15.1}$$

$$A_{mii} = \frac{1}{|l_{ii}(j\omega_{pii})k_{ii}(j\omega_{pii})|} \tag{15.2}$$

$$|l_{ii}(j\omega_{gii})k_{ii}(j\omega_{gii})| = 1 \tag{15.3}$$

$$\varphi_{mii} = \arg[l_{ii}(j\omega_{gii})k_{ii}(j\omega_{gii})] + \pi \tag{15.4}$$

where A_{mii} and φ_{mii} are GM and PM respectively. Here ω_{gii} and ω_{pii} are gain and phase crossover frequencies. The PI controller parameters are given for FOPTD model as in Eq. (15.5).

$$k_{c,ii} = \frac{\omega_{pii}T_{ii}}{A_{mii}k_{p,ii}} \quad T_{I,ii} = \left(2\omega_{pii} - \frac{4\omega_{pii}^2 l_{ii}}{\pi} + \frac{1}{T_{ii}} \right)^{-1} \tag{15.5}$$

Where,

$$\omega_{pii} = \left(\frac{A_{mii}\varphi_{mii} + \frac{1}{2}\pi(1/A_{mii})}{(A_{mii}^2 - 1)l_{ii}} \right) \tag{15.6}$$

PID controller parameters with first order filter are as in Eq. (15.7) and Eq. (15.8).

$$k_{c,ii} = \frac{\omega_{pii}T_{ii}}{A_{mii}k_{p,ii}} \quad T_{Iii} = \left(2\omega_{pii} - \frac{4\omega_{pii}^2 l_{ii}}{\pi} + \frac{1}{T_{ii}} \right)^{-1} \quad T_{Dii} = T_{Iii} \tag{15.7}$$

$$T_{Fii} = T_{Iii} \tag{15.8}$$

It is recommended that a gain margin of 2 and phase margin of 45° can be used.

15.2   Internal Model Control Method

The concept of Internal Model Control (IMC) was introduced by Garcia & Morari (1982) for scalar systems, and further extended to multivariable (linear and non-linear) systems to design a decentralized control system by Garcia and Morari (1985a), Garcia and Morari (1985b), Ravera et al. (1986) and Economou et al. (1986).

Garcia and Morari (1985a) extended the method to MIMO discrete time systems. A model predictive control law formulation is suggested by Garcia and Morari (1985b) for the computation of the IMC and the controller parameters can be calculated explicitly without inversion of the polynomial. In multivariable process with time delays, the Pade approximation is used by Ravera et al. (1986) to get the PI/PID controllers with lead-lag filters. The IMC design approach is further extended to nonlinear process by Economou et al. (1986).

15.2.1 General Steps in IMC Design Procedure

1. Design an IMC controller with set point filter (contains tuning parameter, i.e., closed loop time constant)

$$Q = [\bar{G}_p]^{-1} f \tag{15.9}$$

2. The relationship between the IMC and the classical controller is given in Eq. (15.10).

$$G_c = Q(I - G_p Q)^{-1} \tag{15.10}$$

Generally, to get the classical PI/PID controller form, the Maclaurin series is used. A few researchers recognized the usefulness and the importance of the IMC design procedure and they improved it further by adapting it for the multivariable systems.

Proper input–output pairing is done by using RGA (Bristol, 1966). Then the controllers corresponding to each controlled variable are designed independently by IMC method (Skogestad, 2003) for each system independently. Detuning factor Fd (Chidambaram, 2002) is used for fine-tuning. Thus, the controller parameters are changed as $Kc = kc, ij/Fd$ and $KI = kc, ij/Fd2\tau I, ij$. The range of Fd is from 2 to 5.

Using RGA analysis, if in the system the interaction is not high, then we can use decentralized control structure to get the desired performance.

The diagonal PI controller based on FOPTD process by IMC method (Skogestad, 2003) gives Eq. (15.11) and Eq. (15.12).

$$K_c = \frac{\tau_I}{k(\tau_c + \theta)} \tag{15.11}$$

$$\tau_I = \min\{\tau_I, 4(\tau_c + \theta)\} \tag{15.12}$$

15.3   PI Controllers with No Proportional Kick

This method is discussed in (Chien et al., 1999).

Because the proportional mode is acting on the error signal, there will be a large kick of the controller output during set point changes. If the multi loop control system has large interactions among the loops, the large controller action on the loop will be large for load disturbance rejection. To reject the load disturbance, the corrective action by the controllers in the other loops will be large and the disturbance will go back to the loop with set point changes. The disturbance effect will go back-and-forth among control loops and cause control problem. For this reason, controllers are designed so that there will be no kick at the time of set point changes. This is called PID form with no proportional kick. The proportional mode is acting on the negative sign of the controlled variable alone. The Laplace transform representation of this control is as in Eq. (15.13).

$$u(s) = K_c \left(-y(s) + \frac{1}{\tau_I s} E(s) + \tau_D s [-y(s)] \right) \tag{15.13}$$

Here the derivative mode is also acting on the negative sign of the controlled variable to prevent derivative kick. For PI, tuning rules with no derivative action can be selected.

Controller synthesis method is used to derive the PI tuning set point as in Eq. (15.14).

$$\frac{y}{r} = \frac{\left(\frac{K_c}{\tau_I s} \right) G_p}{1 + \left(\frac{K_c [\tau_I s + 1]}{\tau_I s} \right) G_p} \tag{15.14}$$

Determining the FOPTD process by Taylor approximation, the process model becomes as represented in Eq. (15.15).

$$G_p = \frac{K_p e^{-Ls}}{\tau s + 1} \approx \frac{K_p (1 - Ls)}{\tau s + 1} \tag{15.15}$$

Substituting Eq. (15.15) into Eq. (15.14) and simplifying, we obtain Eq. (15.16).

$$\frac{y}{r} = \frac{1 - Ls}{\left(\frac{\tau_I \tau}{K_c K_p - \tau_I L} \right) s^2 + \left(\frac{\tau_I}{K_c K_p - \tau_I L} + \tau_I - L \right) + 1} \tag{15.16}$$

With the desired closed loop response, the PI tuning rules obtained as Eq. (15.17) and Eq. (15.18).

$$K_c = \frac{1}{K_p} \frac{-\tau^2 cl + 1.414\tau_{cl}\tau + L\tau}{\tau^2 cl + 1.414\tau_{cl}\tau + L^2} \tag{15.17}$$

$$\tau_I = \frac{-\tau^2 cl + 1.414\tau_{cl}\tau + L\tau}{\tau + L} \tag{15.18}$$

15.4 Analytical Method

This method is discussed in (Lee et al., 2004).

$$Y(s) = H(s)r(s) = [I + G(s)K(s)]^{-1}G(s)K(s)r(s) \tag{15.19}$$

Where,

$G(s)$ is the open loop stable process, $K(s)$ is the multi loop controller, $Y(s)$ is the controlled variable vector, and $r(s)$ is the set point vector. According to the synthesis method, the desired closed loop response R_i of ith loop is typically chosen by Eq. (15.20).

$$\frac{y_i}{r_i} = R_i = \frac{G_{ii+}(s)}{(\lambda_i s + 1)^{ni}} \tag{15.20}$$

Where,

G_{ii+} is the non-minimum part of G_{ii}, λ_i is an adjustable constant for system performance and stability, and n_i is chosen such that the IMC controller would be implementable (i.e., the numerator degree in s should be lesser that that in the denominator).

The desired closed loop response matrix would be as in Eq. (15.21).

$$R(s) = \text{diag}[R_1, R_2, \dots R_n] \tag{15.21}$$

Thus, our aim is to design $K(s)$ such that all the diagonal elements of $H(s)$ resemble those of $R(s)$ as close as possible over a frequency range. $K(s)$ can be written in a Maclaurin series as in Eq. (15.22).

$$K(s) = \frac{1}{s}[K_0 + K_1 s + K_2 s^2 + O(s^3)] \tag{15.22}$$

$G(s)$ can be written in a Maclaurin series as in Eq. (15.23).

$$G(s) = G_0 + G_1 s + G_2 s^2 + O(s^3) \tag{15.23}$$

By substituting Eq. (15.23) and Eq. (15.22) into Eq. (15.19) and rearranging we get Eq. (15.24).

$$H(s) = I - (G_0 K_0)^{-1}(I + G_0 K_1 + G_1 K_0)(G_0 K_0)^{-1} s^2$$
$$+ [G_0 K_2 + G_1 K_1 + G_2 K_0 - (I + G_0 K_1 + G_1 K_0)$$
$$(G_0 K_1 + G_1 K_0)(G_0 K_0)^{-1}] s^3 + O(s^4) \tag{15.24}$$

Similarly, $R(s)$ can be written in Maclaurin series as in Eq. (15.25).

$$R(s) = R(0) + R'(0)s + \frac{R''(s)}{2}s^2 + \frac{R'''(s)}{6}s^3 + O(s^4) \tag{15.25}$$

Thus comparing each diagonal element of $H(s)$ and $R(s)$ in Eq. (15.24) and Eq. (15.25) gives K_0, K_1, K_2. Here R', R'', R are respectively the first, second, and third derivative of R with respect to s. But these parameters show severe inaccuracy due to limitations of the Maclaurin series for high frequency region. Proportional (K_1) and derivative (K_2) are predominant at high frequency and insignificant at low frequency. Thus, K_1 and K_2 should be designed based on the process characteristics at high frequencies.

$$K_{c,ij} = f'_{ij}(0); \quad \tau_{I,ij} = \frac{f'_{ij}(0)}{f_{ij}(0)}; \quad \tau_{D,ij} = \frac{f''_{ij}(0)}{2f'_{ij}(0)} \tag{15.26}$$

$|G(j\omega)K(j\omega)| << 1$ at high frequency.

$$H(s) \approx (I + G(s)K(s))^{-1}G(s)K(s) \approx G(s)K(s) \tag{15.27}$$

This indicates that K_1 and K_2 can be designed based on only the diagonal element of $G(s)$. From the IMC method we get Eq. (15.28).

$$K(s) = G^{-1}(s)R \tag{15.28}$$

Where, $G(s) = \text{diag}[G_{11}, G_{22}, \ldots]$

Thus, for i^{th} loop the controller can be derived as in Eq. (15.29) and Eq. (15.30).

$$K_i(s) = \frac{[G_{ii-}(s)]^{-1}}{(\lambda_i s + 1)^{ni} - G_{ii+}(s)} \tag{15.29}$$

$$G_{c,ij}(s) = \frac{1}{s}\left[f_{ij}(0) + f'_{ij}(0)s + \frac{f''_{ij}(0)}{2}s^2 + Os^3\right] \tag{15.30}$$

From Eq. (15.29) and Eq. (15.30), comparing the coefficient of s terms we get Eq. (15.31).

$$f_{ij}(0) = \frac{1}{k_{p,ij}(1 + \lambda_{ic})}$$

$$f'_{ij}(0) = \frac{1}{k_{p,ij}}\left[\frac{\tau_{ij}}{\theta_{ij} + \lambda_{ic}} + \frac{\theta_{ij}^2}{2(1 + \lambda_{ic})^2}\right]$$

$$\frac{f''_{ij}(0)}{2} = \frac{-1}{k_{p,ij}}\left[\frac{\frac{\theta_{ij}^3}{6(\theta_{ij}+\lambda_{ic})} - \frac{\theta_{ij}^4}{4(\theta_{ij}+\lambda_{ic})^2}}{(1 + \lambda_{ic})} - \frac{\tau_{ij}\theta_{ij}^2}{2(\theta_{ij} + \lambda_{ic})^2}\right] \tag{15.31}$$

To obtain a diagonal controller using ETF, there are other different methods available in the literature, reported by Luan and Liu (2014).

15.5 Method of Inequalities

Zakian and Al-Naib (1973) proposed this method for the solution of control problem formulated as a set of inequalities. These inequalities may be in terms of performance specifications like bounds on overshoot, response time, settling time, settling error, sensitivity, stability and integrity, robustness of the system, operational constraints like maximum/minimum limits on the process variables/parameters, and environmental and economic constraints/objectives. The method is computationally involved for solving the set of inequalities. The method is conceptually different from the conventional optimization since the aim is to achieve satisfactory performance rather than optimal performance. The design parameters as well as the physical constraints could be imposed on the optimization problem. The constraints specified as inequalities, should be satisfied by the system parameters, manipulated variables, and outputs. The design problem turns out to be about how to find controller parameters set that satisfies all the inequality constraints. Taiwo (2005) reviews the application of the method of inequalities to the multivariable control of important industrial processes such as distillation columns, a turbofan engine, and power systems. It is shown that the desired performance and stipulated constraints can be expressed in terms of a conjunction of inequalities, any solution of which gives an acceptable design. The method of inequalities is consistently shown to give an efficient solution to the given problem, not only in terms of the ease of problem formulation, but also in terms of the simple and implementable controllers obtained.

15.6 Goal Attainment Method

This technique involves expressing a set of design goals $I^* = [I_1*, I_2*, I_3*, \ldots, I_n*]$. The problem formulation allows the designer to be relatively imprecise about the initial design goals. The relative degree of under- or over-achievement of the goals is controlled by a vector of weighting coefficients, $w = \{w_1, w_2, w_n\}$ and it is expressed as a standard optimization using the formulation given in Eq. (15.32).

$$\text{Min } \gamma \tag{15.32}$$

$$I_1(x) - w_i\gamma \le I_i * \quad i = 1, 2, \ldots, n \tag{15.33}$$

The term $w_i\gamma$ introduces an element of slackness into the problem which otherwise imposes that the goals should be met rigidly. The weighting vector (w_i) enables the designer to express a measure of the relative trade-offs so that the same percentage of under- or over- attainment of the goals $(I*)$ is achieved. Hard constraints can be incorporated into the design by setting the weighting factor to 0 (i.e., $w_i = 0$). The goal attainment method provides a convenient initiative interpretation of the design problem, which is solvable using optimization methods.

The goal attainment formulation as given in Eq. (15.32) can be posed a min–max problem as given in Eq. (15.34).

$$\text{Minimize}\{\max\{\gamma_{ii}\}\} \tag{15.34}$$

Where

$$\gamma_{ii} = [I_i(x) - I_i*]/w_i \quad i = 1, 2, \ldots, n \tag{15.35}$$

15.7 Effective Transfer Function Method

This input–output relation can be expressed as in Eq. (15.36) and Eq. (15.37).

$$Y(s) = G(s)U(s) \tag{15.36}$$

$$Y(s) = \begin{bmatrix} y_1(s) \\ y_2(s) \end{bmatrix}; \quad U(s) = \begin{bmatrix} u_1(s) \\ u_2(s) \end{bmatrix} \tag{15.37}$$

where $Y(s)$ and $U(s)$ are output and input vectors, respectively. The input–output relationship for the TITO system, can be written as (Seborg et al., 2006) given in Eq. (15.38) and Eq. (15.39).

$$y_1(s) = g_{11}(s)u_1 + g_{12}(s)u_2(s) \tag{15.38}$$

$$y_2(s) = g_{21}(s)u_1 + g_{22}(s)u_2(s) \tag{15.39}$$

In the TITO system, when the second loop is closed, the input from u_i to y_i has two transmission paths. The combination of two transmission paths is considered as effective open loop dynamics. If the second feedback controller is in the automatic mode, with $y_{r2} = 0$, then the overall closed loop transfer function between y_1 and u_1 is given by Seborg et al. (2006) as in Eq. (15.40).

$$\frac{y_1}{u_1} = g_{11}\frac{g_{12}g_{21}g_{c2}}{1 + g_{c2}g_{22}} \tag{15.40}$$

This can be written as Eq. (15.41).

$$\frac{y_1}{u_1} = g_{11}\frac{g_{12}g_{21}(g_{c2}g_{22})}{g_{22}(1 + g_{c2}g_{22})} \tag{15.41}$$

Similarly, for the second loop, in Eq. (15.42),

$$\frac{y_2}{u_2} = g_{22} - \frac{g_{21}g_{12}(g_{c1}g_{11})}{g_{11}(1 + g_{c1}g_{11})} \tag{15.42}$$

In MIMO systems, the open loop dynamics between the controlled variable (y_i) and the manipulated variable (u_i) not only depends on the corresponding transfer function model (g_{ii}) but also depend on the other processes and controllers in all the other loops. This implies the tuning of one controller cannot be done independently and it depends on the other controllers. The complicated relations of Eq. (15.41) and Eq. (15.42) can be simplified by making two assumptions (Vu et al., 2010b; Hu et al., 2010).

The first assumption is that the perfect controller approximation for the other loop (the output attains steady state with no transient) was used to simplify the equations Eq. (15.8) and Eq. (15.9), as given in Eq. (15.43).

$$\frac{g_{ci}g_{ii}}{(1 + g_{ci}g_{ii})} = 1 \quad i = 1, 2 \tag{15.43}$$

The second assumption is that the ETFs have the same structure of the corresponding open loop model. By using the perfect controller approximation, Eq. (15.41) and Eq. (15.42) can be approximated as Eq. (15.44) and Eq. (15.45).

$$g_{11}^{eff} = \frac{y_1}{u_1} = g_{11} - \frac{g_{12}g_{21}}{g_{22}} \tag{15.44}$$

$$g_{22}^{eff} = \frac{y_2}{u_1} = g_{22} - \frac{g_{21}g_{12}}{g_{11}} \tag{15.45}$$

Where, g_{11}^{eff} and g_{22}^{eff} are the Effective Open Loop Transfer Functions (EOTF) (Vu et al., 2010b). These EOTFs are complicated transfer function models and it is difficult to directly use them for the controller design. For the purpose of controller design, resulting EOTFs are reduced to FOPTD models using Maclaurin series (Vu et al., 2010b). This method poses complications in higher dimension systems in the formulation of EOTFs and in the model reduction step. In the present example, by using RGA, RNGA and RARTA concepts (Cai et al., 2008; He et al., 2009; Shen et al., 2010; Hu et al., 2010; Vijaykumar et al., 2012), the expression for ETF can be derived easily for higher dimension systems also.

15.8   Model Reference Controller

We consider a MIMO state variable linear system (Chidambaram, 1995) represented by Eq. (15.46) and Eq. (15.47).

$$x" = Ax + Bu + d(t) \tag{15.46}$$

$$y = Cx \tag{15.47}$$

Here x is a state vector of size $n \times 1$, u is a vector of manipulated variables, and y is a vector of output variables which is a linear function of the state vector x. The vector size of both u and y is assumed to be equal and is given by $p \times 1$. Here A is $n \times n$, b is $n \times p$ and C is $q \times n$ size matrices. Let us assume the reference model in the output vector y is given by Eq. (15.48).

$$y'_m = \Lambda_m y_m - \Lambda_m y_r \tag{15.48}$$

where Λ_m is an eigen value diagonal matrix of size $p \times p$ and y_r is the input vector to the reference model. y'_m is the time derivative of y_m. The error vector E is defined by Eq. (15.49)

$$E = y_m - y \tag{15.49}$$

$$E' = y'_m - y' \tag{15.50}$$

On combining Eq. (15.46) to Eq. (15.50), we get Eq. (15.51 to Eq. (15.56).

$$E' = y'_m - CAx - CBu - Cd(t) \tag{15.51}$$

Let,

$$y'_m - CAx - CBu - Cd(t) = -KE - K_I E_I - K_D E_D \tag{15.52}$$

$$K = \text{diag}[k_i], \ K_I = \text{dia}[k_i/\tau_{Ii}] \text{ and } K_D = \text{dia}[k_i\tau_{Di}] \tag{15.53}$$

$$E_I = \left[\int e_1 dt \ \int e_2 dt \ \int e_3 dt \ldots \int e_p dt \right]^T \tag{15.54}$$

$$E_D = [e'_1 \ e'_2 \ldots e'_p]^T \tag{15.55}$$

$$E = [e_1 \ e_2 \ldots e_P]^T \tag{15.56}$$

Here e'_1 refers to the time derivative of e_1. From Eq. (15.52) we get the control law as in Eq. (15.57).

$$CBu = y_m - CAx - Cd(t) + KE + K_I E_I + K_D E_D \tag{15.57}$$

Eq. (15.51) and Eq. (15.57) become Eq. (15.58).

$$E' = -KE - K_I E_I - K_D E_D \tag{15.58}$$

Differentiating once, we get Eq. (15.59).

$$[I + K_D]E'' + KE' + K_I E = 0 \tag{15.59}$$

$''$ refers to the second derivative with respect to time.

The individual error equation can be written as Eq. (15.60).

$$(1 + k_i \tau_{Di})e_1'' + k_I e_i' + (k_i/\tau_{ii})e_i = 0 \qquad (15.60)$$

Eq. (15.60) can be written in the standard form as Eq. (15.61).

$$\tau_i^2 e_i'' + 2\tau_i \zeta_i e_i' + e' = 0 \qquad (15.61)$$

Eq. (15.61) is a second order differential equation in e with respect to time.
Where,

$$\tau_i^2 = (1 + k_i \tau_{Di})/(k_i/\tau_{Ii}) \qquad (15.62)$$

$$\zeta_i = \tau_{Ii}/(2\tau_i) \qquad (15.63)$$

with the restriction given in Eq. (15.64).

$$\zeta_i > 0 \qquad (15.64)$$

Once the damping coefficient (ζ_i) and the effective time constant (τ_i) for the individual error (e_i) is specified, then τ_{Ii} and τ_{Di} are calculated from Eq. (15.64) as Eq. (15.65) and Eq. (15.66).

$$\tau_{Ii} = 2\zeta_i \tau_i \qquad (15.65)$$

$$\tau_{Di} = [(0.5\tau_i/\zeta_i) - (1/k_i)] \qquad (15.66)$$

For a second order system, τ_i and ζ_i are related to the settling time (Cochin, 1980) as in Eq. (15.67).

$$t_{si} = 5\tau_i/\zeta_i \qquad (15.67)$$

Hence, from Eq. (15.66) we get Eq. (15.68) and Eq. (15.69).

$$\tau_{Ii} = 0.4\zeta_i^2 t_{si} \qquad (15.68)$$

$$\tau_{Di} = [0.1t_{si} - (1/k_i)] \qquad (15.69)$$

Once ζ_i and t_{si} for the ith output variable error (e_i) are specified, then τ_{Ii} and τ_{Di} are calculated from Eq. (15.68) and Eq. (15.69) by assuming the values of k_i. For τ_{Di} to be positive, we need to meet the condition in Eq. (15.70).

$$k_i > (10/t_{si}) \qquad (15.70)$$

Hence, u can be calculated.

15.9 Synthesis Method

Lee et al. (2004) extended the generalized synthesis method for SISO systems to MIMO systems by using the frequency properties of the closed loop interactions. Proportional and derivative terms of the PID controllers are determined by using diagonal terms of the process transfer function matrix whereas the integral terms are determined by taking all the elements into account.

The closed loop transfer function matrix $H(s)$ is given by Eq. (15.71).

$$H(s) = [I + G_p(s)G_c(s)]^{-1}G_p(s)G_c(s) \qquad (15.71)$$

The desired closed loop response R_i of the ith loop is chosen by Eq. (15.72).

$$R_i = \frac{g_{p,i+}(s)}{(\tau_{ic}s + 1)^{n_i}} \qquad (15.72)$$

Where,

$g_{p,i+}$ is the non-minimum part of transfer function,

τ_{ic} is an adjustable closed loop time constant for system performance and stability, and

n_i is the order of the processes. The requirement of $g_{p,i+}(0) = 1$ is necessary for the controlled variable to track its set point. The desired closed loop response matrix $R(s)$ for TITO can be as given in Eq. (15.73).

$$R(s) = \begin{bmatrix} R_1 & 0 \\ 0 & R_2 \end{bmatrix} \qquad (15.73)$$

We need to design the multivariable controller $G_c(s)$ such that all the diagonal elements of $H(s)$ resemble those of $R(s)$ as closely as possible over a frequency range relevant to control applications. $G_c(s)$ can be written in a Maclaurin series as in Eq. (15.74).

$$G_c(s) = \frac{K_0 + K_1s + K_2s^2}{s} = \begin{bmatrix} G_{c,1}(s) & 0 \\ 0 & G_{c,2}(s) \end{bmatrix} \qquad (15.74)$$

$Gp(s)$ also can be written in Maclaurin series as in Eq. (15.75).

$$G_p(s) = G_0 + G_1s + G_2s^2 = \begin{bmatrix} G_{p,1} & 0 \\ 0 & G_{p,2} \end{bmatrix} \qquad (15.75)$$

Where,

$G_0 = G(0)$,

$G_1 = G'(0)$, and

$G_2 = G''(0)/2$.

Here G' and G'' are the first and second derivative of G with respect to s. By substituting Eq. (15.75) and Eq. (15.74) into Eq. (15.71) and expanding it by the Maclaurin series, we get Eq. (15.76).

$$H(s) = H_0 + H_1 s + H_2 s^2 + H_3 s^3 + Os^4 = \begin{bmatrix} H_1 & 0 \\ 0 & H_2 \end{bmatrix} \qquad (15.76)$$

Where,

H_0 is full matrix, $H_0 = H(0)$, $H_1 = H'(0)$ and $H_2 = H''(0)/2$.

$R(s)$ can also be expressed in a Maclaurin series as in Eq. (15.77).

$$R(s) = R_0 + R_1 s + R_2 s^2 + R_3 s^3 + Os^4 = \begin{bmatrix} R_1 & 0 \\ 0 & R_2 \end{bmatrix} \qquad (15.77)$$

Where,

$R_0 = R(0)$, $R_1 = R'(0)$, and $R_2 = R''(0)/2$.

By comparing each element of $H(s)$ and $R(s)$ in Eq. (15.76) and Eq. (15.77), for the first three s terms (s, s^2, s^3), we can express K_0, K_1, and K_2 in terms of the process model parameters and the desired closed loop response parameters. Considering the diagonal elements in $Gp(s)$, we can get the ideal multivariable decentralized controller $G_c(s)$ to the desired closed loop responses, and by expanding in the Maclaurin series we get Eq. (15.78).

$$G_{c,ij}(s) = \frac{1}{s}\left[f_{ij}(0) + f'_{ij}(0)s + \frac{f''_{ij}(0)}{2}s^2 + Os^3 \right] \qquad (15.78)$$

By neglecting the higher order terms, the control law for the ideal multivariable controllers can be derived. Hence, K_{cij}, τ_{Iij}, and τ_{Dij} of the decentralized PID controller can be obtained by Eq. (15.79).

$$K_{c,ij} = f'_{ij}(0), \quad \tau_{I,ij} = \frac{f'_{ij}(0)}{f_{ij}(0)}, \quad \tau_{D,ij} = \frac{f''_{ij}(0)}{2f'_{ij}(0)} \qquad (15.79)$$

15.10 Model Reference Robust Control for MIMO Systems

Model Reference Robust Control (MRRC) of SISO systems was introduced as a new means of designing I/O robust control (Qu et al., 1994). This I/O design is an extension of the recursive back-stepping design in the sense that a nonlinear dynamic control (not static) is generated recursively. Back-stepping entails the design of fictitious controls, starting with the output state–space equation and back-stepping until one arrives at the input state–space equation where the actual control can be designed. At each step the system is transformed and a fictitious control is designed to stabilize the transformed state (Naik and Kumar 1992). MRRC of MIMO systems is an extension of MRC of MIMO systems and MRRC of SISO systems. Unwanted coupling exists in many physical MIMO systems. MRRC decouples MIMO systems using only input and output measurements rather than state feedback. This is a very desirable property, because in many instances state information is not available.

A diagonal transfer function matrix is Strictly Positive Real (SPR) if and only if each element on the diagonal is SPR. The fact that complicates the development of robust control laws is that the recursive back-stepping procedure used in non-SPR SISO systems cannot be directly applied to diagonal MIMO non-SPR systems without the introduction of the augmented matrix or a pre-compensator. Ambrose and Qu (1997) proposed a MRRC for MIMO systems.

15.11 Variable Structure Control for Multivariable Systems

The method proposed by Nouri and Abdennour (2007) is discussed here. Let us consider a multivariable linear system described by Eq. (15.80) and Eq. (15.81).

$$\frac{dx}{dt} = Ax + Bu \tag{15.80}$$

$$y = Lx \tag{15.81}$$

Where,

x is a state vector of size $(n \times 1)$, u is the input vector of size $(m \times 1)$, and y is the output vector of size $(p \times 1)$.

The sliding surface chosen linear is given in Eq. (15.82).

$$S = C^T x \tag{15.82}$$

Here, C is $m \times n$, a parameter space.

Assume the control is composed of two terms as in Eq. (15.83).

$$u = u_{eq} + \Delta u \tag{15.83}$$

The equivalent control u_{eq} is given by Eq. (15.84).

$$dS/dt = 0 \tag{15.84}$$

$$= C^T(dx/dt) \tag{15.85}$$

$$= C^T(Ax + Bu) \tag{15.86}$$

Hence

$$u_{eq} = -(C^T B)^{-1} C^T A x, \text{ if}(C^T B)^{-1} \text{ exists.} \tag{15.87}$$

The control law is given in Eq. (15.88).

$$u(t) = -(C^T B)^{-1} C^T A x(t) + \Delta u \tag{15.88}$$

where

$$\Delta u = -k \operatorname{sign}[S(x)] \tag{15.89}$$

and where k is a gain chosen to eliminate the perturbation effect.

Eq. (15.88) requires that $(C^T B)^{-1}$ exists. To overcome this, Nouri and Abdennour (2007) proposed a method based on the decomposition of multivariable system into subsystem controllable by each component of the system.

15.12 Stabilizing Parametric Region of Multi Loop PID Controllers for Multivariable Systems

Despite the rapid development of advanced process control techniques, multi loop PID controller is often used to control MIMO systems (Wang et al., 2008; Nie et al., 2011; Gigi and Tangirala, 2013). The main reason for this popularity is its simple control structure and the ease of handling loop failures. Compared with single loop controller design, the tuning of multi loop controllers to meet performance requirements is much more difficult because of the interactions among the loops. Much attention has been devoted to achieve good performance in view of the limitation imposed by process interactions, such as the commonly used independent design method (Wu and Lee, 2010b; Huang et al., 2003; He et al., 2005).

Although great progress on multi loop PID control has been achieved, some fundamental issues remain to be addressed for better understanding and more effective applications of multi loop PID control to MIMO multi-delayed plants. Since the essential requirement imposed on the PID controller is to ascertain the stability of the resulting closed loop system, the very first task when using PID control is to get a stabilizing PID controller for a given process with time delay (Wang et al., 2008). This problem is of great importance in both theoretical and practical aspects and it is also related to the problems of stability or robustness. Such a stabilizing set can not only be useful to design PID controllers satisfying defined performance requirements, but also avoid the time-consuming stability analysis during the controller tuning stage.

Therefore, approaches to design stabilizing PID controllers have attracted much attention in the process control field. In particular, FOPTD processes (Silva et al., 2001; Silva et al., 2002; Xu et al., 2003; Ho et al., 2004; Ho et al., 2005) have been studied. It has been shown that this kind of stability analysis is also useful in the design context (Silva et al., 2002; Kim et al., 2003; Keel and Bhattacharyya, 2008). A solution for linear time-invariant systems using low-order controller has been proposed in Ou et al. (2009). Explicit solutions have been also introduced for the case of integral processes (Wang et al., 2006; Ou et al., 2005; Ou et al., 2006). Unfortunately, the aforementioned approaches are still limited only to SISO systems. Luan et al. (2014) tried to provide a complete solution to the problem of characterizing the set of all multi loop PID parameters stabilizing multivariable systems with time delay.

Luan et al. (2014) proposed a method to determine the stabilizing PID parametric region for multivariable systems. First, a general equivalent transfer function parameterization method is proposed to construct the multi loop equivalent process for multivariable systems. Then, based on the equivalent single loops, a model-based method is presented to derive the stabilizing PID parametric region by using the generalized Hermite–Biehler theorem. By sweeping over the entire ranges of feasible proportional gains and determining the stabilizing regions in the space of integral and derivative gains, a complete set of stabilizing PID controllers can be determined. The robustness of the design procedure against the approximation in getting the SISO plants is analysed. Finally, simulation of a practical model is carried out to illustrate the effectiveness of the proposed technique.

Lin et al. (2008) discuss the computation of the maximum ranges of stabilizing PI controllers for MIMO systems. A time-domain scheme is proposed by converting the considered problem to a robust stability problem for a polytopic system. An algorithm based on Linear Matrix Inequality (LMI) is established to find the maximum ranges.

Gundes (2008) analysed the simultaneous stabilization with asymptotic tracking of step input references for linear, time-invariant MIMO stable plants. Necessary conditions are presented for existence of simultaneous integral action controllers and existence of simultaneous PID controllers. A systematic simultaneous PID controller synthesis method is proposed under a sufficient condition.

Summary

In this chapter, a brief review has been given on some of the newer methods of designing multivariable control systems.

Problems

1. Jin et al. (2017) presented centralized IMC–PID controller design for multivariable processes with multiple time delays. Read the paper and give a summary of the work.

2. Rajapandiyan and Chidambaram (2012) presented a simple decoupled equivalent transfer function s and simplified decoupler for MIMO systems. Read the paper and give a summary of the work.

3. Cui and Jacobsen (2002) made a presentation on the performance limitations in decentralized control. Read the paper and give a summary of the work.

4. Reddy et al. (2006) presented a comparison of multivariable controllers for non minimum phase systems. Read the paper and a give a summary of the work.

5. Shaji and Chidambaram (1996) presented robust decentralized PI controllers for crude unit product quality using the method of inequalities. Read the paper and present a summary of the work.

6. Balachandran and Chidambaram (1996) compared the performances of decentralized controllers. Read the paper and present a summary of the work.

7. Sadasivarao and Chidambaram (2014) presented multivariable controllers tuning by GAs. Read the paper and present a summary of the paper.

8. Bhat et al. (1991) presented model reference control of mixed culture bioreactor with competition and external inhibition. Read the paper and give a summary of the work.

Appendix A

Identification of Unstable Second Order Transfer Function Model with a Zero by Optimization Method

---

A technique is presented to identify a transfer function of an unstable SOPTD system with a 0. An optimization method is used to estimate the model parameters to match the closed loop responses. A method is given for the initial guess values of the model parameters (time delay, steady state gain and the values of poles, and 0). The method is applied to three simulation examples. The presented method gives good results. The measurement noise and disturbance effects on the model identification are also reported.

A.1   Introduction

Transfer function models are required to design PID controllers. Closed loop identification method is not sensitive to disturbances and is essential to identify the transfer function models for unstable systems. Kavdia and Chidambaram (1996) and Srinivas and Chidambaram (1996) have extended the method proposed by Yuwana and Seborg (1982) for stable systems to identify an unstable FOPTD model. The method uses a proportional controller, and a closed loop response for a step change in the set point is used to identify the model. Harini and Chidambaram (2005) have extended the method to identify an unstable SOPTD model. Since only the proportional controller is used, an offset in the response is present and the method may not be employed in chemical plants. In addition, in case of certain parameter values of unstable FOPTD systems (for example, when the ratio of time delay to time constant is greater than 0.7), a proportional controller alone cannot stabilize the process. Ananth and Chidambaram (1999), Cheres (2006), and Padmasree and Chidambaram (2006) have proposed methods for identifying an unstable FOPTD model using a closed loop response of the system under the PID control mode. All the above methods have used analytically derived nonlinear algebraic equations relating an open loop and closed loop model parameters.

Pramod and Chidambaram (2000, 2001) have proposed an optimization method for identifying an unstable FOPTD model using the step response of a PID-controlled system's response. Padmasree and Chidambaram (2002) have extended the method for identifying an unstable FOPTD model with a 0.

For stable systems, there are several methods available for closed loop identification of SOPTD systems under PID control mode (Suganda et al., 1998; Cheres and Eydelzon, 2000; Sung et al., 2009; Dhanya Ram and Chidambaram, 2015). However, there are systems (Padmasree and Chidambaram, 2003, 2006; Rao and Chidambaram, 2006a, 2012) which are to be modelled as an unstable SOPTD with a 0, for example, $k_p(1+ps)\exp(-\theta s)/(a_1 s^2 + a_2 s + 1)$. Here k_p is process gain, p is system 0, θ is time delay, a_1 and a_2 are coefficients of the denominator. Methods of designing PID controllers are given by synthesis method and IMC method. Liu and Gao (2010) have proposed a method of identifying unstable FOPTD and SOPTD systems using a closed loop response under the PID control mode. Unstable systems with one stable pole and one unstable pole only are considered by Liu and Gao (2010) and the system does not have any 0. For accurate results, the method also needs an iterative scheme. In the present example, the optimization method proposed by Padmasree and Chidambaram (2002) will be extended to the closed loop identification of parameters of unstable SOPTD systems with a 0.

A.2 Proposed Method

The procedure involved is basically the same as given by Padmasree and Chidambaram (2002). The response (y_{pc}) of the closed system using a PI or PID

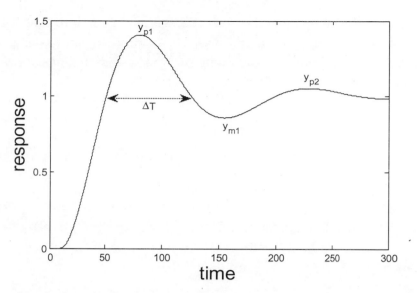

FIGURE A.1   Typical closed loop response for a step change in set point and associated parameters to be noted

controller is obtained (Fig. A.1 gives a typical response) for a known magnitude of step change in the set point. The transfer function of the open loop process is assumed as $k_p(1 + ps)\exp(-\theta s)/(a_1 s^2 + a_2 s + 1)$. From the closed loop response, the initial guess values for the model parameters $(k_p, p, a_1, a_2, \theta)$ are calculated (see next section). Using this model and the same controller settings, the closed loop step response (y_{mc}) is obtained by simulation. The final model parameters are obtained by minimizing the sum of the squared difference between y_{pc} and y_{mc} over several points of the response. The Matlab optimization routine *lsqnonlin* is used in the present example. This method employs the Trust–region–reflective algorithm.

If we know the change in the steady state value of u_s reached in order for the output y to reach the set point value $(y_r = 1)$, then the process steady state gain can be calculated as y_r/u_s. Even the exact value of u is not recorded and if we know the sign of u (deviation value) at steady state, that will be enough to detect if there is one unstable pole or two unstable poles present in the system. If the sign of final steady state u (deviation value) is positive and the sign of k_p (calculated from the initial guess value from the controller setting as given later in this book) is positive, then there are two unstable poles present. If the sign of k_p is negative, then there is one stable pole and one unstable pole present in the system. Similarly, if the sign of the final steady state value of u (deviation value) is negative and the sign of k_p is positive, then there is one unstable pole and one stable pole present in the system. In the present example, we assume that the value of u_s is not noted but only the sign of the final value of u reached is known. Even if the sign of u reached is not known, the optimization method selects the time constant as negative so that the reformatted model of the system will give the system that has appropriate one stable or two unstable poles.

The following procedures are suggested in the present example for the guess values of the model parameters $(k_p, p, a_1, a_2, \theta)$:

(1) The guess value of the time delay is assumed to be that of the closed loop response.

(2) The guess value for k_p is calculated from the controller design formula (Pramod and Chidambaram, 2001) $k_p = (\tau/\tau_d)^{0.5}/k_c$. This equation is valid only for FOPTD systems. The guess value of k_p for SOPTD system is considered as 110% of k_p (from FOPTD).

(3) Overshoot, peak value (y_p), peak time (t_p), time delay, and period of oscillation are measured from the closed loop response for calculating the damping coefficient (ζ) and effective time constant (τ_e) of the closed loop system

(4) Since the open loop system is unstable, the initial guess values of a_1 and a_2 are calculated from $a_1 = \tau_{eo}^2$, $a_2 = -2\zeta\tau_{eo}$, where τ_{eo} is the open loop effective time constant to be assumed from τ_e (as 2 to 5 times that of τ_e). For getting the initial guess values of open loop model parameters, we assume that the open loop system is similar to the closed loop system but with unstable poles. Let us discuss the method of calculating the initial guess value for the 0 of the system.

A.2.1   Initial Guess Value for the System Zero

The closed loop system model is assumed to be a stable SOPTD with a 0 as given in Eq. (A.1).

$$y(s)/y_r(s) = (1 + ps)\exp(-\theta s)/(\tau_e^2 s^2 + 2\tau_e \zeta s + 1) \qquad (A.1)$$

The expression for step response of the closed loop system can be obtained (Seborg et al., 1996) given in Eq. (A.2).

$$y(\varphi) = \{[pq/\tau_e]\sin(b)\exp(-\zeta\varphi/\tau_e)\} + \{1 - \exp(-\zeta\varphi/\tau_e)[q\zeta\sin(b) + \cos(b)]\} \quad (A.2)$$

Where,

$$\varphi = t - \theta \tag{A.3}$$

$$b = [(1 - \zeta^2)^{0.5} \varphi / \tau_e] \tag{A.4}$$

$$q = 1/(1 - \zeta^2)^{0.5} \tag{A.5}$$

The expressions for the effective time constant (τ_e) and for the damping coefficient (ζ) can be derived (Ananth and Chidambaram, 1999) given in Eq. (A.6) and Eq. (A.7).

$$\tau_e = [\Delta T/(\pi)](1 - \zeta^2)^{0.5} \tag{A.6}$$

$$\zeta = 0.5(\zeta_1 + \zeta_2) \tag{A.7}$$

$$\zeta_1 = -\ln(v_1)/\{\pi^2 + [\ln(v_1)]^2\}^{0.5} \tag{A.8}$$

$$\zeta_2 = -\ln(v_2)/\{4\pi^2 + [\ln(v_2)]^2\}^{0.5} \tag{A.9}$$

$$v_1 = (y_\infty - y_{m1})/(y_{p1} - y_\infty) \tag{A.10}$$

$$v_2 = (y_{p2} - y_\infty)/(y_{p1} - y_\infty) \tag{A.11}$$

If the response shows only one peak, then we consider $\zeta = \zeta_1$.

By using equations Eq. (A.6) to Eq. (A.9), we can calculate ζ and τ_e. Substituting τ_e, ζ, y_p, $t_p(\varphi_p)$ in Eq. (A.2), we can get the value for p (system 0). The guess values of a_1 and a_2 of the open loop unstable system can be calculated from Eq. (A.12).

$$a_1 = \tau_{eo}^2, \quad a_2 = -2\zeta\tau_{eo} \tag{A.12}$$

A transfer function model with known model parameters is considered with suitable PID controller settings and the closed loop step response is obtained. The closed loop response of the identified model is compared with that of the original system using the same PID settings. For the system whose transfer function is not known, a PID controller is designed for the identified model. This PID controller will be implemented on the original system and also on the identified model and the servo responses can be compared.

A.3   Simulation Results

Case Study 1: Transfer Function Model with Two Real Unstable Poles and a Negative Zero

The open loop transfer function model was considered as $k_p(1 + ps)\exp(-\theta s)/(a_1 s^2 + a_2 s + 1)$ with $k_p = 2$, $p = 5$, $a_1 = 3$, $a_2 = -4$ and $\theta = 0.3$. This system has two unstable poles at 1.0 and 0.3333. The PID controller settings by the synthesis method were obtained as $k_c = 1.33$, $\tau_I = 1.4865$, $\tau_D = 0.1349$ with the filter time constant $\alpha = 0.02252$. The closed loop response of the system gave $y_{p1} = 2.69$, $y_{m1} = 0.257$, $y_{p2} = 1.53$, $y_\infty = 1$, $t_p = 0.86$, $\Delta T = (1.2184 - 0.5069) = 0.7115$, $T = 1.423$. Using Eq. (A.7) to Eq. (A.9), ζ is obtained as 0.0527. Using Eq. (A.6), the closed loop τ_e is calculated as 0.22. The guess value for $\tau_{e,o}$ was taken as $5\tau_e$. The other parameters were calculated as $k_P = (1.1/0.3)^{0.5}/1.33 = 1.4432$, $p = 6.53$, [substitute τ, ζ, y_p, t_p in Eq. (A.2) to get p], $a_1 = (5\tau_{eo})^2 = 1.2217$, $a_2 = -2\zeta(5\tau_{eo}) = 0.4807$. The value of T (period of oscillation) was determined as 1.4230. The equation for calculating the initial guess value of k_p is valid for the first order system. The guess value of k_p for SOPTD system was considered as 110% of k_P. Hence, k_P was calculated as 1.5875. The Matlab routine *lsqnonlin* was used. The limits of the constraints are given as 0 to infinity for k_P, p, a_1, and θ. For a_2, the constraints are given as $-\infty$ to 0. The optimization method converged to the actual values of the parameters $k_P = 2.0$, $p = 5.0$, $a_1 = 3.0$, $a_2 = -4.0$, and $\theta = 0.3$. The computational time (using Pentium Dual Core) was 40.5233 seconds and the number of iterations was 78. Fig. A.2 shows that the actual response and identified responses are nearly the same. When a +10% or −10% of perturbation in the initial guess values was given separately, the optimization converged to the values of the actual model. The computational times were 41.30 seconds and 32.93 seconds, respectively.

If the initial guess is selected arbitrarily ($k_P = 3.175$, $p = 13.06$, $a_1 = 2.44$, $a_2 = -0.96$, $\theta = 0.6$), then it is found that the optimization routine does not converge, giving the values as $k_P = 3.7276$, $p = 11.0765$, $a_1 = 17,5715$, $a_2 = -0.8886$, and $\theta = 0.325$. These values are different from the actual parameters. With the initial guess for p as 0 and values of other parameters kept as given in the earlier, the optimization method does not converge to true values. The method gives the values as $k_P = 1.8271$, $p = 2.0522$, $a_1 = 1.2254$, $a_2 = -0.9551$, and $\theta = 0.3051$, which are different from the actual parameters. When the initial guess values are given as $k_P = 1$, $p = 1$, $a_1 = 1$, $a_2 = -1$ and $\theta = 0.6$, the optimization method does not converge and gives different values of the parameters ($k_P = 1.321$, $p = 7.2756$,

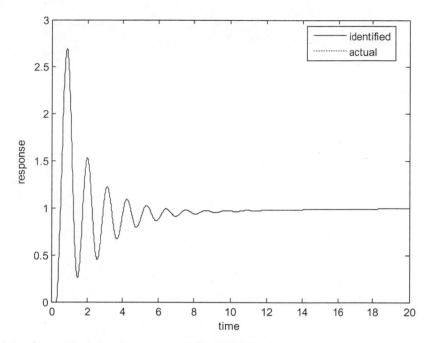

FIGURE A.2 Closed loop servo response of the Case Study 1
Identified model response coincides with that of the actual system (same controller settings).

$a_1 = 3.555$, $a_2 = -0.6957$, and $\theta = 0.3351$). Hence, the proposed method of getting the guess values for the model parameters is recommended.

The effect of load on the model parameters was evaluated by adding a step disturbance of 0.02 to the system along with the input variable. The closed loop response (similar to Fig. A.1) of the system gave $y_{p1} = 2.7238$, $y_{p2} = 1.5403$, $y_{m1} = 0.2494$, $y_\infty = 1$, $t_p = 0.86$. Eq. (A.7) was used for calculating the damping coefficient ζ as 0.2187. Time period of the response was obtained as 1.4284. Using Eq. (A.6), τ_e was calculated from the time period as 0.2203. For the optimization method, the guess value for the open loop $\tau_{e.o}$ was taken as $5\tau_e$. The other parameters were calculated as $k_P = (0.2203/0.1)^{0.5}/1.33 = 1.4457$, $p = 6.6639$ [substitute τ, ζ, y_p, t_p in Eq. (A.2) to get p], $a_1 = \tau_{eo}^2 = 1.2302$, $a_2 = -2\xi\tau_{eo} = -0.4852$. The guess value of k_p was calculated as 1.5903. The optimization method gives the converged parameters as $k_P = 2.076$, $p = 5.0698$, $a_1 = 3.1539$, $a_2 = -4.3496$, $\theta = 0.2985$. The computational time was 81.28401 seconds (Pentium Dual Core) and the number of iterations was 71. Fig. A.3 shows that the actual response and identified response are nearly the same.

The effect of measurement noise on the model parameters was evaluated by adding a white noise (of power 0.00001 and sample time 0.01) to the system.

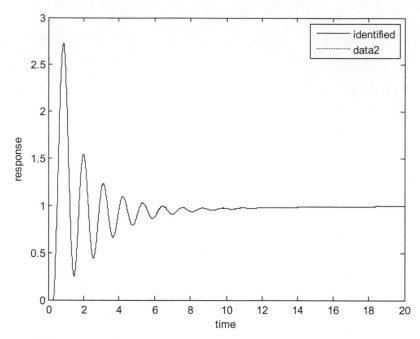

FIGURE A.3 Effect of load for Case Study 1
The response of the actual system and the identified model coincide (same controller settings).

The servo response was evaluated to get the initial guess values. The closed loop response of the system gave $y_{p1} = 2.71$, $y_{p2} = 1.54$, $y_{m1} = 0.24$, $y_\infty = 1$, $t_p = 0.85$. Eq. (A.7) was used for calculating the damping coefficient ζ as 0.2152. The time period of response was obtained as 1.4012. Using Eq. (A.6), τ_e was calculated from the time period as 0.2178. For the optimization method, the guess value of (open loop) $\tau_{e,o}$ was taken as $5\tau_e$. The other parameters were calculated as $k_p = (0.2178/0.1)^{0.5}/1.33 = 1.4325$, $p = 6.4938$ (substitute τ, ξ, y_p, t_p in Eq. (A.1) to get p), $a_1 = \tau_{eo}^2 = 1.1857$, $a_2 = -2\zeta\tau_{eo} = -0.4686$. The initial guess of k_p was calculated as 1.5757. The optimization method converged closer to the actual parameters $k_P = 2.0085$, $p = 4.9948$, $a_1 = 3.0068$, $a_2 = -4.0347$, $\theta = 0.2997$. The computational time was 35.908 seconds (Pentium Dual Core) and the number of iterations was 70. Fig. A.4 shows the actual response and identified response of the closed loop system.

Case Study 2: Transfer Function Model with Two Unstable Conjugate Poles and a Negative Zero

The open loop transfer function of a Continuous Stirred Tank Reactor (CSTR) was considered as $k_p(1 + ps)\exp(-\theta s)/(a_1 s^2 + a_2 s + 1)$ with $k_P = 3.87$, $p = 0.5283$, $a_1 =$

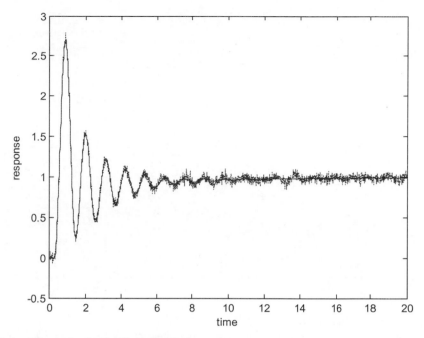

FIGURE A.4 Effect of noise in Case Study 1
Solid line: identified model; oscillatory line: noise corrupted signal (same controller settings)

0.4769, $a_2 = -0.348$ and $\theta = 0.1$. This system has two unstable complex conjugate poles at $0.3649 \pm 1.4013i$. The PID controller settings designed by the synthesis method (Rao and Chidambaram, 2012) were $k_c = 2.32$, $\tau_I = 0.704$ and $\tau_D = 0.0464$ and filter time constant $\alpha = 0.013$. The closed loop response of the system gave $y_{p1} = 1.8195$, $y_{m1} = 0.996$, $y_\infty = 1$, $t_p = 0.308$. Eq. (A.7) to Eq. (A.9) are used for calculating the damping coefficient ζ as 0.86124. The time period of response was obtained as 0.6756. Using Eq. (A.6), τ_e was calculated from the time period as 0.0547. For the optimization method, τ_{eo} was taken as $5\tau_e$. The guess values for the other parameters were calculated as $k_p = (0.2733/0.1)^{0.5}/2.32 = 0.7126$, $p = 1.1574$ (substitute τ, ξ, y_p, t_p in Eq. (A.1) to get p), $a_1 = \tau_{eo}^2 = 0.0747$, $a_2 = -2\zeta\tau_{eo} = -0.4707$. The initial guess of k_P was calculated as 0.7838. The optimization method converged with the values of the actual parameters $k_P = 3.87$, $p = 0.5283$, $a_1 = 0.4769$, $a_2 = -0.348$, $\theta = 0.1$. The computational time (Pentium Dual Core) was 39.0431 seconds and the number of iterations was 57. Fig. A.5 shows that the actual response and the identified response are nearly the same. Even after giving $+10\%$, -10% of perturbation in the each of the initial guess values, the same parameters were obtained. The computational times were 43.30 seconds and 35.64 seconds, respectively.

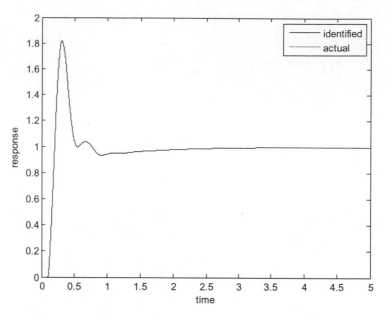

FIGURE A.5 Closed loop servo response for Case Study 2
Identified model and actual system coincide (same controller settings).

If the initial guess is selected arbitrarily, ($k_P = 3.1352$, $p = 4.6296$, $a_1 = 0.2988$, $a_2 = -1.8828$, $\theta = 0.4$), it is found that the optimization routine does not converge. The method gives values as $k_P = 2.4627$, $p = 15.6918$, $a_1 = 11.9561$, $a_2 = -0.8581$, and $\theta = 0.1085$, which are different from the actual parameters. Hence, the proposed method for the initial guess values of the model parameters is recommended.

Case Study 3: Transfer Function Model with Two Unstable Conjugate Poles, a Stable Pole and a Negative Zero

The open loop transfer function model was given as $k_p(1 + ps)\exp(-\theta s)/[(a_1 s^2 + a_2 s + 1)(a_3 s + 1)]$ with $k_p = 3.87$, $p = 0.5283$, $a_1 = 0.4769$, $a_2 = -0.348$, $a_3 = 0.1$, and $\theta = 0.1$. This system had two unstable complex conjugate poles at $0.3649 \pm 1.4013i$ and a stable pole -9.9981. The parameters for PID controller were $k_c = 0.4229$, $\tau_I = 0.3$, $\tau_D = 1.5121$, with the filter time constant $\alpha = 0.1$ and $\beta = 0.4811$. The closed loop response of the system gave $y_{p1} = 1.5467$, $y_{m1} = 0.7275$, $y_\infty = 1$, $t_p = 1.78$. Eq. (A.7) to Eq. (A.9) were used for calculating the damping coefficient ζ as 0.2166. The time period of response was obtained as 3.7146. Using Eq. (A.6), τ_e was calculated from the time period as 0.5772. The initial guess values were calculated as $k_P = (0.5772/0.1)^{0.5}/0.4299 = 5.6808$, $p = 0.2617$ (substitute τ_e, ζ, y_p, t_p in Eq.

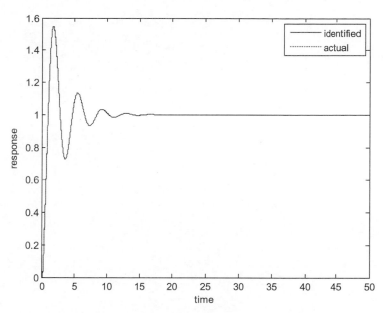

FIGURE A.6 Closed loop servo response for Case Study 3
Identified model and the actual system coincide (same controller settings).

(A.2) to get p), $a_1 = \tau_{eo}^2 = 0.3331$, $a_2 = -2\zeta\tau_{eo} = -0.25$. The guess value of k_P was calculated as 6.2489. The optimization method converged closer to the actual parameters $k_P = 3.9269$, $p = 0.4863$, $a_1 = 0.482$, $a_2 = -0.3376$, and $\theta = 0.1743$. The computational time (Pentium Dual Core) was 12.153 seconds and the number of iterations was 13. Fig. A.6 shows that the actual and identified responses are nearly the same. Even for giving +10% or −10% of perturbation in the initial guess values, the optimization method converged to values closer to the actual parameters.

If the initial guess is selected arbitrarily ($k_P = 13$, $p = 0.5$, $a_1 = 0.7$, $a_2 = -0.5$, $\theta = 0.2$), it is found that the optimization routine does not converge. The method gives the values of the parameters as $k_P = 12.5012$, $p = 0.524$, $a_1 = 0.8849$, $a_2 = -0.4954$, $\theta = 0.1934$. These values are different from the actual parameters. Hence, the proposed method for the initial guess values of the model parameters is recommended.

To conclude, an unstable SOPTD model with a 0 is identified using the step response of the PID-controlled system. A method is given for calculating the initial guess values of the model parameters. The optimization method (*lsqnonlin*) is used. Simulation studies on three unstable transfer function models show a good agreement with the corresponding original transfer function model. The noise in the measurement of the output or the disturbance entering the process does not affect the identified model parameters.

Appendix B

B.1 Details of the Relay Auto-Tuning Method for TITO Systems

The method proposed by Palmor et al. (1995) concentrated on the identification of DCP (Desired Critical Point) along with the steady state gains. The usual method involves the open loop step or pulse response test. Palmor et al. (1995) recommend a closed loop method. The process output y(t) and manipulated variable u(t) are determined by two experiments conducted on the system with two different reference (that are not both 0 mean) signals along with the relays. The steady state gain of the system is then calculated as (Palmor et al., 1995):

$$[p_1 \quad p_2] = G_p(0)[q_1 \quad q_2] \tag{B.1}$$

where p_1 and p_2 (vectors) are calculated from the area under the corresponding waveform for one period duration in the responses and interactions in the output (Palmor et al., 1995). Similarly, q_1 and q_2 (vectors) are calculated from the area under the corresponding waveform for one period duration in the manipulated variables. $G_P(0)$ which is the SSGM of the system can be easily calculated as discussed in Chapter 11.

To identify DCP, relative importance is given to one loop over another using a weighting factor C, as given in Eq. (B.2).

$$C = \frac{K_{2c} g_{p22}(0)}{K_{1c} g_{p11}(0)} \tag{B.2}$$

K_{1c} and $K_{2,c}$ are the critical gains of the controllers. Once the steady state gains and the resultant error amplitudes are known, the relay height ratio (h_1 and h_2) can be calculated as in Eq. (B.3).

$$\frac{h_1}{h_2} = \frac{1}{C_d} * \left(\frac{a_1}{a_2}\right) \left|\frac{g_{p22}(0)}{g_{p11}(0)}\right| \tag{B.3}$$

Here a_1 and a_2 are the amplitude of output oscillations obtained and C_d indicates the desired quantity of C. None of these quantities in Eq. (B.3) is available initially to the auto tuner. The relay ratio may be calculated by a series of limit cycle experiments. The algorithm is repeated till the convergence criterion is satisfied $|\varphi - \varphi_d| < \varepsilon$. Once the optimum relay height is known, the DCP can be obtained by the relay feedback test.

From the Fourier series expansion, the amplitude of the relay response is assumed to be the result of the primary harmonic output. Therefore the ultimate gain can then be calculated (Yu, 2006; Astrom and Hagglund, 1995) from Eq. (B.4).

$$K_{u,i} = \frac{4h_i}{\pi a_i} \quad (i = 1, 2) \tag{B.4}$$

Atherton (1975), Zhuang and Atherton (1994), Wadey and Atherton (1982) proposed a method to analyse the limit cycles in multivariable systems with the decentralized relays. In order to design the PID controller for a MIMO system, the identification of critical points is essential. By varying the magnitudes of the relay (i.e., relay heights) different critical points forming the stability limit can be identified (Palmor et al. 1995; Wang et al., 2003). Each critical point results in different controller settings, hence in different performance. It is necessary to define the desired critical point (DCP) based on the relative importance of the loops.

B.2   Derivation of Eq. (3.50a) and Eq. (3.50b)

Consider a stable FOPTD system as in Eq. (B.5).

$$\frac{y(s)}{u(s)} = \frac{k_p \exp(-Ds)}{\tau s + 1} \tag{B.5}$$

The input to the system is the output from the relay. Let the system response be as given in Eq. (B.5).

$$y(t) = \left[a_1 \sin(\omega t + \varphi_1) + \frac{1}{3} a_3 \sin(3\omega t + \varphi_3) + \frac{1}{5} a_5 \sin(5\omega t + \varphi_5) + \cdots \right] \quad \text{(B.6)}$$

$$a_1 = \frac{1}{\sqrt{\tau^2 \omega^2 + 1}}; \quad a_3 = \frac{1}{\sqrt{9\tau^2 \omega^2 + 1}}; \quad a_5 = \frac{1}{\sqrt{25\tau^2 \omega^2 + 1}} \quad \text{(B.7)}$$

$$\varphi_1 = -D\omega - \tan^{-1} \tau\omega \quad \text{(B.8)}$$

$$\varphi_3 = -3D\omega - \tan^{-1} 3\tau\omega \quad \text{(B.9)}$$

Eq. (B.5) can be written as Eq. (B.9).

$$y(t) = a_1 \left[\sin(\omega t + \varphi_1) + \frac{1}{3} b_3 \sin(3\omega t + \varphi_3) + \frac{1}{5} b_5 \sin(5\omega t + \varphi_5) + \cdots \right] \quad \text{(B.10)}$$

Where,

$$b_3 = \frac{a_3}{a_1}; \quad b_5 = \frac{a_5}{a_1} \quad \text{(B.11)}$$

For the limiting case we have:

Case 1

$\tau\omega \ll 1$, $\tan^{-1}(.)$ is negligible.
Equation (B.7) leads to Eq. (B.11).

$$\varphi_1 = -D\omega_u = -\pi \quad \text{(B.12)}$$

$$D\omega_u = \pi \quad \text{(B.13)}$$

Eq. (B.8) will result in Eq. (B.13).

$$\varphi_3 = -3\pi \quad \text{(B.14)}$$

In general,

$$\varphi_n = -n\pi \quad \text{(B.15)}$$

For smaller value of $\tau\omega$, the value of $\tau\omega$ can be neglected when compared to the value of one. The values of b_3, b_5 can be approximated as given in Eq. (B.15).

$$b_3 = b_5 = \cdots = b_j = 1 \quad \text{(B.16)}$$

Substituting values of $\varphi's$ and b's in Eq. (B.9), we get Eq. (B.16).

$$y(t) = |-a_1|[\sin(\omega t) - \frac{1}{3}\sin(3\omega t) + \frac{1}{5}\sin(5\omega t)\cdots] \tag{B.17}$$

$$y(t^*) = |a_1| \left[\sum_1^\infty \frac{\sin i\omega t^*}{i} \right] \tag{B.18}$$

Here the value of a_1 is to be calculated. Consider Eq. (B.16) when $\omega = \omega_u$. Substituting $\omega_u t^* = 0.5\pi$ in Eq. (B.5), we will get Eq. (B.18).

$$y(t^*) = |a_1| \left[1 - \frac{1}{3} + \frac{1}{5} - \frac{1}{7} + \frac{1}{9} + \cdots \frac{1}{N} \right] \tag{B.19}$$

The time at which $y(t^*)$ is noted from the graph when $t^* = (0.5\pi)/\omega_u$.

Case 2

$\tau\omega \gg 1$, $\tan^{-1}(.)$ is approximated to $\pi/2$. Eq. (B.7) is derived to Eq. (B.19).

$$\varphi_1 = -D\omega_u - \frac{\pi}{2} = -\pi \tag{B.20}$$

$$D\omega_u = \frac{\pi}{2} \tag{B.21}$$

Eq. (B.8) leads to Eq. (B.21).

$$\varphi_3 = -\frac{4\pi}{2} \tag{B.22}$$

In general,

$$\varphi_n = \frac{(n+1)\pi}{2} \tag{B.23}$$

For smaller value of $\tau\omega$, the value of 1 is neglected when compared to the value of $\tau\omega$. So values of b_3, b_5 can be approximated as in Eq. (B.23).

$$b_3 = \frac{1}{3}; \quad b_5 = \frac{1}{5}; \cdots ; b_j = \frac{1}{j} \tag{B.24}$$

Substituting the values of φ and b in Eq. (B.5), we get Eq. (B.24).

$$y(t) = |-a_1|[\sin(\omega t) - \frac{1}{3^2}\sin(3\omega t) + \frac{1}{5^2}\sin(5\omega t) - \cdots] \tag{B.25}$$

$$y(t^*) = |a_1| \left[\sum_1^\infty \frac{\sin i\omega t^*}{i} \right] \tag{B.26}$$

Here the value of a_1 is to be calculated. Let $\omega_u t^* = 0.5\pi$ and substituting in Eq. (B.25), we get Eq. (B.26).

$$y\left(t^*\right) = |a_1| \left[1 + \frac{1}{9} + \frac{1}{25} + \frac{1}{49} + \frac{1}{81} + \cdots \frac{1}{N*N}\right] \qquad \text{(B.27)}$$

Appendix C

C.1 Simulink Block for WB Distillation Column Process

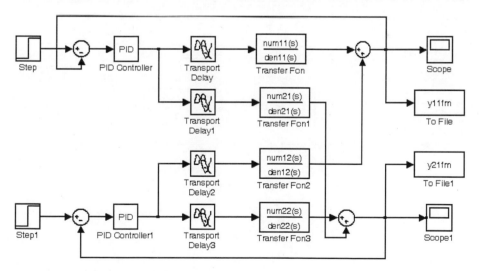

C.2 Simulink Block for OR Distillation Column for Response

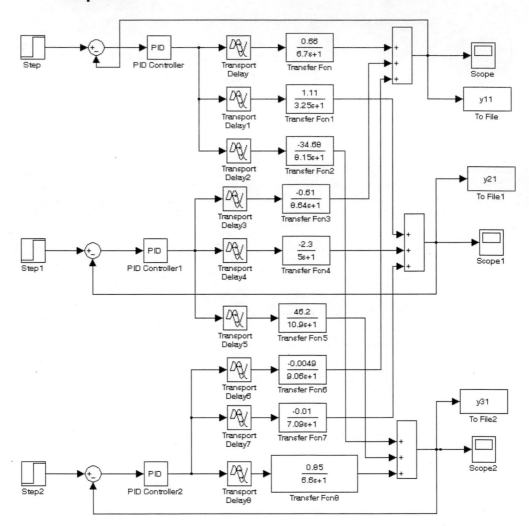

C.3 Simulink Block for OR Distillation Column Model for Identification

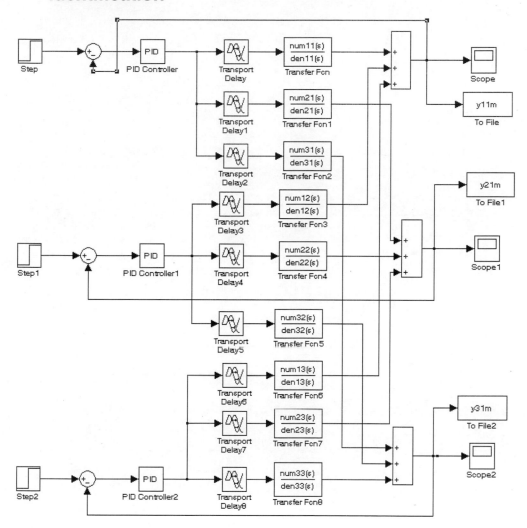

C.4   Matlab Code for Identification of Wood and Berry (WB) Column Transfer Function Matrix (Example 1)

```
clc;
clear all;
global pros22 pros21 pros12 pros11 num11 den11 del11 num12
  den12 del12 num21 den21 del21 num22 den22 del22 f11
  f21  f12 f22

%% step  change in loop_11 to get (for guess) response G11
 and interaction G21
sim('Ftito11',100);

%% step change in loop_22 to get (for guess) response-G22
 and interaction-G12
sim('Ftito22',100);

%% load output response values for each step change

% Process set point change in yr1
load y11f.mat;
load y21f.mat;

% Process set point change in yr2
load y12f.mat;
load y22f.mat;

xx = 0;
step_size = 1;
pros11 = (y11f(:,2)-xx)/step_size;
pros21 = (y21f(:,2)-xx)/step_size;
pros12 = (y12f(:,2)-xx)/step_size;
pros22 = (y22f(:,2)-xx)/step_size;
t = y11f(:,1);
```

```
%% Guess value calculation (area under the curve for
   non-diagonal process gains)
x0=[5.0505 6.25 1 5.2391 8.75 7 -7.245 6.25 3 -16.129 7.5 3];
lb=[0.2525 0.3125 0.25 0.2619 0.4375 1.75 -144.9 0.3125
    0.75 -322.58 0.375 0.75];
ub=[101.01 31.25 4 104.782 43.75 28 -0.3623 31.25 12
    -0.8065 37.5 12];

%% function call for nonlinear least square analysis

tic;
options=optimset('Largescale','on','Diagnostics','on',
        'Display','Iter','TolX',1e-3)
[x_sol,resnorm,residual,exitflag,output]=lsqnonlin('Ftito',
    x0,lb,ub,options);
toc;

%% To plot final converged values response and real response

f = Ftito(x_sol);

predt11 = pros11+f11;
predt21 = pros21+f21;
predt12 = pros12+f12;
predt22 = pros22+f22;
res11 = [pros11; predt11];
res21 = [pros21; predt21];
res12 = [pros12; predt12];
res22 = [pros22; predt22];

subplot(2,2,1);plot(t,predt11,'-',t,pros11,':');
xlabel('time'), ylabel('y11');
subplot(2,2,3);plot(t,predt21,'-',t,pros21,':');
xlabel('time'), ylabel('y21');
subplot(2,2,2);plot(t,predt12,'-',t,pros12,':');
xlabel('time'), ylabel('y12');
```

```matlab
subplot(2,2,4);plot(t,predt22,'-',t,pros22,':');
xlabel('time'), ylabel('y22');

%% Calculation of IAE values

iae11 = 0;
iae21 = 0;
iae12 = 0;
iae22 = 0;


for i = 1:10000
iae11 = iae11 + ((abs(pros11(i)-predt11(i)))*(y11f(i+1,1)
        -y11f(i,1)));
iae21 = iae21 + ((abs(pros21(i)-predt21(i)))*(y21f(i+1,1)
        -y21f(i,1)));
iae12 = iae12 + ((abs(pros12(i)-predt12(i)))*(y12f(i+1,1)
        -y12f(i,1)));
iae22 = iae22 + ((abs(pros22(i)-predt22(i)))*(y22f(i+1,1)
        -y22f(i,1)));
end

iae11
iae21
iae12
iae22

IAE = iae11+iae21+iae12+iae22

%% Calculation of ISE values

ise11 = 0;
ise21 = 0;
ise12 = 0;
ise22 = 0;


for i = 1:10000
ise11 = ise11 + ((abs(pros11(i)-predt11(i))^2)*(y11f(i+1,1)
```

```
            -y11f(i,1)));
ise21 = ise21 + ((abs(pros21(i)-predt21(i))^2)*(y21f(i+1,1)
            -y21f(i,1)));
ise12 = ise12 + ((abs(pros12(i)-predt12(i))^2)*(y12f(i+1,1)
            -y12f(i,1)));
ise22 = ise22 + ((abs(pros22(i)-predt22(i))^2)*(y22f(i+1,1)
            -y22f(i,1)));
end

ise11
ise21
ise12
ise22

ISE = ise11+ise21+ise12+ise22


%% FUNCTION FILE

function f=Ftito(x)

global pros22 pros21 pros12 pros11 num11 den11 del11
 num12 den12 del12 num21 den21 del21 num22 den22 del22
 f11 f21 f12 f22

%% assigning parameters to each block

step_size1 = 1;
num11 = x(1);
den11 = [x(2) 1];
del11 = x(3);
num21 = x(4);
den21 = [x(5) 1];
del21 = x(6);
num12 = x(7);
den12 = [x(8) 1];
del12 = x(9);
```

```
num22 = x(10);
den22 = [x(11) 1];
del22 = x(12);

%% assigning of parameters in first loop model transfer
    function & delay blocks

set_param('Ftito11_model/Transport Delay','delaytime','del11')
set_param('Ftito11_model/Transfer Fcn','numerator','num11',
          'denominator','den11')
set_param('Ftito11_model/Transport Delay1','delaytime',
          'del21')
set_param('Ftito11_model/Transfer Fcn1','numerator','num21',
          'denominator','den21')
set_param('Ftito11_model/Transport Delay2','delaytime',
          'del12')
set_param('Ftito11_model/Transfer Fcn2','numerator',
          'num12','denominator','den12')
set_param('Ftito11_model/Transport Delay3','delaytime',
          'del22')
set_param('Ftito11_model/Transfer Fcn3','numerator','num22',
          'denominator','den22')

%% assigning of parameters in second loop model transfer
    function & delay blocks

set_param('Ftito22_model/Transport Delay','delaytime',
          'del11')
set_param('Ftito22_model/Transfer Fcn','numerator','num11',
          'denominator','den11')
set_param('Ftito22_model/Transport Delay1','delaytime',
          'del21')
set_param('Ftito22_model/Transfer Fcn1','numerator','num21',
          'denominator','den21')
set_param('Ftito22_model/Transport Delay2','delaytime',
          'del12')
set_param('Ftito22_model/Transfer Fcn2','numerator','num12',
```

```
            'denominator','den12')
set_param('Ftito22_model/Transport Delay3','delaytime',
            'del22')
set_param('Ftito22_model/Transfer Fcn3','numerator','num22',
            'denominator','den22')


%% step change in loop_11 to get model main response-G11 and
   interaction-G21
sim('Ftito11_model',100);


%% step change in loop_22 to get model main response-G22 and
   interaction-G12
sim('Ftito22_model',100);


%% load output response values for model

% Model set point change in yr1
load y11fm.mat;
load y21fm.mat;

% Model set point change in yr2
load y12fm.mat;
load y22fm.mat;


guess11 = y11fm(:,2)/step_size1;
guess21 = y21fm(:,2)/step_size1;
guess12 = y12fm(:,2)/step_size1;
guess22 = y22fm(:,2)/step_size1;


%calculation of error vectors from model and real system
 response

f11 = (guess11-pros11);
f21 = (guess21-pros21);
f12 = (guess12-pros12);
f22 = (guess22-pros22);
```

```
f = [f11; f21; f12; f22];

%% CALCULATION OF MODLE PARAMETERS GUESS VALUES
   (another script file)

clc;
clear all;

%% step change in loop_11 to get model main response-G11
   and interaction-G21

sim('Ftito11',70);

%% step change in loop_22 to get model main response-G22
   and interaction-G12

sim('Ftito22',70);

%% Assigning values to the parameters

s11 = 0;
s21 = 0.1143;
s12 = 0.16;
s22 = 0;

kc11 = 0.396;
ki11 = 0.06682;
kd11 = 0.4237;
kc22 = -0.124;
ki22 = -0.01748;
kd22 = -0.2292;
d11 = 1; t11 = 6.25;
d21 = 7; t21 = 8.75;
d12 = 3; t12 = 6.25;
d22 = 3; t22 = 7.5;

% guess values calculation for main loop process gains
```

```
kp11 = 2/kc11
kp22 = 2/kc22

%% guess values calculation for non-diagonal process gains

% evaluation of integral of response
A = y11f(:,2).*(exp(-s11*y11f(:,1)));
B = y21f(:,2).*(exp(-s21*y21f(:,1)));
C = y12f(:,2).*(exp(-s12*y12f(:,1)));
D = y22f(:,2).*(exp(-s22*y22f(:,1)));
E = trapz (y11f(:,1),A);
F = trapz (y21f(:,1),B)
G = trapz (y12f(:,1),C)
H = trapz (y22f(:,1),D);

% calculation non-diagonal process gain (kp21) from closed
  loop transfer function

Gc2=(kc22+(ki22/s21)+(kd22*s21))
Gp22=(kp22*exp(-d22*s21)/(t22*s21+1))
CG22=(1+(Gc2*Gp22))
Gc1=(kc11+(ki11/s21)+(kd11*s21))
Gp11=(kp11*(exp(-d11*s21))/(t11*s21+1))
CG11=(1+(Gc1*Gp11))
J21=F*CG22*CG11
kk21=(1/s21)*Gc1*exp(-d21*s21)/((t21*s21)+1)
kp21=J21/kk21

% calculation non-diagonal process gain (kp12) from closed
  loop transfer function

iGc2=(kc22+(ki22/s12)+(kd22*s12))
iGp22=(kp22*exp(-d22*s12)/((t22*s12)+1))
iCG22=(1+(iGc2*iGp22))
iGc1=(kc11+(ki11/s12)+(kd11*s12))
iGp11=(kp11*(exp(-d11*s12))/((t11*s12)+1))
```

```
iCG11=(1+(iGc1*iGp11))
J22=G*iCG22*iCG11
kk22=(1/s12)*iGc2*exp(-d12*s12)/((t12*s12)+1)
kp12=J22/kk22
```

C.5 Code for Plotting Nyquist Diagram of Weischedel and McAvoy (1980) System

```
clc;
clear all;


%TITO SOPTD ACTUAL

g11 = tf([0.58],[12.35 8.26 1],'iodelay',1);
g12 = tf([-0.45],[17.04 8.69 1],'iodelay',1);
g21 = tf([0.35],[3.35 5.67 1],'iodelay',1.28);
g22 = tf([-0.48],[1.69 5.06 1],'iodelay',1);
G = [g11 g12;g21 g22];


%TITO SOPTD SUSD response model

h11 = tf([0.5798],[12.4199 8.2506 1],'iodelay',0.9939);
h12 = tf([-0.4497],[17.0478 8.633 1],'iodelay',0.9982);
h21 = tf([0.3498],[3.5608 5.684 1],'iodelay',1.2439);
h22 = tf([-0.4797],[1.8274 5.0714 1],'iodelay',0.9748);
H = [h11 h12;h21 h22];


%TITO SOPTD step response model

j11 = tf([0.5141],[4.9173 7.0361 1],'iodelay',1.9342);
j12 = tf([-0.3869],[13.2227 8.0013 1],'iodelay',0.9982);
j21 = tf([0.3038],[1.1868 4.6517 1],'iodelay',1.2439);
j22 = tf([-0.4318],[2.4171 4.6287 1],'iodelay',0.9748);
J = [j11 j12;j21 j22];


figure(1),subplot(2,2,1), nyquist(g11,h11,'r+',j11,':');
figure(1),subplot(2,2,2), nyquist(g12,h12,'r+',j12,':');
```

```
figure(1),subplot(2,2,3), nyquist(g21,h21,'r+',j21,':');
figure(1),subplot(2,2,4), nyquist(g22,h22,'r+',j22,':');

%figure(2),nyquist(G,H,'r+',J,':');
*********************************************************
```

Bibliography

Agrawal, P. and Lim, H. (1984). Analyses of various control schemes for continuous bioreactors. *Advances in Biochemical Engineering Biotechnology*, **30**, 61–90.

Ambrose, H. and Qu, Z. (1997). Model reference robust control for MIMO systems. *International Journal of Control*, **68 (3)**, 599–623.

Ananth, I. and Chidambaram, M. (1999). Closed-loop identification of transfer function model for unstable systems. *Journal of the Franklin Institute*, **336**(7), 1055–1061.

Ananth, I. and Chidambaram, M. (2000). Closed loop identification of transfer function model for stable systems. *Indian Chemical Engineer*, **42**, 5–10.

Åström, K. J. and Hägglund, T. (1984). Automatic tuning of simple regulators with specifications on phase and amplitude margins. *Automatica*, **20**(5), 645–651.

Astrom, K. J. and Hagglund, T. (1995). PID *Controllers Theory Design and Tuning*, 2nd Edition, New Jersey: Instrument Society of America.

Atherton, D. P. (1975). *Non Linear Control Engineering*, London: Van Nostrand Reinhold Co.

Babji, B. S. and Saraf, D. N. (1991). Dynamics and control of a pilot scale distillation column. PhD diss., IIT Kanpur, *Report of Department of Chemical Engineering, IIT Kanpur*.

Balachandar, R. and Chidambaram, M. (1996). Comparison of performances of decentralized controllers, *Process Control and Quality*, **8**, 123–131.

Bhat, J., Chidambaram, M. and Madhavan, K. P. (1991). Robust multivariable control of a mixed culture bioreactor with competition and external inhibition.*Indian Chemical Engineer*, **2**, 29–34.

Binder, Z., Magalhaes, M. F. and Rey, D. (1984). Multivariable control of distillation columns. *IFAC Proceedings*, **17**(2), 1707–1712.

Bogere, M. N. and Ozgen, C. (1989). On-line controller tuning of second order dead time processes. *Chemical Engineering Research and Design*, **67**(6), 555–561.

Bristol, E. (1966). On a new measure of interaction for multivariable process control. *IEEE Transactions on Automatic Control*, **11**, 133–134.

Cai, W. J., Ni, W., He, M. J. and Ni, C. Y. (2008). Normalized decoupling: a new approach for MIMO process control system design. *Industrial and Engineering Chemistry Research*, **47**(19), 7347–7356.

Chen, C. L. (1989). Simple method for on-line identification and controller tuning. *AIChE Journal*, **35**(12), 2037–2039.

Chen, D. and Seborg, D. E. (2002). Multiloop PI/PID Controller design based on Gershgovin Bands. *IEE process Control Theory and Applications*, **149**(1), 68–73.

Chen, D. and Seborg, D. E. (2003). Design of decentralized PI control system based on Nyquist stability analysis. *Journal of Process Control*, **13(1)**, 27–39.

Chen, L., Li, J. and Ding, R. (2011). Identification for the second-order systems based on the step response. *Mathematical and Computer Modelling*, **53**(5–6), 1074–1083.

Cheres, E. (2006). Parameter estimation of an unstable system with a PID controller in a closed loop configuration. *Journal of the Franklin Institute*, **343**(2), 204–209.

Cheres, E. and Eydelzon, A. (2000). Parameter estimation of a second order model in the frequency domain from closed loop data. *Transactions of the Institution of Chemical Engineers*, **78**(2), 293–298.

Chidambaram, M. (1995). *Nonlinear Process Control*, New Delhi: Wiley Eastern Limited.

Chidambaram, M. (1998). *Applied Process Control*, New Delhi: Allied Publishers.

Chidambaram, M. and Nikita, S. (2018). *Relay Tuning of PID Controllers for Unstable MIMO Systems*, New Delhi: Springer.

Chidambaram, M. and Padmasree, R. (2006). *Control of Unstable Systems*, 1st ed., Chennai: Narosa Publishing House.

Chidambaram, M. and Padmasree, R. (2016). *Control of Unstable Systems*, 2nd ed., Chennai: Narosa Publishing House.

Chidambaram M. and Sathe, S. (2014). *Relay Auto Tuning for Identification and Control*, Cambridge: Cambridge University Press.

Chien, I., Huang, H. and Yang, J. (1999). A simple multiloop tuning method for PID controllers with no proportional kick. *Industrial and Engineering Chemistry Research*, **38**(4), 1456–1468.

Chien, I. and Fruehauf, P. S. (1990). Consider IMC tuning to improve controller performance. *Chemical Engineering Progress*, **10**, 33–41.

Chiu, M. S. and Arkun, Y. (1990). Decentralized control structure selection based on integrity considerations. *Industrial and Engineering Chemistry Research*, **29**(3), 269–273.

Chipperfield, A., Fleming, P. J., Pohlheim, H. and Fonseca, C. M. (1994). *Genetic Algorithm Toolbox: User's Guide*, Sheffield, UK: Department of Automatic Control and Systems Engineering, University of Sheffield.

Clark, R. (2005). *Control System Dynamics*, London: Cambridge University Press.

Cochin, I. (1980). *Analysis and Design of Dynamic Systems*, New York: Harper and Row.

Cordons, B. (2005). *Process Modelling for Control: A Unified Approach Using Standard Black- Box Techniques*, London: Springer-Verlog, 48.

Cui, H. and Jacobsen, M. A. (2002). Performance limitations in decentralized control. *Journal of Process Control*, **12**(4), 485–494.

Davison, E. (1976). Multivariable tuning regulators: the feed-forward and robust control of general servo- mechanism problem. *IEEE Transactions on Automatic Control*, **21**(1), 35–41.

Davison, E. J. (1983). Some properties of minimum phase systems and 'Squared-Down' systems. *IEEE Transactions on Automatic Control*, **28**(2), 221–222.

De Paor, A. M. and O'Malley, M. (1989). Controllers of Ziegler-Nichols type for unstable process with time delay. *International Journal of Control*, **49**(4), 1273–1284.

Dhanya Ram, V. and Chidambaram, M. (2014). Closed loop reaction curve method for identification of TITO systems, *IFAC Proceedings Volumes* (IFAC-PapersOnline), **47**(1), 989–996.

Dhanya Ram, V. D. and Chidambaram, M. (2015). On-line controller tuning for critically damped SOPTD systems. *Chemical Engineering Communications*, **202**(1), 48–58.

Dhanya Ram, V. and Chidambaram M. (2015). Simple method of designing centralized PI controllers for multivariable systems based on SSGM. *ISA Transactions*, **56**, 252–260.

Dhanya Ram V. and Chidambaram, M. (2016). Identification of centralized controlled multivariable systems. *Indian Chemical Engineer*, **58**(3), 240–254.

Dhanya Ram, V. and Chidambaram, M. (2017). Closed loop identification of two input two output critically damped second order systems with delay. *Indian Chemical Engineer*, **57**(2), 79–100.

Dhanya Ram, V. and Chidambaram, M. (2018). SSGM based multivariable control of unstable non-square systems. *Chemical Product and Process Modelling*, **13**(1), 1–14.

Dhanya Ram, V., Karlmarx, A. and Chidambaram, M. (2014). Identification of unstable second order transfer function model with a zero by optimization method. *Indian Chemical Engineer*, **58**(1), 29–39.

Dhanya Ram, V. D., Rajapandiyan, C. and Chidambaram, M. (2015). Steady-state gain identification and control of multivariable unstable systems. *Chemical Engineering Communications*, **202**(2), 151–162.

Dinesh, P., Chidambaram, M. and Pandit, M. (2003). Comparison of multivariable controllers for non minimum phase systems.*International Symposium on Process Systems Engineering and Control*, January 3–4, Mumbai, India, 386–391.

Economou, C. G., Morari, M. and Palsson, B. (1986). Internal model control. 5. Extension to nonlinear systems. *Industrial and Engineering Chemistry Process Design and Development*, **25**(2), 403–411.

Franklin, G. F., Powell, J. D. and Emami-Naeini, A. (1989). *Feedback Control of Dynamic Systems*, New Jersey: Pearson.

Friman, M. and Waller, K. V. (1994). Autotuning of multiloop control systems. *Industrial and Engineering Chemistry Research*, **33**(7), 1708–1717.

Fruehauf, P. S., Chien, I. and Lauritsen, M. D. (1994). Simplified IMC–PID tuning rules. *ISA transactions*, **33**(1), 43–59.

Ganesh, P. and Chidambaram, M. (2010). Multivariable controller tuning for nonsquare systems with RHP zeros by genetic algorithm. *Chemical and Biochemical Engineering Quarterly*, **24**(1), 17–22.

Garcia, C. E. and Morari, M. (1982). Internal model control. 1. A unifying review and some new results. *Industrial and Engineering Chemistry Process Design and Development*, **21**(2), 308–323.

Garcia, C. E. and Morari, M. (1985a). Internal model control. 2. Design procedure for multivariable systems. *Industrial and Engineering Chemistry Process Design and Development*, **24**(2), 472–484.

Garcia, C. E. and Morari, M. (1985b). Internal Model Control. 3. Multivariable Control Law Computation and Tuning Guidelines. *Industrial and Engineering Chemistry Process Design and Development*, **24**(2), 484–494.

Garcia, D., Karimi, A. and Longchamp, R. (2005). PID controller design for multivariable systems using Gershgorin bands. *IFAC Proceedings Volumes* (IFAC Papers Online), **V38**(1), 183–188.

Gigi, S. and Tangirala, A. K. (2013). Quantification of interaction in multiloop control systems using directed spectral decomposition. *Automatica*, **49**(5), 1174–1183.

Gopal, M. (2014). *Modern Control System Theory*, New Delhi: New Age International.

Govindakannan, G. and Chidambaram, M. (1997). Multivariable PI control of unstable systems. *Process Control and Quality*, **10**, 319–329.

Govindakannan, G. and Chidambaram, M. (2000). Two stage multivariable controllers for unstable plus time delay systems. *Indian Chemical Engineer*, **42**, 34–38.

Grosdidier, P., Morari, M. and Holt, B. R. (1985). Closed-loop properties from steady state gain information. *Industrial Engineering Chemistry Fundamentals*, **24**(2), 221–235.

Gu, K., Chen, J. and Kharitonov, V. (2003). *Stability of Time-Delay Systems*, Berlin: Springer Science, Business Media.

Gupta, S. K. (1995). *Numerical Methods for Engineers*, New Delhi: Wiley Easter Ltd.

Gundes, A. N. (2008). Simultaneous stabilization of MIMO systems with integral action controllers. *Automatica*, **44**(4), 1156–1160.

Gustavsson, I., Ljung, L. and Soderstrom, T. (1997). Identification of processes in closed loop identifiability and accuracy aspects. *Automatica*, **13**(1), 59–75.

Halevi, Y., Palmor, Z. J. and Efrati, T. (1997). Automatic tuning of decentralized PID controllers for MIMO processes. *Journal of Process Control*, **7**(2), 119–128.

Ham, T. W. and Kim, Y. H. (1998). Process identification and PID controller tuning in multivariable systems. *Journal of Chemical Engineering of Japan*, **31**(6), 941–949.

Harini, S. and Chidambaram, M. (2005). Identification of unstable systems. *Indian Chemical Engineer*, **47**(2), 88–94.

Haritha, P., Dhanya Ram, V. and Chidambaram, M. (2021). *Identification of CSOPT System by Optimization Method*, Research Report, Department of Chemical Engineering, NIT-Calicut.

Harriott, P. (1964). *Process Control*, New York: McGraw-Hill Education.

He, M. J., Cai, W. J., Ni, W. and Xie, L. H. (2009). RNGA based control system configuration for multivariable processes. *Journal of Process Control*, **19**(6), 1036–1042.

He, M. J., Cai, W. J, Wu, B. F. and He, M. (2005). Simple decentralized PID controller design method based on dynamic relative interaction analysis. *Industrial and Engineering Chemistry Research*, **44**(22), 8334–8344.

Ho, W. K, Hang, C. C, Cao, L. S. (1995). Tuning of PID controller based on gain and phase margin specifications, *Automatica*, **31**(3) 497–502.

Ho, M. T., Silva, G. J., Datta, A. and Bhattacharyya, S. P. (2005). *PID Controllers for Time Delay Systems*, Boston, Mass, USA: Birkhauser.

Hovd, M. and Skogestad, S. (1993). Improved independent design of robust decentralized controllers. *Journal of Process Control*, **3**(1), 43–51.

Hovd, M. and Skogestad, S. (1994). Pairing criteria for decentralized control of unstable plants. *Industrial and Engineering Chemistry Research*, **33**(9), 134–2139.

Hu, W., Xiao, G., Li, X. (2011). An analytical method for PID controller tuning with specified gain and phase margin for integral plus time delay processes. *ISA Transactions*, **50**(2), 268–276.

Huang, C. and Chou, C. (1994). Estimation of the underdamped second-order parameters from the system transient. *Industrial Engineering Chemistry Research*, **33**(1), 174–176.

Huang, C. and Clements Jr., W. C. (1982). Parameter estimation for the second-order-plus-dead-time model. *Industrial Engineering Chemistry Process Design and Development*, **21**(4), 601–603.

Huang, C. and Huang, M. (1993). Estimation of the second-order parameters from the process transient by simple calculation. *Industrial and Engineering Chemistry Research*, **32**(1), 228–230.

Huang, H. P., Jeng, J. C., Chiang, C. H. and Pan, W. (2003). A direct method for multi-loop PI/PID controller design. *Journal of Process Control*, **13**(8), 769–786.

Hu, W., Cai, W. and Xiao, G. (2010). Decentralized control system design for MIMO processes with integrators/differentiators. *Industrial and Engineering Chemistry Research*, **49**(2), 12521–12528.

Hwang, S. (1995). Closed-loop automatic tuning of single-input-single-output systems. *Industrial and Engineering Chemistry Research*, **34**(7), 2406–2417.

Ikonen, E. and Najim, K. (2001). *Advanced Process Identification and Control*, CRC Press.

Ivezic, D. D. and Petrovic, T. B. (2003). New approach to milling circuit control robust inverse Nyquist array design. *International Journal of Mineral Processing*, **70**, 171–182.

Jacob, E. F. and Chidambaram, M. (1996). Design of controllers for unstable first order plus time delay systems. *Computers and Chemical Engineering*, **20**(5), 579–584.

Jevtovic, B. T. and Matauek, M. R. (2010). PID controller design based on ideal decoupler. *Journal of Process Control*, **20**, 869–876.

Jin et al. (2017). Novel centralized IMC–PID controller design for multivariable processes with multiple time delays, *Industrial and Engineering Chemistry Research*. **56**, 4431–4445.

Johansson, K. H. (2000). The quadratic-tank process: a multivariable laboratory process with an adjustable zero. *IEEE Transactions on Control Systems Technology*, **8**, 456–465.

Johnson, M. and Morari, M. (2006). *PID Control: New Identification and Design Methods*, London: Springer–Verlog.

Jutan, A. (1989). Comparison of three closed-loop PID tuning algorithms. *AIChE Journal*, **35**(11), 1912–1914.

Jutan, A. and Rodriguez II, E. S. (1983). Extension of a new method for on-line controller tuning. *Canadian Journal of Chemical Engineering*, **62**(6), 802–807.

Katebi, R. (2012). Robust multivariable tuning methods. In R. Vilanova and A. Visioli, eds., *PID Control in the third Millennium*, London: Springer–Verlag, 225–280.

Kavdia, M. and Chidambaram, M. (1996). On-line controller tuning for unstable systems. *Computers and Chemical Engineering*, **20**(3), 301–305.

Kaya, I. (2004). Two degree of freedom IMC structure and controller design for integrating processes based on gain and phase margin specifications, *IEE Proceedings: Control Theory and Applications*, **151** (4), 481–487.

Keel, L. H. and Bhattacharyya, S. P. (2008). Controller synthesis free of analytical models: three term controllers. *IEEE Transactions on Automatic Control*, **53**(6), 1353–1369.

Keesman, K. (2011). *System Identification: An Introduction*, Springer.

Kumar, N., Pandit, M. and Chidambaram, M. (2004). Multivariable control of four tank systems. *Indian Chemical Engineer*, **46**(4), 216–221.

Kurniawan, A. M., Handogo, R., Sutikno, J. P. and Lee, H. Y. (2018). Stability criterion of modified inverse Nyquist array on a simple non-square MIMO process. *International Journal of Applied Chemistry*, **14**(4), 293–310.

Lee, C.H. (2004). A survey of PID controller design based on gain and phase margin. *International Journal of Computational Cognition*, **2**, 63–100.

Lee, J. (1989). On-line PI controller tuning from a single closed-loop test. *AICHE Journal*, **35**, 329–331.

Lee, M., Lee, K., Kim, C. and Lee, J. (2004). Analytical design of multiloop PID controllers for desired closed loop responses. *AIChE Journal*, **50**(7), 1631–1635.

Levien, K. and Morari, M. (1987). Internal model control of coupled distillation columns. *AIChE Journal*, **33**(1), 83–98.

Lieslehto, J., Tanttu, J. T. and Koivo, H. N. (1993). An expert system for multivariable controller design. *Automatica*, **29**(4), 953–968.

Lin, Chong, Wang, Q. G., He, Y., Wen, G., Han, X., Guangyao, L. and Zhong, Z. H. (2008). On stabilizing PI controller ranges for multivariable systems. *Chaos, Solitons and Fractals*, **35**, 620–625.

Liu, T. and Gao, F. (2010). Closed Loop Step Response Identification of Integrating and Unstable Processes, *Chemical Engineering Science*, **65**, 2884–2895.

Liu, T. and Gao, F. (2011a). Enhanced IMC design of load disturbance rejection for integrating and unstable processes with slow dynamics. *ISA Transactions*, **50**(2), 239–248.

Liu, T. and Gao, F. (2011b). *Industrial Process Identification and Control Design: Step- Test and Relay-Experiment-Based Methods*, London: Springer–Verlag.

Liu, T. and Gao, F. (2012). *Industrial Process Identification and Control Design: Step Test and Relay Experiment Based Methods*, London: Springer–Verlag.

Liu, T., Gao, F. and Wang, Y. (2008). A systematic approach for on-line identification of second-order process model from relay feedback test. *AIChE Journal*, **54**(6), 1560–1578.

Liu, T., Wang, Q. and Huang, H. (2013). A tutorial review on process identification from step or relay feedback test. *Journal of Process Control*, **23**(10), 1597–1623.

Ljung, L. (1998). *System Identification: Theory for the User*, Pearson Education.

Loh, E. J. and Chiu, M. (1997). Robust decentralized control of non-square systems. *Chemical Engineering Communications*, **158**, 157–180.

Loh, P. A., Hang, C. C., Quek, K. C. and Vasani, U. V. (1993). Auto tuning of multi-loop proportional–integral controllers using relay feedback. *Industrial and Engineering Chemistry Research*, **33**, 1102–1107.

Luan, X., Chen, Q. and Liu, F. (2014). Centralized PI control for high dimensional multivariable systems based on equivalent transfer function. *ISA Transactions*.

Luyben, M. and Luyben, W. (1997). *Essentials of process control*, McGraw-Hill.

Luyben, W. L. (1986). Simple method for tuning SISO controllers in multivariable systems. *Industrial and Engineering Chemistry Research*, **25**(3), 654–660.

Luyben, W. L. and Vinante, C. D. (1972). Experimental studies of distillation decoupling. *Kem. Teollisuus*, **29**(8), 499–514.

Lyzell, C. (2009). Initialization methods for system identification. PhD diss., Linkoping University, Sweden. *Linköping studies in science and technology*, Thesis No. 1426.

Maciejowski, J. (1989). *Multivariable Feedback Design*, Workingham, England: Addison.

Maghade, D. K. and Patre, B. M. (2012). Decentralized PI/PID controllers based on gain and phase margin specifications for TITO processes. *ISA Transactions*, **51**(4), 550–558.

Mamat, R. and Fleming, P. J. (1995). Method for on-line identification of a first order plus dead-time process model. *Electronics Letters*, **31**(15), 1297–1298.

Marchetti, G., Scali, C. and Romagnoli, J. R. (2002). Relay auto tuning of multivariable systems: application to an experimental pilot scale column. *IFAC Proceedings*, **35**(1), 127–132.

Mei, H., Li, S., Cai, W. and Xiong, Q. (2005). Decentralized closed-loop parameter identification for multivariable processes from step responses. *Mathematics and Computers in Simulation*, **68**(2), 171–192.

Melo, D. L. and Friedly, J. C. (1992). On-line, closed-loop identification of multivariable systems. *Industrial Engineering Chemistry Research*, **31**(1), 274–281.

Morari, M. (1982). Flexibility and resiliency of process systems. In Proceedings of the International Symposium on Process Systems Engineering.

Morari, M. and Zafiriou, E. (1989). *Robust Process Control*, New Jersey: Prentice Hall.

Munro, N. (1972). Multivariable systems design using the inverse Nyquist array. *Computer Aided Design*, **4**(5), 222–227.

Muske, K., Young, J., Grosdidier, P. and Tani, S. (1991). Crude unit product quality control. *Computers and Chemical Engineering*, **15**(9), 629–638.

Mutalib, A. (2014). System identification for a pilot plant scale acetone and isopropyl alcohol distillation column. Graduate thesis, University of Petronas, Malaysia.

Naik, S. and Kumar, P. (1992). Robust continuous-time adaptive control by parameter projection. *IEEE Transactions on Automatic Control*, **37**(2), 182–197.

Narasimha Reddy, S. and Chidambaram, M. (2020). Model identification of critically damped second order plus time delay systems. *Indian Chemical Engineer*, **62**, 67–77.

Niederlinski, A. (1971). A heuristic approach to the design of linear multivariable interacting control systems. *Automatica*, **7**(6), 691–701.

Nikita, S. and Chidambaram, M. (2018). Case studies of improved relay auto tuning of PID controllers for SISO systems. *Indian Chemical Engineer*, **60**(4), 438–456.

Nie, Z. Y., Wang, Q. G., Wu, M. and He, Y. (2011). Tuning of multi-loop PI controllers based on gain and phase margin specifications. *Journal of Process Control*, **21**(9), 1287–1295.

Nouri, A. S. and Abdennour, R. B. (2007). A sliding mode control for multivariable systems. *International Journal on Sciences and Techniques of Automatic Control*, **1**(1).

Ogata, K. (2017). *Modern Control Engineering*, New Delhi: Prentice Hall.

Ogunnaike, B. A., Lemaire, J. P., Morari, M. and Ray, W. H. (1983). Advanced multivariable control of a pilot-plant distillation column. *AIChE Journal*, **29**(4), 632–640.

Ogunnaike, B. A. and Ray, W. H. (1994). *Process Dynamics, Modelling and Control*, New York: Oxford University Press.

Okiy, S. (2015). Transfer function modelling: a literature survey research. *Journal of Applied Sciences, Engineering and Technology*, **11**, 1265–1279.

Ou, L., Tang, Y., Gu, D. and Zhang, W. (2005). Stability analysis of PID controllers for integral processes with time delay. *Proceedings of the American Control Conference* (ACC 2005), 4247–4252.

Ou, L., Zhang, W. and Gu, D. (2006). Sets of stabilising PID controllers for second-order integrating processes with time delay. *IEE Proceedings: Control Theory and Applications*, **153**(5), 607–614.

Padma Sree, R. and Chidambaram, M. (2002). Identification of unstable transfer model with a zero by optimization method. *Journal of the Indian Institute of Science*, **82**(5–6), 219–225.

Padmasree, R. and Chidambaram, M. (2003). IMC controller design for unstable with a zero. In Proceedings of the International Conference on Chemical and BioChem. Engineers University Malaysia Sabah, Kota Kinabalu.

Padmasree, R. and Chidambaram, M. (2006). *Control of Unstable Systems*, New Delhi: Narosa.

Padma Sree, R. and Chidambaram, M. (2006). Improved closed loop identification of transfer function model for unstable systems. *Journal of the Franklin Institute*, **343**(2), 152–160.

Padmasree, R. P. and Chidambaram, M. (2008). Identification of unstable transfer model with zero by optimization method. *Journal of Indian Institute of Science*, **82**, 219–225.

Palmor, Z. J., Halevi, Y. and Krasney, N. (1995). Automatic tuning of decentralized PID controllers for TITO processes. *Automatica*, **31**(7), 1001–1010.

Papastathopoulo, H. S. and Luyben, W. L. (1990). A new method for the derivation of steady-state gains for multivariable processes. *Industrial and Engineering Chemistry Research*, **29**(3), 366–369.

Park, H. I., Sung, S. W., Lee, I. and Lee, J. (1997). On-line process identification using the Laguerre series for automatic tuning of the Proportional-Integral-Derivative controller. *Industrial and Engineering Chemistry Research*, **36**(1), 101–111.

Park, J. H., Park, H. I. and Lee, I. (1998). Closed-loop on-line process identification using a proportional controller. *Chemical Engineering Science*, **53**(9), 1713–1724.

Pavkovic, D., Polak, S. and Zorc, D. (2014). PID controller auto-tuning based on process step response and damping optimum criterion. *ISA transactions*, **53**(1), 85–96.

Pensar, J. A. and Waller, K. V. (1993). Steady-state-gain identification for multivariable process control. *Industrial and Engineering Chemistry Research*, **32**(9), 2012–2016. 244.

Pintelon, R. and Schoukens, J. (2001). *System Identification: A Frequency Domain Approach*, John Wiley & Sons.

Postlethwaite, I. (1977). A generalized inverse Nyquist stability criterion. *International Journal of Control*, **26**(3), 325–340.

Pouliquen, M., Gehan, O. and Pigeon, E. (2014). Bounded-error identification for closed-loop systems. *Automatica*, **50**(7), 1884–1890.

Pramod, S. and Chidambaram, M. (2000). Closed loop identification of transfer function model for unstable bioreactors for tuning PID controllers. *Bioprocess Engineering*, **22**(2), 185–188.

Pramod, S. and Chidambaram, M. (2001). Closed loop identification by optimization method for tuning PID controllers. *Indian Chemical Engineer*, **43**(2), 90–94.

Prett, D. M. and Morari, M. (1987). *The Shell Process Control Workshop*, London: Butterworths.

Qu, Z., Dorsey, J. and Dawson, D. (1994). Model reference robust control of a class of SISO systems. *IEEE Transactions on Automatic Control*, **39**(1), 2219–2234.

Rajapandiyan, C. and Chidambaram, M. (2012a). Closed-loop identification of multivariable systems by optimization method. *Industrial and Engineering Chemistry Research*, **51**(3), 1324–1336.

Rajapandiyan, C. and Chidambaram M. (2012b). Closed-loop identification of second-order plus time delay (SOPTD) model of multivariable systems by optimization method. *Industrial and Engineering Chemistry Research*, **51**(28), 9620–9633.

Rajapandiyan, C. and Chidambaram M. (2012c). Controller design for MIMO processes based on simple decoupled equivalent transfer functions and simplified decoupler. *Industrial and Engineering Chemistry Research*, **51**(38), 12398–12410.

Ramachandiran, R., Lakshimanarayanan, S., Rangaiah, G. P. (2005). Process identification using open-loop and closed-loop step responses. *Journal of the Institution of Engineers, Singapore*, **45**(6).

Rangaiah, G. P. and Krishnaswamy, P. R. (1994). Estimating second-order plus dead time model parameters. *Industrial and Engineering Chemistry Research*, **33**(7), 1867–1871.

Rangaiah, G. P. and Krishnaswamy, P. R. (1996). Estimating second-order dead time parameters from underdamped process transients. *Chemical Engineering Science*, **51**(7), 1149–1155.

Rao, A. S. and Chidambaram, M. (2006a). Enhanced two-degrees-of-freedom control strategy for second-order unstable processes with time delay. *Industrial and Engineering Chemistry Research*, **45**(10), 3604–3614.

Rao, A. S. and Chidambaram, M. (2006b). Smith delay compensator for multivariable non-square systems with multiple time delays. *Computers and Chemical Engineering*, **30**(8), 1243–1255.

Rao, A. S. and Chidambaram, M. (2012). PI/PID controllers design for integrating and unstable systems. In R. Vilanova and A. Visioli, eds., *PID Control in the Third Millennium*, London: Springer–Verlag, 75–111.

Reddy, B. C., Chidambaram, M. and Al-Gobaisi, D. M. K. (1997). Design of centralized controllers for a MSF desalination plant. *Desalination*, **113**(1), 27–38.

Reddy, P.D.S., Pandit. M. and Chidambaram, M. (2006).Comparison of multivariable controllers for non minimum phase systems. *International Journal of Modelling and Simulation*, **26**, 237–243.

Reeves, D. E. and Arkun, Y. (1989). Interaction measures for non square decentralized control structures, *AIChE Journal*, **135**, 603–613.

Revelo, J., Herrera, M., Camacho, O. and Alvarez, H. (2020). Nonsquare multivariable chemical processes: a hybrid centralized control proposal. *Industrial & Engineering Chemistry Research*, **59**, 14410–14422.

Rivera, D. E., Morari, M. and Skogestad, S. (1986). Internal Model Control. 4. PID Controller Design. *Industrial and Engineering Chemistry, Process Design and Development*, **25**(1), 252–265.

Rosenbrock, H. H. (1969). Design of multivariable control systems using the inverse Nyquist array. *Proceedings of the Institution of Electrical Engineers*, **116**(11), 1929–1936.

Sadasivarao, M. V. and Chidambaram, M. (2006). PID controller tuning of cascade control systems using genetic algorithm. *Journal of Indian Institute of Science*, **86**, 343–354.

Sarma, K. L. N. and Chidambaram M. (2005). Centralized PI/PID controllers for nonsquare systems with RHP zeros. *Journal of the Indian Institute of Science*, **85**(4), 201–214.

Savitzky, A., Golay, M. J. E. (1964). Smoothing and differentiation of data by simplified least squares procedures, *Analytical Chemistry*, **36**(8), 1627–1639.

Schwanke, C. O., Edgar, T. F. and Hougen, J. O. (1977). Development of multivariable control strategy for distillation columns. *ISA Transactions*, **16**, 69–81.

Seborg, D., Edgar, T. and Mellichamp, D. (2006). *Process Dynamics Control*, 2nd ed. New York: John Wiley & Sons.

Semino, D., Scali, C. (1998). Improved identification and autotuning of PI controllers for MIMO processes by relay techniques. *Journal of Process Control*. **8**(3), 219–227.

Shaji, C. and Chidambaram, M. (1996). Robust decentralized PI controllers for crude product quality. *Process Control & Quality*, **8**, 167–175.

Shen, Y., Cai, W. and Li, S. (2010). Multivariable process control: decentralized, decoupling, or sparse? *Industrial and Engineering Chemistry Research*, **49**(2), 761–771.

Shen, Y., Sun, Y. and Xu, W. (2014). Centralized PI/PID controller design for multivariable processes. *Industrial and Engineering Chemistry Research*, **53**(25), 10439–10447.

Silva, R., Sbarbaro, D. and Barra, B. D. L. (2006). Closed-loop process identification under PI control: a time domain approach. *Industrial and Engineering Chemistry Research*, **45**(13), 4671–4678.

Silva, G. J., Datta, A. and Bhattacharyya, S. P. (2001). PI stabilization of first-order systems with time delay. *Automatica*, **37**(12), 2025–2031.

Silva, G. J., Datta, A. and Bhattacharyya, S. P. (2002). New results on the synthesis of PID controllers. *IEEE Transactions on Automatic Control*, **47**(2), 241–252.

Silva, G. J., Datta, A. and Bhattacharyya, S. P. (2002). New results on the synthesis of PID controllers. *IEEE Transactions on Automatic Control*, **47**(2), 241–252.

Simi, S. and Chidambaram, M. (2015). Tuning of proportional integral-derivative controllers for critically damped second order plus time delay systems. *Indian Chemical Engineer*, **57**, 32–51.

Skogestad, S. (2003). Simple analytic rules for model reduction and PID controller tuning. *Journal of Process Control*, **13**(4), 291–309.

Skogestad, S. (2006). Tuning for smooth PID control with acceptable disturbance rejection. *Industrial and Engineering Chemistry Research*, **45**(23), 7817–7822.

Skogestad, S. and Morari, M. (1989). Robust performance of decentralized control systems by independent designs. *Automatica*, **25**(1), 119–125.

Skogestad, S. and Postlethwaite, I. (2005). *Multivariable Feedback Control: Analysis and Design*, 1st ed., New York: John Wiley & Sons.

Skogestad, S. and Postlethwaite, I. (2014). *Multivariable Feedback Control: Analysis and Design*, 2nd ed., New York: John Wiley & Sons.

Smith, C. and Corripio, A. (1985). *Principles and Practice of Automatic Process Control*, New York: John Wiley & Sons.

Soderstrom, T. and Stoica, P. (1989). *System Identification*. In *Prentice Hall International Series in Systems and Control Engineering*, Prentice Hall.

Srinivas, N. and Chidambaram, M. (1996). Experimental comparisons of on-line identification methods for controller Tuning, *Process Control and Quality*, **8**, 85–89.

Srinivas, N. and Chidambaram, M. (1996). Comparisons of on-line controller tuning methods for unstable systems. *Process Control and Quality*, **8**, 177–183.

Srividya, R. and Chidambaram, M. (1997). On-line controllers tuning for integrator plus time delay systems. *Process Control and Quality*, **9**, 59–66.

Stephanopoulos, G. (1984). *Chemical Process Control: An Introduction to Theory and Practice*. Prentice-Hall.

Subramaniam, M., Rajkumar, A. and Chidambaram, M. (1996). Multivariable PI controllers for a nuclear reactor. Proc. Adv. Chem. Eng. ICAChE, Allied Publishers, Madras, India.

Suganda, P., Krishnaswamy, P. R. and Rangaiah, G. P. (1998). On-line process Identification from closed-loop tests under PI control. *Transactions of the Institute of Chemical Engineers*, **76**, 451–457.

Sujatha, V. and Panda, R. C. (2013). Control configuration selection for multi input multi output processes. *Journal of Process Control*, **23**(10), 1567–1574.

Sundaresan, K. R., Chandra Prasad, C. and Krishnaswamy, P. R. (1978). Evaluating parameters from process transients. *Industrial & Engineering Chemistry Process Design and Development*, **17**(3), 237–241.

Sundaresan, K. R. and Krishnaswamy, P. R. (1978). Estimation of time delay time constant parameters in time, frequency and Laplace domains. *Canadian Journal of Chemical Engineering*, **56**(2), 257–262.

Sung, S., Lee, J. and Lee, I. (2009). *Process Identification and PID Control*, John Wiley & Sons.

Taiwo, O. (1993). Comparison of four methods of on-line identification and controller tuning. *IEE Proceedings D: Control Theory and Applications*, **140**(5), 323–327.

Taiwo, O. (2005). Design of multivariable industrial control system by the method of inequalities. In V. Zakian (Ed.) *Control System Design, a New Frame Work*, Berlin: Springer, 251–285.

Tanttu, J. T. and Lieslehto, J. (1991). A comparative study of some multivariable PI controller tuning methods. In R. Devanathan, ed., *Intelligent Tuning and Adaptive Control: IFAC Symposia Series*, Oxford: Pergamon Press.

Treiber, S, (1984). Multivariable control of non-square systems, *Industrial Engineering and Chemistry Research, Process Design and Development*, **23**, 854–857.

Tung, L. S. and Edgar, T. F. (1981). Analysis of control-output interactions in dynamic systems. *AIChE Journal*, **27**(4), 690–693.

Tyreus, B. D. (1979). Multivariable control system design for an industrial distillation column, *Industrial & Engineering Chemistry Process Design and Development*, **19**(1) 177–182.

Venkatashankar, A. and Chidambaram, M. (1994). Control of PI controller for unstable first order plus time delay processes. *International Journal of Control*, **60**, 137–144.

Vijaykumar, V., Rao, V. S. R. and Chidambaram, M. (2012). Centralized PI controllers for interacting multivariable processes by synthesis method. *ISA Transactions*, **51**(3), 400–409.

Viswanathan, P. and Rangaiah, G. (2000). Process identification from closed-loop response using optimization methods. *Chemical Engineering Research and Design*, **78**(4), 528–541.

Viswanathan, P. K., Toh, W. K. and Rangaiah, G. P. (2001). Closed-loop identification of TITO processes using time-domain curve fitting and genetic algorithms. *Industrial and Engineering Chemistry Research*, **40**(13), 2818–2826.

Vivek, S. and Chidambaram, M. (2012). An improved relay auto tuning of PID controllers for critically damped SOPTD systems. *Chemical Engineering Communications*, **199**(11), 1437–1462.

Vlachos, C., Williams, D. and Gomm, J. B. (1999). Genetic approach to decentralized PI controller tuning for multivariable processes. *IEEE: Process Control Theory Applications*, **146**, 58–64.

Vlachos, C., Williams, D. and Gomm, J. B. (2002). Solution to shell control problem using genetically tuned PID controllers. *Control Engineering Practice*, **10**, 151–163.

Vu, T. N. L., Hong, S. and Lee, M. (2009). Analytical design of robust multi-loop PI controller for multivariable processes, *ICROS-SICE International Joint Conference, Fukuoka International Congress Center*, August 18–21, 2961–2966.

Vu, T. N. L. and Lee, M. (2010a). Multi-loop PI controller design based on the direct synthesis for interacting multi-time delay processes. *ISA Transactions*, **49**(1), 79–86

Vu, T. N. L. and Lee, M. (2010b). Independent design of multi-loop PI/PID controllers for interacting multivariable processes, *Journal of Process Control*, **20**(8), 922–933.

Wang, B., Rees, D. and Zhong, Q. C. (2006). Control of integral processes with dead time. Part IV: various issues about PI controllers, *IEE Proceedings: Control Theory and Applications*, **153**(3), 302–306. Mathematical Problems in Engineering. 7

Wang, Q. G. et al., (1997). Auto-tuning of multivariable PID controllers from decentralized relay feedback. *Automatica*, **33**(3), 319–330.

Wang, Q. and Nie, Z. (2012). PID control for MIMO systems. In R. Vilanova and A. Visioli, eds., *PID Control in the Third Millennium*, London: Springer–Verlag, 117–204.

Wang, Q. G., Lee, T. H. and Lim, C. (2003) *Relay Feedback Analysis, Identification and Control*, London: Springer–Verlag.

Wang, Q. G., Ye, Z., Cai, W. J. and Hang, C. C. (2008). *PID Control for Multivariable Processes*. Vol. 373 of *Lecture Notes in Control and Information Sciences*, London: Springer.

Wang, Y. (2008). Closed-loop multivariable process identification in the frequency domain. *Proceedings of the 27th Chinese Control Conference, CCC*.

Wood, R. K. and Berry, M. W. (1973). Terminal composition control of a binary distillation column. *Chemical Engineering Science*, **28**(9), 1707–1717.

Weischedel, K. and McAvoy, T. J. (1980). Feasibility of decoupling in conventionally controlled distillation columns. *Industrial & Engineering Chemistry Fundamentals*, **19**(4), 379–384.

Xiong, Q., Cai, W. and He, M. (2007). Equivalent transfer function method for PI/PID controller design of MIMO processes. *Journal of Process Control*, **17**(8), 665–673.

Xu, C. and Shin, Y. C. (2011). A self-tuning fuzzy controller for a class of multi-input multi-output nonlinear systems. *Engineering Applications of Artificial Intelligence*, **24**(2), 238–250.

Xu, H., Datta, A. and Bhattacharyya, S. P. (2003). PID stabilization of LTI plants with time delay. *Proceedings of 42nd IEEE International Conference on Decision and Control*, 4038–4043.

Yu, C.C. (2006). *Autotuning of PID Controllers: A Relay Feedback Approach*, 2nd ed., London: Springer–Verlag.

Yu, C. and Luyben, W. L. (1986). Design of multiloop SISO controllers in multivariable processes. *Industrial Engineering Chemistry Process Design and Development*, **25**(2), 498–503.

Yuwana, M. and Seborg, D. E. (1982). New method for on-line controller tuning. *AIChE Journal*, **28**(3), 434–440.

Zarei, M. (2018). Multi point kinetics based MIMO PI control of PWR reactors, *Nuclear Engineering and Design*, **328**, 283–291.

Zakian, V. and Al-Naib, U. (1973). Design of dynamical and control systems by the method of inequalities. *Proceedings of the Institution of Electrical Engineers*, **120**, 1421–1427.

Zhang, W. (2012). *Quantitative Process Control Theory (Automation and Control Engineering)*, Florida: CRC Press.

Zhao, Y. M., Xie, W. F. and Tu, X. W. (2012). Performance-based parameter tuning method of model-driven PID control systems. *ISA Transactions*, **51**(3), 393–399.

Zhuang, M. and Atherton, D. P. (1994). PID Controllers for a TITO System, *IEE Proceedings: Control Theory Technology Applications*, **141**, 111–120.

Zhu, Y. (2001). *Multivariable System Identification for Process Control*, London: Elsevier.

Zhu, Z. X. and Jutan, A. (1996) Loop decomposition and dynamic interactive single loop and overall stability of decentralized control systems, **51**(12), 3325–3335.

Ziegler, J. and Nichols, N. (1942). Optimum settings for automatic controllers. *Transactions of the A.S.M.E.*, **64**, 759–768.

Index